Animal Cell Biotechnology

Volume 3

Contributors

G. J. Berg
B. G. D. Bödeker
P. C. Brown
B. J. Bulbulian
H. Büntemeyer
M. Butler
S. R. Cernek
M. A. C. Costello
N. A. de Bruyne
C. Figueroa
J. B. Griffiths
H. Katinger
N. F. Kirkby
C. Lambe
J. Lehman
A. S. Lubiniecki
S. M. Maciukas
B. Z. Menken
O.-W. Merten
A. D. Murdin
T. J. Murphy
R. Oakley
J. C. Petricciani
C. P. Prior
A. Rosevear
P. W. Runstadler
W. Scheirer
O. T. Schönherr
R. E. Spier
W. R. Srigley
W. R. Tolbert
M. A. Tyo
P. J. T. A. van Gelder
J. Vorlop
R. Wilson

Animal Cell Biotechnology

Volume 3

Edited by

R. E. SPIER

Department of Microbiology
University of Surrey
Guildford, Surrey
United Kingdom

J. B. GRIFFITHS

Vaccine Research and Production Laboratory
Public Health Laboratory Service
Centre for Applied Microbiology and Research
Salisbury, Wiltshire
United Kingdom

1988

ACADEMIC PRESS

Harcourt Brace Jovanovich, Publishers

London San Diego New York
Boston Sydney Tokyo Toronto

ACADEMIC PRESS LIMITED
24–28 Oval Road, London NW1 7DX

United States Edition published by
ACADEMIC PRESS INC.
San Diego, CA 92101

British Library Cataloguing in Publication Data
Animal cell biotechnology.
Vol. 3
1. Cell culture
I. Spier, R. E. II. Griffiths, J. B.
(John Bryan)
591′.0724 QH585

ISBN 0-12-657553-3

Phototypeset by
Paston Press, Loddon, Norfolk
Printed by The Alden Press, Oxford

Contents

Part II Environmental Factors

3 Environmental Factors: Medium and Growth Factors

R. E. SPIER

4 Immobilized Hybridomas: Oxygen Diffusion

A. D. MURDIN, N. F. KIRKBY, R. WILSON, AND R. E. SPIER

5 Sensors for the Control of Mammalian Cell Processes

O.-W. MERTEN

6 The Design of Bench-scale Reactors

N. A. DE BRUYNE

Part III Culture Systems

7 Overview of Cell Culture Systems and Their Scale-up

J. B. GRIFFITHS

8 Bubble-free Reactors and Their Development for Continuous Culture with Cell Recycle

J. LEHMANN, J. VORLOP AND H. BÜNTEMEYER

9 A Tubular Biological Film Reactor Concept for the Cultivation and Treatment of Mammalian Cells

H. KATINGER

10 Protein Production from Mammalian Cells Grown on Glass Beads

P. C. BROWN, C. FIGUEROA, M. A. C. COSTELLO, R. OAKLEY, AND S. M. MACIUKAS

11 High-density Growth of Animal Cells within Cell Retention Fermenters Equipped with Membranes

W. SCHEIRER

12 A Comparative Review of Microcarriers Available for the Growth of Anchorage-dependent Animal Cells

M. BUTLER

13 Large-scale Fluidized-Bed, Immobilized Cultivation of Animal Cells at High Densities

P. W. RUNSTADLER AND S. R. CERNEK

14 Employing a Ceramic Matrix for the Immobilization of Mammalian Cells in Culture

G. J. BERG AND B. G. D. BÖDEKER

Part IV Downstream Processing

18 Downstream Processing of Animal Cell Culture Products—Recent Developments

A. ROSEVEAR AND C. LAMBE

Contributors

Numbers in parentheses indicate the pages on which the authors' contributions begin.

G. J. Berg[3] (*321*), Charles River Biotechnological Services, D-6200 Wiesbaden, Federal Republic of Germany

B. G. D. Bödeker (*321*), Bayer AG, Wuppertal, Federal Republic of Germany

P. C. Brown (*251*), Bio-Response Inc., 1978 West Winton Avenue, Hayward, CA 94545, United States

B. J. Bulbulian (*357*), Endotronics Inc., 8500 Evergreen Boulevard NW, Coon Rapids, MN 55433, United States

H. Büntemeyer (*221*), GBF mbH, Mascheroder Weg 1, D-3300 Braunschweig, Federal Republic of Germany

M. Butler (*283*), Department of Biological Sciences, Manchester Polytechnic, John Dalton Building, Chester Street, Manchester M1 5GD, United Kingdom

S. R. Cernek (*305*), Verax Corporation, Etna Road, HC-61, Box 6, Lebanon, NH 03766, United States

M. A. C. Costello (*251*), Bio-Response Inc., 1978 West Winton Avenue, Hayward, CA 94545, United States

N. A. de Bruyne[5] (*141*), Pyne's House, 8 Chapel Street, Duxford, Cambridge CB2 4RJ, United Kingdom

C. Figueroa (*251*), Bio-Response Inc., 1978 West Winton Avenue, Hayward, CA 94545, United States

J. B. Griffiths (*179*), PHLS CAMR, Porton Down, Salisbury SP4 0JG, United Kingdom

H. Katinger (*239*), Institute of Applied Microbiology, Faculty of Food and Biotechnology, University of Agriculture, Peter Jordan Strasse 82, A-1182 Vienna, Austria

N. F. Kirkby (*55*), Department of Chemical and Process Engineering, University of Surrey, Guildford GU2 5XH, United Kingdom

C. Lambe (*397*), Biotechnology Group, Harwell Laboratory, AERE, Didcot OX11 0RA, United Kingdom

J. Lehmann (*221*), GBF mbH, Mascheroder Weg 1, D-3300 Braunschweig, Federal Republic of Germany

A. S. Lubiniecki (*3*), Genentech Inc., 460 Point San Bruno Boulevard, South San Francisco, CA 94080, United States

S. M. Maciukas (*251*), Bio-Response Inc., 1978 West Winton Avenue, Hayward, CA 94545, United States

B. Z. Menken (*357*), Endotronics Inc., 8500 Evergreen Boulevard NW, Coon Rapids, MN 55433, United States
O.-W. Merten (*75*), Institut Pasteur, Technologie Cellulaire, 28 rue du Docteur Roux, 75724 Paris cedex 15, France
A. D. Murdin[1] (*55*), Department of Microbiology, University of Surrey, Guildford GU2 5XH, United Kingdom
T. J. Murphy (*357*), Endotronics Inc., 8500 Evergreen Boulevard NW, Coon Rapids, MN 55433, United States
R. Oakley (*251*), Bio-Response, 1978 West Winton Avenue, Hayward, CA 94545, United States
J. C. Petricciani (*13*), World Health Organization, Biologicals, CH-1221 Geneva 27, Switzerland
C. P. Prior (*373*), Invitron Corporation, 4766 La Guardia Drive, Berkeley, MO 63134, United States
A. Rosevear (*397*), Biotechnology Group, Harwell Labs, AERE, Didcot OX11 0RA, United Kingdom
P. W. Runstadler (*305*), Verax Corporation, Etna Road, HC-61, Box 6, Lebanon, NH 03766, United States
W. Scheirer (*263*), Sandoz Forschungsinstitut mbH, A-1253 Vienna, Austria
O. T. Schönherr[4] (*337*), Microbiological RDL, Organon, PO Box 20, 5340 BH Oss, The Netherlands
R. E. Spier (*29, 55*), Department of Microbiology, University of Surrey, Guildford GU2 5XH, United Kingdom
W. R. Srigley (*373*), Invitron Corporation, 4766 La Guardia Drive, Berkeley, MO 63134, United States
W. R. Tolbert (*373*), Invitron Corporation, 4766 La Guardia Drive, Berkeley, MO 63134, United States
M. A. Tyo (*357*), Endotronics Inc., 8500 Evergreen Boulevard NW, Coon Rapids, MN 55433, United States
P. J. T. A. van Gelder (*337*), Microbiological RDL, Organon, PO Box 20, 5340 BH Oss, The Netherlands
J. Vorlop (*221*), GBF mbH, Mascheroder Weg 1, D-3300 Braunschweig, Federal Republic of Germany
R. Wilson[2] (*55*), Department of Microbiology, University of Surrey, Guildford GU2 5XH, United Kingdom

[1] Present address: Department of Microbiology, SUNY at Stony Brook, Stony Brook, NY 11794-8621, United States.
[2] Present address: Department of Biological Sciences, University of Stirling, Stirling, United Kingdom.
[3] Present address: Chemap AG, Hölzilwisenstrasse 5 CH-8604 Volketswill, Switzerland.
[4] Present address: Diosynth BV, P.O. Box 20, 5340 BH Oss, Holland.
[5] Present address: Techne Inc., 3700 Brunswick Pike, Princeton, NJ 08540, United States.

Preface

Animal cell biotechnology is a rapidly developing and expanding subject and this means that many books on the subject are already dated on publication. When Volumes 1 and 2 of this series were published, the Editors were aware that although they provided a comprehensive introduction to the whole subject of animal cell biotechnology, the subject had progressed. The rationale for this volume is to review the advances that have been made in all facets of the subject and to make an in-depth assessment of one particular area—the bioreactor. It is felt that the number and diversity of bioreactors now commercially available for cell culture is causing confusion. The choice is so wide that many companies are delaying investment because they are spoilt for choice and lack of comparative information. In reviewing the different concepts and designs of cell reactor, one has to be cautious, because there are no bad ones, nor are there any obviously superior ones at the moment. Different products have different process requirements, varying from their biological properties to the required annual product yield. Thus, as far as possible, innovators, or investigators with a substantial experience, of a particular reactor have been asked to contribute a chapter in such a manner that comparisons can be made. An overview is added to include bioreactors not individually described, and to help the comparative process. The area of bioreactor development has been highlighted because it has moved rapidly over the past few years, there is a wide range of concepts being used, and because we hope that it will allow the more rapid evolution of one or two methods that will become the dominant culture technologies of the future.

The primary reasons for this rapid development in animal cell technology are, firstly, that it is now recognized that only animal cells can produce many of the required biomolecules in the correct configurations and with the necessary post-translational modifications, and, secondly, that developments in cellular engineering, media and bioreactor development have closed the productivity, or cost, gap with prokaryotic production systems. The fact that the use of animal cell cultures is no longer just one of the possible options but the method that has to be used, has meant a greater input of effort and resources into its development. Monoclonal antibodies were responsible for the initial surge of investment in the early 1980s, but as this was mainly to meet the requirements of the diagnostics industry, production targets were not too difficult to achieve. Nevertheless, it was the birth of animal cell biotechnology as we now know it, increasing the scope of products from the traditional narrow confines of virus vaccines, and more recently interferon. It also proved that cell technology would lead to a multibillion-dollar industry. However, viral vaccine manufacturers, in particular, have been slow to take advantage of newly developed processes, needing the tests of time and experience to prove the reliability and consistency of such systems before moving from such safety-first techniques as roller culture. Also, the possibility of using recombinant bacteria delayed any progress in this direction for a while. Thus, the efforts of academic research laboratories in which the different aspects of cell technology continued to be developed (derivation of recombinant cell lines, serum-free media, culture reactors capable of supporting high-density growth, and more efficient purification systems) have largely been confined to the laboratory scale.

The realization that cell culture had a commercial future has overcome this hesitation, and rapid progress is now being made in two important areas. Firstly, the scope of potential products has greatly increased. In health care alone, a wide range of hormones, immunoregulators, new viral vaccines, and enzymes has been added to the classical list of vaccines, interferon and antibodies, all as a result of recombinant-DNA technology. Many of the products require 10^5–10^{11} cells per clinical dose and to be commercially viable not only does the scale of production need to be increased enormously but the cost-efficiency of the process also has to be significantly improved. Secondly, this has provided the necessary impetus and funding for many of the laboratory-scale systems to be scaled-up and used for the commercial manufacture of products, many of which are in various stages of clinical trial.

Despite the breakthrough in the use of animal cells described above, it is still only a beginning. What we now have is a range of tools, techniques, and goals. What is needed is the blending operation in order to assemble these new, but disparate, forms of expertise into a more efficient and productive *process*. The nature of the developmental work has been such that different components of the production process have been developed largely in isolation. Cell lines have been derived and selected on the basis of product expression and stability. The fact that these lines may be nutritionally extremely fastidious, or sensitive to "shear" damage in stirred tanks, was not considered. Similarly, reactors have been designed on sound engineering principles but using the most robust model cell that could be found. There is also the problem of quantifying the productivity of processes in meaningful and uniform terms, so that everyone can comprehend the performance and capability levels of the bioreactor, and of the process as a whole. It is for these, and similar, reasons that scale-up from laboratory to industrial scale has been so slow, and why many companies have kept with the well-tried and proven methods.

When Volumes 1 and 2 were prepared, it was not intended as the beginning of a series, but as a background text to the subject. However, this intention has been overtaken by events. One cannot keep updating old chapters, because the scope of the subject is widening and some areas move faster than others. It is also essential to present publications that are balanced for the subject as a whole rather than to publish books on specialized areas of the topic. The answer, we feel, is to have an in-depth treatment of one particular aspect, but to accompany this with state-of-the-art reviews on all the other components of animal cell technology. In this volume, for reasons stated above, the bioreactor has been chosen as the special subject, and the accompanying collection of reviews on cell and product quality control, on media development and the use of growth factors, on biosensors and mixing technology, and on downstream processing, serve not only to update the reader on all these important aspects of the subject, but also as a reminder that they are all important components which have to be optimally blended together to achieve a successful process. Future volumes will follow this pattern as we can foresee the need for detailed sections on cell products, cell physiology, and product purification as these rapidly developing areas progress.

R. E. Spier
J. B. Griffiths

PART I

CELL SUBSTRATES

1

Safety of Recombinant Biologics: Issues and Emerging Answers

A. S. LUBINIECKI
Genentech Inc.,
South San Francisco,
California, U.S.A.

1. INTRODUCTION

As the 1980s began, animal cell culture was viewed by some in industrial and academic circles as a dying technology. Soon the molecular biologists would teach bacteria how to make interferons and even virus vaccines. Then the

ANIMAL CELL BIOTECHNOLOGY VOL. 3
ISBN 0-12-657553 3

remaining few practitioners of industrial cell culture might be forced to put down the tools of black magic developed over the decades and finally seek honest work. As the same decade now approaches its close, the earlier prediction of the death of industrial cell culture appears premature. Recombinant DNA (rDNA) technology has proved practical to practise in animal cells. This is fortunate, since many genes are now known to be expressed more optimally in mammalian cells than in microbes. Combined with broad interest in large-scale hybridoma culture, rDNA-modified mammalian cells have rekindled industrial enthusiasm for cell-culture technology and stimulated considerable technological development as well.

The products which will soon emerge from rDNA cell culture technology, such as tissue plasminogen activator and erythropoietin and others never before available, are significant additions to medical practice. In addition to the clear prospects of the benefits of products from rDNA-modified mammalian cells, the theoretical prospect of risks is also present. This chapter will describe some of these theoretical risk factors and how manufacturers and regulators have begun to approach them. The result of these efforts is likely to become a steady stream of new recombinant protein products whose benefits clearly outweigh their risks, and which are significantly safer than previously known conventional cell-culture biologics.

2. NATURE OF THE RISKS

The theoretical risk factors fall into two principal categories: risk of tumour formation and risk of toxic response in recipients of rDNA products. Current rDNA pharmaceutical technology is practised in continuous cell lines capable of immortal growth *in vitro*. Their biological behaviour is reminiscent of tumour cells and, hence, there is some concern that some tumorigenic factor might be transferred from the product-expressing cell to the recipient of the product. There are four theoretical mechanisms which might be responsible for this hypothetical risk: transmission of intact cells, contamination by oncogenic DNA, contamination by oncogenic proteins, and contamination by tumour viruses (*17*).

The second broad category of potential risk factors is toxicity. This might manifest itself as response to the innate pharmacological activity or immunogenicity of the product. Either the active ingredient or its derivatives or contaminating substances might be responsible. The possibility of contamination by adventitious agents also exists, although this is not different from that associated with conventional cell culture biologics.

The relative importance of these putative risk factors varies according to the intended use of the product in terms of route, dose, and frequency of administration, as well as the health status of the recipient. In addition, different manufacturing processes will tend to have different spectra of impurities. As such, these risks must be evaluated on a case-by-case basis for each product.

These risks present a complex picture which seems difficult to resolve or analyse. Earlier attempts in the 1950s to use continuous cells as vaccine substrates were unsuccessful, in part owing to the overlapping layers of complexity of these multiple-risk factors. Another contributing cause was the lack of scientific tools for evaluating the risk factors. In the 1980s the march of science has significantly progressed beyond that earlier point. The same advances which have enabled rDNA technology to be implemented also permit dissection and analysis of the risk factors. Part of this improvement lies in the multitude and resolving power of modern experimental techniques, and part in the considerable improvement in product purity. The rest of this chapter will examine how modern scientific technologies can be employed to assess these potential risk factors and assure their elimination from products of rDNA-modified mammalian cells. Additional information on mammalian cell culture rDNA products is available in the "Points to Consider" documents formulated by the Office of Biologics, US Food and Drug Administration (*3*, *15*).

3. CELL LINE CHARACTERIZATION

Freedom from adventitious agents for conventional biologics is assured primarily by rigorous testing of master cell banks (*5*). Extensive searches are made for evidence of contamination by inoculation of bank cells into intact animals and indicator cell lines. Recombinant biologics have borrowed these principles for detection of exogenous agents and added thorough testing for endogenous agents such as retroviruses (*11*). Retroviruses can be elusive and available tests are often insensitive. In some cases, only virus-like particles or retroviral components are present, but no infectious virus can be detected (*7*, *11*). The types used to detect retroviruses are summarized in Table I.

Cytogenetic evaluation is employed to assure the safety of conventional products prepared in diploid cells which normally show stable karyological patterns (*8*). In contrast, continuous cell lines frequently exhibit unstable karyology, whether modified by rDNA technology or not. Therefore, cytogenetic evaluation is not expected to provide evidence of cell substrate safety for rDNA cell lines. Despite this, banding cytogenetic methods (*19*) can serve as a test of identity for the recombinant cell lines, confirming the lineage of the construction and verifying the absence of contaminants such as HeLa cells (*11*). Isoenzyme analysis by electrophoresis may be employed to provide

TABLE I Some Tests Used to Detect Retroviruses

Type	Test	Reference
Morphogenetic	Transmission electron microscopy of bank cells	(*13*)
Biochemical	Reverse transcriptase[a]	(*2*)
	Tritiated uridine labelling at appropriate isopyknic density	(*12*)
Immunochemical	Competitive radioimmunoassay for p30 core protein	(*7*)
Biological	Plaque and focus formation	(*18*)
	Co-cultivation	(*9*)
	Induction of halogenated pyrimidines	(*10*)
Molecular biological	Hybridization with common retroviral sequences	(*1*)

[a] Positive results should be confirmed by simultaneous detection assay (*20*) to assure that the activity detected is producing a retrovirus-like 70S RNA.

additional confirmation of identity and lack of contaminants such as HeLa cells (*14*).

Bank cells are inoculated into suitably immunosuppressed rodents to assess tumorigenicity. Either nu/nu congenitally athymic (nude) mice (*21*) or neonatal rats treated with anti-thymocyte serum (*22*) are acceptable. Continuous cell lines are frequently positive in these tumorigenicity assays. In addition to testing cells, nucleic acids and/or nucleohistones may be tested for tumorigenicity. While being criticized as insensitive and of uncertain relevance to humans, these assay systems are useful in providing clues to the risk of tumorigenicity associated with cells and their components. These assays can also provide evidence of reduced risk following processing, e.g. inoculation of intact cells may lead to nodule formation but DNA, nucleohistone, or purified product from these cells might uniformly fail to form nodules (*16*).

At the conclusion of these studies, the master cell bank system (usually a seed bank with multiple working banks) is known to be free of risks from adventitious agents. Its identity is confirmed. Indications of the risk of tumorigenesis are available from these studies to guide the remainder of the strategy.

Although employed in several classic studies of viral tumorigenicity (*6*), the experimental use of "cell-free filtrates" has dubious value. The difficulty of proving the integrity of membranes employed for such studies (usually due to the use of small-scale units), the viscosity and particulate nature of the "cell-free" homogenates, and the substantial pressure needed to filter these preparations invite membrane leaks, which may allow a few intact cells to contaminate the filtrate. In view of these technical difficulties and of our current heightened understanding of tumorigenesis and process validation, the filterability experiments are not recommended, unless rigorous confirmation of membrane integrity is documentable.

4. PROCESS VALIDATION

During this phase, the design of the manufacturing process is examined critically for its ability to allow reduction of risk factors. The first activity is to list the risk factors, drawing upon research knowledge and the cell bank characterization studies. For example, it may be known that the product-secreting cell is tumorigenic in nude mice, that it contains retrovirus-like particles, and that some cells die during the product-collection phase. It is reasonable to hypothesize that, in the worst case, the harvested culture fluids might contain live tumorigenic cells, potentially tumorigenic retroviruses, and potentially tumorigenic cell DNA, in addition to the protein of interest.

The next phase involves removing the risk factors by designing specific process steps to achieve these purposes (or taking advantage of fortuitous steps which unintentionally accomplish the same task). This procedure also requires an understanding of the quantitative goal or target to be achieved. Combined with actual or worst-case estimates of the starting concentration of each risk factor, it becomes possible to estimate how much needs to be removed. The process is designed accordingly to achieve the necessary level of removal.

In the final phase, the performance of the process is evaluated under relevant process conditions. For the example mentioned above, the manufacturer might demonstrate that the process removes live cells, retroviruses, and cellular nucleic acid. Often this is done by means of "spiking" experiments, in which suitably labelled or recoverable preparations of each risk factor are added to specific process steps (*4, 23*). Their removal or inactivation can be accurately quantified. Two points are emphasized: first, these studies are meticulously planned and executed to represent faithfully the actual events during full-scale manufacturing, so that the data and conclusions are above reproach. Indeed, much of the perception of safety of the product rests on these studies. Second, the validation studies need to demonstrate the phenomenon of removal or inactivation reliably and reproducibly in repeated trials. Besides cells, viruses, and nucleic acids, other candidates for this exercise are noxious fermentation or purification chemicals, and potential allergens.

In summary, validation is the identification, quantification, and removal of known risks in a reliable manner. The unknown risk cannot be removed, since it may have properties which allow its escape. In this way, millions of people received live tumorigenic SV40 virus in early polio vaccines, owing to the resistance of the undetected SV40 contaminants to formaldehyde used to inactivate the poliovirus. The risk must be quantified, otherwise it is difficult to determine the adequacy of the removal procedure. Finally, removal or inactivation must be demonstrated reliably under conditions relevant to full-scale manufacturing. From a regulatory viewpoint, the product is now inseparably linked to the process used to prepare it. A different process might yield a different product, whether in terms of measurable test results or hypothetical contaminants. Process changes usually require repetition of affected validation and characterization efforts at least, and therefore are not undertaken lightly.

5. PRODUCT CHARACTERIZATION

Characterization studies are usually done only once, or a very limited number of times. Recombinant pharmaceuticals are proteins, and those from cell culture are usually glycoproteins. It is important for manufacturers to demonstrate that the molecule of interest reliably manifests important structural and functional properties. Tests for these might include amino acid analysis, amino acid sequence, carbohydrate analysis, lipid analysis (for lipoproteins), circular dichroism, tryptic peptide mapping, high-resolution chromatography and/or electrophoresis, epitope mapping with monoclonal antibodies, functional tests for enzymatic activity or receptor binding, and tests for aggregate formation, denaturation, or chemical modification. Other specialized tests may be appropriate, depending on the specific molecule and the limit of resolution for a given technique. The stability of the molecule must be demonstrated reliably in the exact formulation and container system intended for market use.

Another set of assays is generally devised for contaminants. Some of these might be detected by the assays listed above for the molecule of interest, but more specific assays for contaminants are desirable. One of the most popular immunochemical methods is based on polyvalent antiserum prepared against process fluid that lacks the molecule of interest. This immunogen consists of product stream depleted of the molecule of interest, or product stream from a recombinant expression system lacking the product gene. The resulting antiserum should be capable of detecting and quantitating antigenic impurities in the final product in an appropriate immunoassay.

Biological characterization may also be utilized. Purified mammalian cell products may be administered to immunosuppressed rodents to determine risk of tumorigenicity (*16*). This kind of assay may be more important for products chronically administered than for products given acutely or occasionally. The performance of preclinical pharmacological and toxicological testing in animals is another form of biological characterization of the molecule and of the manufacturing process used to prepare it. Clinical trials in humans might include evaluation of pharmacological, toxicological, and immunological effects on recipients of its product and its contaminants.

Another type of biological characterization may be required for specialized recombinant DNA products. For example, live vaccinia virus modified by rDNA technology to contain the immunogenic envelope glycoproteins of other viruses might have altered tissue specificity. These possibilities should be approached by suitable studies *in vitro* and in animal models. Protein chemists and biologists have an almost bewildering array of tools at present to employ in the characterization of recombinant molecules. Many of these were not available to previous generations of scientists. They should permit eloquent description of molecular secrets, resulting in products superior to any of those in the past.

At the completion of these studies, the nature of the molecule of interest and its contaminants are known quantitatively. This information is evaluated to decide what level of purity will be necessary for a specific application of a product.

6. PRODUCT QUALITY CONTROL TESTING

By law, in the United States and many other countries, cell culture biologics must be proven safe, effective, pure, potent, stable and reliable. Manufacturers must perform assays on final product, and perhaps some intermediate forms, which help assure that each lot of product possesses those six quality attributes. Some of the tests are standard pharmacopoeial tests, such as those for sterility and general safety, which are done for all products. Others are tailored to the specific protein, and are selected from those assays developed during the previous characterization phase by virtue of their information value. In general, glycoprotein products from cell cultures might use quality control tests to determine molecular size, molecular homogeneity and composition, and biochemical activity. The appropriate tests from among the characterization assays are validated for use in lot-release protocols.

Another set of release tests might involve screening for adventitious agents. These might include bacteria, fungi, viruses and mycoplasma. They do not substantially differ from tests performed for conventional cell-culture products such as virus vaccines. Similarly, a sample of final product vials is examined by classical chemical technology to confirm that the contents of the vial (excipient concentration, residual moisture, and so forth) are within specifications.

It is arguable that final product testing for recombinant cell culture products differs very little from that for conventional products. When they exist the main differences are in the greater resolution afforded by the greatly increased purity and state-of-the-art tools for products of rDNA technology.

7. OTHER TOPICS

7.1. Applicability to Other Products

Although the above is written from the perspective of recombinant products, most of these ideas also apply to conventional products of continuous cell lines. Throughout the world, a variety of products such as monoclonal antibodies, lymphoblastoid interferons, virus vaccines, and other products of continuous cells are being developed and used in clinical trials. A few have been approved for use in selected countries (lymphoblastoid interferon in the United Kingdom, Germany, and Austria; poliovirus vaccine from Vero cells in France and Belgium; OKT3 monoclonal antibody in the United States). Developers of those products have applied some of the concepts described here, and indeed, their successes in those efforts lighted the trail for the subsequent development of recombinant products (*4*).

7.2. Stability of Inserted Genes

The requirement for stability of inserted genes is unique to recombinant products, including those prepared in cell culture. It is incumbent upon manufacturers of cell-culture biologics to establish the reproducibility of each manufacturing system employed, i.e., to ensure that the same product is made every time. This is difficult to do with a single study, and usually requires a multifaceted approach to demonstrate consistency. The methods available include: cloning and sequencing of the inserted plasmid or its messenger RNA, measurement of specific production rates (amount of product per cell per unit time) throughout the manufacturing cycle, and assessment of molecular quality throughout the manufacturing cycle. Each has its strengths and weaknesses; all are laborious. Each manufacturer must decide the nature of the appropriate approach in consultation with national control authorities.

7.3. Case-by-Case Analysis

It is difficult to be definitive on the special regulatory requirements for recombinant cell culture products, owing to their great diversity. Different constructions are prepared by competing molecular biologists, which are superimposed upon existing differences among cell lines. A good deal of diversity of thought and approach exists among process development scientists, which adds yet another dimension of complexity. The different recombinant proteins themselves have specialized structural, functional, clinical and quality control attributes. Different governments may also emphasize different aspects of the risks and benefits of each product. While this may seem to promote disharmony, it seems to work reasonably well in getting products approved. As more history is acquired with recombinant products, additional unity on requirements will likely emerge. It is unlikely that case-by-case analysis will disappear entirely.

8. SUMMARY

In this chapter I have attempted to discuss some of the regulatory considerations facing cell culture biologics prepared from cells modified by recombinant DNA technology. The putative risks identified include tumorigenic factors, allergic reactions, and toxic reactions. Tools available to analyse these risks include cell bank characterization, process validation, product characterization and quality control testing. Completion of these studies permits risk associated with the product to be defined. Clinical trials delineate the benefit of the product. When both sets of data become available, assessment of the risk–benefit relationship is usually straightforward, if all factors have been defined in empirical terms.

The result of these efforts will be the availability of glycoprotein products unlike any available in the past. The broad approval of tissue plasminogen activator from recombinant CHO cells is expected. Other rDNA cell culture products used in advanced clinical trials include erythropoietin and hepatitis B surface-antigen vaccine. Others beginning clinical trials are Factor VIII and various haematological growth factors. Many others are under development in pharmaceutical companies throughout the world. For a technology which seemed moribund less than 10 years ago, cell culture has experienced a phenomenal rebirth thanks to the march of molecular technology. The remarkable purity and safety of these products are major reasons for this resurgence.

REFERENCES

1. Benveniste, R. E., and Todaro, G. J. (1973). Homology between type-C viruses of various species as determined by molecular hybridization. *Proc. Natl. Acad. Sci. U.S.A.* **70**, 3316–3320.
2. Baltimore, D., and Smoler, S. (1971). Primer requirement and template specificity of the DNA polymerase of RNA tumor viruses. *Proc. Natl. Acad. Sci. U.S.A.* **68**, 1507–1511.
3. Division of Blood and Blood Products, U.S. Food and Drug Administration (1984). Points to consider in the characterization of cell lines used to produce biologicals. Rockville.
4. Finter, N. B., and Fantes, K. H. (1980). The purity and safety of interferons prepared for

clinical use: The case for lymphoblastoid interferons. *In* "Interferon 1980" (I. Gresser, ed.), pp. 65–80. Academic Press, New York.
5. Hayflick, L., Plotkin, S. A., Norton, T. W., and Koprowski, H. (1962). Preparation of poliovirus vaccines in a human fetal diploid cell strain. *Am. J. Hyg.* **75**, 240–258.
6. Harvey, J. J. (1964). An unidentified virus which causes the rapid production of tumors in mice. *Nature (London)* **204**, 1104–1105.
7. Jackson, M. L., Nakamura, G. R., Lubiniecki, A. S., and Patzer, E. J. (1987). Attempts to detect retroviruses in continuous cell lines: Radioimmunoassays for hamster P30 protein. *Dev. Biol. Stand.* **66**, 541–553.
8. Jacobs, J. P., Magrath, D. I., Garrett, A. J., and Schild, G. C. (1981). Guidelines for the acceptability, management, and testing of serially propagated human diploid cells for the production of live virus vaccines for use in man. *Dev. Biol. Stand.* **9**, 331–342.
9. Kaplan, H. S., Goodenow, R. S., Epstein, A. L., Gartner, S., Decleve, A., and Rosenthal, P. N. (1977). Isolation of a type C RNA virus from an established human histiocytic lymphoma cell line. *Proc. Natl. Acad. Sci. U.S.A.* **74**, 2564–2568.
10. Lieber, M. M., Benveniste, R. E., Livingston, D. M., and Todaro, G. J. (1973). Mammalian cells in culture frequently release type C viruses. *Science* **182**, 565–69.
11. Lubiniecki, A. S., and May, L. H. (1985). Cell bank characterization for recombinant DNA mammalian cell lines. *Dev. Biol. Stand.* **60**, 141–146.
12. Manley, K. F., Givens, J. F., Taber, R. L., and Ziegel, R. F. (1978). Characterization of virus-like particles released from the hamster cell lines CHO-K1 after treatment with 5-bromodeoxyuridine. *J. Gen. Virol.* **39**, 505–517.
13. Michaides, R., Schlom, J., Dahlberg, J., and Perk, J. (1975). Biochemical properties of the bromodeoxyuridine-induced guinea pig virus. *J. Virol.* **16**, 1039–1050.
14. O'Brien, S. J., Shannon, J. E., and Gail, M. H. (1980). A molecular approach to the identification and individualization of human and mammalian cells in culture: Isoenzyme and allozyme genetic signatures. *In Vitro* **16**, 119–135.
15. Office of Biologics Research and Review, U.S. Food and Drug Administration (1985). Points to consider in the production and testing of new drugs and biologicals produced by recombinant DNA technology. Rockville.
16. Palladino, M. A., Levinson, A. D., Svedersky, L. P., and Obijeski, J. F. (1987). Safety issues related to the use of recombinant DNA-derived cell culture products. I. Cellular components. *Dev. Biol. Stand.* **66**, 13–22.
17. Petricciani, J. C. (1985). Regulatory considerations for products derived from the new biotechnology. *Pharmaceut. Manufac.* **5**, 31–34.
18. Rowe, W. P., Pugh, W. E., and Hartley, J. W. (1970). Plaque assay techniques for murine leukemia viruses. *Virology* **42**, 1136–1138.
19. Seabright, M. (1971). Rapid banding technique for human chromosomes. *Lancet* **2**, 971–972.
20. Schlom, J., and Spiegelman, S. (1971). Simultaneous detection of reverse transcriptase and high molecular weight RNA unique to oncogenic RNA viruses. *Science* **174**, 840–843.
21. Shin, S. I. (1979). Use of nude mice for tumorigenicity testing and mass propogation. *In* "Methods in Enzymology", Vol. 58 (W. B. Jakoby and I. H. Pastan, eds.), pp. 370–379. Academic Press, New York.
22. van Steenis, G., and van Wezel, A. L. (1982). Use of ATG-treated newborn rat for *in vivo* tumorigenicity testing of cell substrates. *Dev. Biol. Stand.* **50**, 37–46.
23. van Wezel, A. D., van der Marel, A., van Bevern, C. P., Verma, I., Salk, P., and Salk, J. (1982). Detection and elimination of cellular nucleic acids in biologicals produced on continuous cell lines. *Dev. Biol. Stand.* **50**, 59–69.

2

Changing Attitudes and Actions Governing the Use of Continuous Cell Lines for the Production of Biologicals

JOHN C. PETRICCIANI
World Health Organization,
Geneva, Switzerland

1. INTRODUCTION

Over the past decade, a number of conferences and discussions have been held on the possible benefits and the potential problems associated with the use of continuous cell lines (CCLs) such as Vero, Namalwa, and CHO for the production of various types of biological products (*1–3*). The purpose of this chapter is to review the events which have resulted in the widening use of CCLs to manufacture products of significant value in preventing and curing human disease, and to draw some tentative conclusions about what the immediate future may hold in this regard.

As in the case of any evolutionary change in attitudes, the acceptance of CCLs has been slow and difficult; nevertheless, it is now clear that during this past decade there has been a great deal of progress. One might even say that we are approaching the conclusion of the era of controversy and entering into a new stage where there are no longer significant reservations about the use of well-characterized cell lines of various types to produce biologicals when the manufacturing process can be shown to eliminate and/or inactivate specific contaminants of concern.

1.1. Initial Considerations

The initial discussions and decisions associated with the use of abnormal mammalian cells for the production of human biological products have been described by others (*4, 5*), and they will not be discussed in detail here. It is important to note, however, that the initial decision made in 1954 by the United States Armed Forces Board was an extremely important one, because it set the tone for attitudes towards cell cultures in vaccine production for the following three decades. When that group met to consider what cells would be an acceptable substrate to use for the production of an experimental live adenovirus vaccine, they basically had two choices: (i) HeLa, a human cancer cell line, or (ii) primary monkey kidney cells. Based on very straightforward pragmatic reasoning that there would be less risk in using a cell culture derived from the tissues of normal animals than if cells derived from a human tumour were used, they decided to approve primary monkey kidney cell cultures. The significance of that decision cannot be overstated, because it essentially ruled out as unsafe the use of cells other than those derived from normal tissues of animals in good health.

1.2. Human Diploid Cells

It was not until the suggestion by Hayflick and Morehead in 1961 that human diploid cells (HDCs) could serve as a useful and well-standardized cell system for the production of biological products that any serious reconsideration of the traditional concept of using only primary cells was undertaken. As reviewed by others (*4–6*), intense debate, disagreement, and dogmatism surrounded the question of the safety of products derived from HDCs. Eventually, vaccines derived from such cells were used in clinical trials, and ultimately they were licensed by national control authorities in many different countries.

1.3. Continuous Cell Lines

It is interesting to note, that during the HDC debates, Dr. Robert Hull of the Eli Lilly Company suggested that CCLs could be used for the production of vaccines; he pointed out that such cells had a significant number of advantages over both primary tissues and HDCs. Although data were presented (*7*) in support of the safety of vaccines derived from cell lines such as LLC-MK2, no serious attention was paid to that proposal, since the focus of controversy and interest was at that time on the HDC issue.

The question of heteroploid cell lines (e.g. CCLs) for vaccine production was essentially lost in the turmoil surrounding HDCs. The very few proponents of CCLs understood very clearly that if HDCs were having such a difficult time gaining acceptance, there was little hope for CCLs.

1.4. Lymphoblastoid Interferon

It was not until the mid-1970s that the next serious proposal to use other than primary or diploid cells was made. At that time, there was a great deal of interest in the use of interferon (IFN) in various clinical studies. The major source of IFN at that time was from primary human leukocyte cultures, although some was also available from human fibroblast cultures. In both cases, however, the amount of IFN which could be produced was very limited. An alternative source was developed by the Wellcome Research Laboratories using the Namalwa human lymphoblastoid cell line. Rather large amounts of IFN could be produced, and such cells were proposed as a source of IFN for human clinical studies (*8*). As in the case of considerations in 1954 of the safety of adenovirus vaccine which could be produced in human cancer cells, IFN from malignant human lymphoid tissue posed serious questions of safety. However, there were two major differences between lymphoblastoid IFN and live adenovirus vaccine: (i) interferon was not a replicating agent; and (ii) the purification and validation procedures which were available in the mid-1970s were significantly better than those of the mid-1950s.

In an effort to initiate a dialogue within the scientific community on the acceptability of lymphoblastoid cells as a source for IFN, a conference was held at Lake Placid in 1978 to discuss the issues and to open a reconsideration of the entire area of cell substrates (*1*). The conclusion of that meeting was a very cautious "yes" to the continuing exploration of lymphoblastoid IFN as an experimental biological product.

A decade later, it is easy to appreciate how important that research and development effort was, because it laid the groundwork for approaches which might be taken to ensure the safety of products derived from a variety of other CCLs, including those now being used in biotechnology. Following the Lake Placid meeting, interest began to develop in the use of other CCLs for manufacturing products. Chief among those CCLs was the possibility of using Vero, a continuous monkey cell line, for the production of inactivated polio vaccine.

1.5. Biotechnology

Recombinant DNA technology then emerged on the scientific landscape, and soon made its way into the industrial setting with the possibility that a wide range of biological products might be produced in bacterial cells. It quickly became evident, however, that bacterial cells were not able to synthesize adequate amounts of certain products. An example is the hepatitis B surface antigen which bacterial cells were unable to accumulate in sufficient amounts to make them reasonable as an expression system. In addition, the inability of bacterial cells to glycosylate was thought to be a potential drawback at least for some products in which glycosylation sites might be important for their proper biological functioning.

The use of recombinant yeasts and mammalian cells as expression systems for some products then loomed as the next step in product development, and various manufacturers began to explore the use of yeasts and several different mammalian cells for the production of these new biologicals. Among major determinants in the choice of the cell expression system were the following: rapid growth, high level of expression, and the ability to grow to high cell density in suspension culture. Those requirements meant that, in the case of mammalian cells, the systems of choice would be CCLs. The development of hybridoma technology and the potential usefulness of monoclonal antibodies also led to heightened interest in the use of CCLs.

Once again, however, there was a great deal of concern relating to the safety of products derived from these various abnormal mammalian cells. A number of meetings and discussions addressed the safety issues, and a consensus began to grow that CCLs would be acceptable under certain conditions. In an effort to develop a more formal consensus on the use of CCLs, the United States

Public Health Service sponsored a workshop in July 1984 at the National Institutes of Health in Bethesda (*3*). The focus of attention at that meeting was on the risks that might be associated with the following cellular contaminants in a biological product: (i) transforming proteins; (ii) viruses; and (iii) DNA. Those three contaminants have continued to be at the centre of debate regarding the safety of products derived from CCLs.

2. MAJOR ISSUES

2.1. Transforming Proteins

A number of proteins have now been identified which can induce the proliferation of many different types of cells, but most of them are active only intracellularly. Such proteins are encoded by cellular transforming genes or by an altered form of those genes.

Although transforming proteins have involved safety considerations, they have not loomed large as realistic problems. The major reason for the lesser importance attached to oncogene-encoded proteins is that only those which act extracellularly pose even a theoretical risk. That in turn means the consideration is restricted to growth factors which might be secreted by cells (*9*). But, even then, several aspects concerning the known growth factors lead to the conclusion that they are trivial considerations with respect to safety of products derived from CCLs.

First of all, they do not appear to be oncogenic in and of themselves, and their growth-promoting effects are transient and reversible. In addition, they do not replicate, and many of them are rapidly inactivated when administered *in vivo*. Finally, they are secreted in such small quantities that it would be necessary to concentrate them along with the product during the manufacturing process in order for them to begin to express any biological activity *in vivo*. In view of data *in vivo* (*10*) in which 0.3 $\mu g\ g^{-1}$ body weight of transforming growth factor alpha was required to induce a biological effect in highly susceptible mice, it is reasonable to assume that even microgram amounts of a growth factor would be needed to begin to produce observable effects in humans. Such large amounts of a contaminating protein are extraordinarily unlikely in biotechnology products.

2.2. Viruses

Viral contaminants from cells and tissues used to produce biologicals have been a problem for manufacturers, the medical profession, and the recipients of biologicals ever since the early days of vaccine production in cell cultures; and viral agents continue to be discovered. It is interesting to note, however, that in each case in which a virus of potential pathogenicity for humans was discovered (e.g. SV40), it was primary cell cultures or tissues in which the viruses were found. In contrast, no endogenous virus has ever been isolated from any of the HDCs used for vaccine production. This fact underscores the point that cells drawn from a well-characterized cell bank offer great advantages, because with careful studies it should be possible to establish whether a viral agent is present in the cells, and if so whether it carries any risk for humans. In this regard, potential viral contaminants derived from human and non-human primates clearly pose the greatest risk, while avian viruses are probably the least worrisome.

Fortunately, most of the viruses associated with cells in culture have turned out not to have caused diseases in humans. The presence of avian leukosis in chickens and in hens' eggs is a specific case relevant to the consideration of whether or not cells containing endogenous viruses might be acceptable. For many years, all live yellow fever vaccines were produced in such eggs, and the vaccines contained live avian leukosis. Because of uncertainty regarding the long-term health implications of having administered live avian leukosis to humans, a large retrospective study was undertaken, and the results showed no increased risk for cancer in the recipients of the vaccine (*11*). Nevertheless, even though safety was not an issue, the current yellow fever seed virus is avian-leukosis-free more as a matter of purity than of safety. Although many of the yellow fever vaccines are now being produced in leukosis-free eggs in various countries, such eggs are not available or are too expensive to be considered in many parts of the world where yellow fever is endemic. As a result, a large number of doses of yellow fever vaccine now being produced and used contain avian leukosis virus.

The situation is quite similar for a killed vaccine. Influenza vaccine is also produced in fertilized hens' eggs which contain avian leukosis viruses. But most influenza vaccines are inactivated, in contrast to the yellow fever vaccine which is live. Even in highly developed countries, no attempts have been made to eliminate the use of avian-leukosis-containing hens' eggs in the large-scale production of inactivated influenza vaccine because there is no perceived safety issue, especially since the avian leukosis is killed along with the influenza virus before being administered to humans in the form of a vaccine. The alternatives of using avian-leukosis-free eggs for influenza vaccine or avian-leukosis-free chickens for food, are not viable options, because of the difficulty and expense associated with providing the very large quantities which would be needed.

These two cases illustrate the point that the presence of a virus in cells need not preclude the use of such cells in the production of vaccines or other biologicals. The safety of the final product is the appropriate endpoint in deciding on the acceptability of cell substrates.

2.3. DNA

In evaluating the theoretical risk of cellular DNA in products administered to humans, it is important to keep in mind the fact that normal cellular DNA has already been accepted as a contaminant in the sense that all products produced from cell-culture systems up to the present have to some degree contained DNA, and it has neither frightened anyone nor apparently caused any adverse health effects. Concern about DNA from CCLs should therefore concentrate on those portions of the genome of CCLs which are different from the genome of normal cells. Only recently, with the discovery of cellular oncogenes (c-onc), has there been significant progress in identifying whatever DNA differences do exist between normal and abnormal cells. Certainly, oncogenes do not lead to a direct explanation for cancer or even to a solid understanding of the genetic changes which occur as a cell progresses from normal to abnormal to overtly tumorigenic. For example, strong promoters and gene amplification may be important in some cases, and chromosome rearrangements with subsequent activation of a normal unexpressed gene or inactivation of a gene which normally acts to suppress an oncogene (or other genes) are possibilities worth consideration. But even though the precise role of cellular oncogenes in human cancer remains to be determined, they are useful models to consider in trying to come to grips with the biological meaning of contaminating DNA, because they represent special sequences of some tumour cell DNAs which have transforming potential, and experiments using cellular oncogenes can be helpful in making judgements about the significance of various levels of DNA which might be present in products.

The amount of DNA in many biologicals can now be reduced to 10 pg per dose, or less. This is equivalent to about one mammalian cell genome. Although specific data are not available, it is reasonable to assume that such DNA is heterogeneous and contains random portions of the genome, and that various sizes of DNA are present without selective elimination or concentration of any specific gene sequences.

Various approaches have been taken to calculate risk. One is to assume that at least some contaminating DNA sequences are large enough to retain the potential for biological activity, and that a single c-onc is present in the 10 pg of DNA. Although the sizes of c-oncs vary, the smallest are in the range of 1 kb which serves to illustrate the point. One copy of a 1 kb c-onc is equivalent to about 1×10^{-6} pg of DNA. A variety of studies have shown that under optimal experimental conditions *in vitro*, purified c-onc DNA has a transformation efficiency of about 10^4 focus-forming units per microgram. Converted to picograms, that means 100 pg of purified c-onc DNA are needed to give one focus of transformation in the 3T3 system. Going back to the cellular DNA with one c-onc copy, it becomes clear that there is a factor of about 10^{-8} (10^{-6} pg/100 pg) between a minimally effective DNA transformation dose *in vitro* and the amount of c-onc DNA which theoretically might be in a product. If more realistic assumptions were made which relate to the *in vivo* situation, the safety factor could only become larger by at least several orders of magnitude. For example, if one does similar calculations based on recent *in vivo* data showing that 2 μg of cloned viral oncogene DNA was required to induce tumours in chickens, the 10^{-8} safety factor just mentioned becomes 10^{-12}. Those and similar calculations by others (*12, 13*) have suggested that if the DNA in a product is in the range of 10–100 pg per dose, then there is a safety factor of at least 10^9.

3. WHO ACTIVITIES

The World Health Organization (WHO) has from the very beginning been closely involved with developments in biotechnology, and has sponsored many meetings to address both general and specific issues.

3.1. Inactivated Polio Vaccine

The 1980s have been a time of intensive scientific activity and product development. During this period, WHO has taken actions, after consultation with experts, to facilitate the acceptance of CCLs as substrates for biological production. In 1981, for example, WHO made a major revision of the requirements for inactivated polio vaccine by incorporating provisions for the use of non-tumorigenic and virus-free CCLs as substrates for vaccine production (*14*).

Four major points were taken into consideration before the concept of CCLs for inactivated virus vaccines was accepted:

1. Purification procedures could reduce contaminating DNA to the 10 pg range.
2. Formalin was used to inactivate the virus.
3. Poliovirus is an RNA virus which is not known to incorporate DNA into the virion.
4. The absence of any detectable endogenous virus in Vero cells which were being proposed as the cell substrate.

3.2. Characterization of Continuous Cell Lines

As a follow-up to that step, WHO convened a group of consultants in early 1985 to reconsider those requirements in light of the advances which had been made since 1981. The result of that meeting was a more general set of requirements, which was considered and accepted by the Expert Committee on Biological Standardization (ECBS) in November 1985 (*15*). That document established generally applicable requirements for the characterization of CCLs used in the preparation of biological products. Those general CCL requirements provided WHO acceptance of the general principle of using CCLs, and they gave guidance on what needs to be done to prepare a well-characterized and standard cell bank. The acceptance of CCLs for any given product is dealt with in the requirements for that product as was done with inactivated polio vaccine.

3.3. Study Group on Biologicals

Concerns remained, however, regarding the safety of products derived from CCLs, and such reservations have impeded the more general acceptance of CCLs as a substrate for biologicals administered to humans. It was, in fact, because of such concerns that a group of consultants which met at WHO in September 1985 to consider the development of requirements for hepatitis B vaccine produced by recombinant DNA techniques in CCLs recommended that WHO convene a special group in order to provide an independent international evaluation of the safety issues.

In response to that recommendation, the Director-General convened the Study Group on Biologicals. The purpose of the Study Group was to undertake a general review of the acceptability of CCLs as substrates for biological products administered to humans, and to provide advice to WHO on certain specific points of concern.

Six somewhat overlapping disciplines were represented among the members of the Study Group: general biology and medicine, immunology, cancer cell biology, virology clinical oncology, and molecular biology and pharmacology. In addition, several consultants in very specialized areas were invited to assist in the discussions.

Background documents providing both experimental data and theoretical considerations were discussed and led to the formulation of the summary statements and recommendations in their report (*16*). A major concern was the long-term risk of malignancy represented by heterogeneous contaminating DNA, especially if it were to contain potentially oncogenic or regulatory sequences.

The group noted that there is already some evidence for the safety of certain CCLs as substrates for the production of biologicals. For example, inactivated foot-and-mouth disease vaccine has been prepared in the BHK-21 cell line, and it is estimated that more than 10^8 doses have been administered to cattle over a 20-year period. Carcass inspection has failed to discover any ill-effects attributable to the vaccine, suggesting that in the short term (2–4 years) this vaccine has been safe. Although human clinical experience with biologicals produced in CCLs has been more recent, and therefore more limited than with veterinary products, it also was noted that over 10^7 doses of inactivated polio vaccine produced in Vero cells have already been administered to children since 1983.

There was agreement that the major potential risks associated with the use of biologicals produced in CCLs fall into three categories: transforming proteins, viruses, and heterogeneous contaminating DNA.

The Study Group concluded that, in general, CCLs are acceptable as substrates for the production of biological products, but that differences in the nature of the products derived from CCLs and the specifics of the manufacturing processes must be taken into account in making a decision on the acceptability of each given product. There is, therefore, no reason to exclude CCLs from consideration as substrates for biological products. In this regard, there was agreement with the actions taken to date in approving the use of products derived from CCLs, after having been satisfied that a given manufacturing process yields a product with no detectable risk attributable to the cell substrate.

The importance of validating the efficiency with which various steps in a manufacturing process inactivate and/or eliminate unwanted material such as cellular DNA and viruses was emphasized. Validating the ability of a process to yield a product with certain specifications and to establish the consistency of that process are essential elements in providing the basis for an acceptable biological derived from CCLs. Once a process has been validated and consistency of production has been established, limited tests appropriate for each product should suffice, as has been the usual practice with biologicals in the past.

The Study Group concluded that, based on the experimental data available to date, the probability of risk associated with heterogeneous contaminating DNA in a product derived from a CCL is so small as to be negligible when the amount of such DNA is 100 pg or less in a single dose given parenterally. The assessment of the safety of any product with respect to DNA should take into consideration:

1. the elimination of the biological activity of DNA by various steps in the manufacturing process.
2. the reduction in the amount of DNA during the purification of the product in the manufacturing process.

A given product may be considered safe on the basis of reliable data from either of those two elements, or their combination.

The use of special DNA sequences such as viral regulatory sequences in the construction of recombinant cells is considered acceptable because there is no evidence to suggest that such sequences would impose any additional risk beyond that of heterogeneous contaminating DNA in general. Nevertheless, the manufacturing process should be validated to show that they are not concentrated in any detectable contaminating DNA.

In considering the potential for infecting man with viruses, the Study Group agreed that cells may be divided into three risk categories with respect to their potential for carrying viral agents pathogenic for man:

High risk: Any material derived from either human or primate blood or bone marrow cells, caprine or ovine cells, or hybridomas when at least one fusion partner is of human or non-human primate origin.

Less risk: Any material derived from mammalian non-hematogenous cells such as fibroblasts or epithelial cells.

Low risk: Any material derived from either human diploid lines or avian tissues.

Taking into account the above classification, the Study Group agreed that different degrees of concern, and therefore testing, were appropriate for products manufactured from the various types of cells mentioned. Nevertheless, it was emphasized that when either diploid or continuous cell lines are used for production, a cell seed lot system must be used and the cell seed must be characterized as already provided in WHO Requirements.

The Study Group stressed the importance of validating the ability of a manufacturing process to eliminate and to inactivate those viruses which may pose a risk to humans when cells carrying viruses are proposed for use in the manufacture of human biologicals. In addition, cells from humans and animals with diseases of unknown origin should not be used, nor should animal cells which may contain "slow viruses" be used to produce biologicals for human administration.

With regard to transforming proteins, the Study Group did not consider that the presence of contaminating known growth factors in the concentration at which they are ordinarily to be found constitutes a serious risk in the preparation of biologicals.

4. SUMMARY AND CONCLUSIONS

Looking back at the history of issues and decisions that were made regarding the acceptability of various cell substrates leads one to wonder how much more rapidly products might have been developed if different decisions had been made. It is also strikingly evident that primary tissues and primary cells have been the source of whatever problems have existed in the past. This is not to say that everything is known about CCLs or that there are no surprises in store as more experience is gained with their use in biological production. It is difficult to argue, however, that one is not better off with a very well-tested cell system and a manufacturing process which can cope with even theoretically worrisome contaminants.

The prevention and treatment of human diseases will be greatly facilitated by products derived from CCLs in the developing world as well as in technologically advanced countries. The use of CCLs in the manufacture of rabies vaccine, for example, opens the door to developing countries for the replacement of sheep brain vaccine with a much safer and more readily available cell culture vaccine at low cost.

Based on the conclusions and recommendations of the Study Group, WHO plans to pursue the development of a limited number of CCL cell seed banks which will be characterized according to current requirements. These banks will then be available, upon request, to serve as source material for biologicals manufacturers. They should be especially useful to developing countries with limited resources and which might otherwise be unable to establish a bank of characterized cells. Developed countries may also find the cell seeds useful in the same way that virus seeds for polio and yellow fever have provided basic starting material for vaccine production.

In addition, WHO will attempt to stimulate the development of data on the effect of commonly used viral inactivating agents such as Beta-propriolactone and formalin on the biological activity of DNA. Results from such studies will be helpful in assessing the biological significance of contaminating DNA in products such as rabies and inactivated polio even when such DNA is present in amounts greater than 100 pg/dose.

As pointed out by the Study Group, a number of questions remain unanswered about viruses, DNA, and protein as potential contaminants in biological products, and one must be alert to new information and new technology so as to reassess the subject from time to time. Nevertheless, it was also clear to the Study Group that under certain conditions the use of CCLs was reasonable. The next steps in the process which began in 1978 will be for national control authorities, industry, and the general scientific community to consider the report of the Study Group as new products are proposed for clinical trials and for approval in various countries.

REFERENCES

1. Hopps, H. E., and Petricciani, J. C. (eds.) (1985). "Abnormal Cells, New Products, and Risk." Tissue Culture Association, Gaithersburg.

2. Bar-Sagi, D., and Feramisco, J. R. (1986). Induction of membrane ruffling and fluid-phase pinocytosis in quiescent fibroblasts by *ras* proteins. *Science* **233**, 1061–1068.
3. Beale, A. J. (1979). Choice of cell substrate for biological products. *Adv. Exp. Biol. Med.* **118**, 83–97.
4. Petricciani, J. C., Hopps, H. E., and Chapple, P. J. (eds.) (1979). "Cell Substrates." Plenum, New York.
5. Hayflick, L., Plotkin, S., and Stevenson, R. E. (1987). History of the acceptance of human diploid cell strains as substrates for human virus vaccine production. *Develop. Biol. Standard* **68**, 9–17.
6. Hillenam, M. R. (1979). Cell line saga: an argument in favor of production of biologicals in cancer cells. *Adv. Exp. Biol. Med.* **118**, 47–58.
7. Hull, R. N. (1968). Immunization of experimental animals with vaccines produced in known oncogenic cell lines. *Natl. Cancer Inst. Monog*, **29**, 503–509.
8. Lowy, D. R. (1985). Potential hazards from contaminating DNA that contains oncogenes. *In* "Abnormal Cells, New Products, and Risk" (H. E. Hopps and J. C. Petricciani, eds.), pp. 36–40. Tissue Culture Association, Gaithersburg.
9. Report of the meeting of the study group on biologicals (1987). WHO Technical Report Series, No. 747.
10. Salk, J. (1979). The spector of malignancy and criteria for cell lines as substrates for vaccines. *Adv. Exp. Biol. Med.* **118**, 107–113.
11. Tam, J. P. (1985). Physiological effects of transformating growth factor in the newborn mouse. *Science* **229**, 673–675.
12. Perkins, F. T., and Hennessen, W. (eds.) (1982). "Use of Heteroploid and Other Cell Substrates for the Production of Biologicals." Karger, Basel.
13. Wahl, G. (1985). Detection of adventitious agents and sensitivity of methods. *In* "Abnormal Cells, New Products, and Risk" (H. E. Hopps and J. C. Petricciani, eds.), pp. 50–56. Tissue Culture Association, Gaithersburg.
14. Waters, T. D., Anderson, Jr., P. S., Beebe, G. W., and Miller, R. W. (1972). Yellow fever vaccination, avian leukosis virus, and cancer risk in man. *Science* **177**, 76–77.
15. WHO (1982). Requirements of poliomyelitis vaccine (inactivated). World Health Organization Technical Report Series, No. 673, Annex 2 (1982).
16. WHO (1982). Requirements of continuous cell lines used for biological production. World Health Organization Technical Report Series, No. 747, Annex 3 (1987).

PART II

ENVIRONMENTAL FACTORS

3

Environmental Factors: Medium and Growth Factors

R. E. SPIER
Department of Microbiology,
University of Surrey,
Guildford, U.K.

ANIMAL CELL BIOTECHNOLOGY VOL. 3
ISBN 0-12-657553 3

1. INTRODUCTION

The objectives of the animal cell biotechnologist are:

1. To establish and use the necessary science and technology to grow or make available large quantities of animal cells in an *in vitro* environment.
2. To generate materials from such cells which are of benefit to man and/or animals.
3. To achieve (1) and (2) above in a manner which is (a) safe, both to the technologists themselves and to the people or animals who use the products, and (b) so efficient as to make the products widely available while providing the manufacturer with an adequate financial return on his investment so that he may be encouraged to do the necessary research and development for the establishment of further such products.

Such objectives need to be brought to the fore in any discussion of those environmental factors which are designed to provide for the nutrition of animal cells in culture, because, today, we are faced with the enormous opportunities afforded by the welter of discoveries in the area of the factors which can control the proliferation and differentiation of animal cells *in vitro*. In the use of such factors, while providing the technologist with unbounded possibilities for making cells grow where they once would not grow (no matter what medium formulation was assayed) there are also possibilities of accentuating the degree of differentiation of cells such that they either produce the materials representative of their tissue of origin, or they may be induced to produce to excess materials which are of value irrespective of their original programming.

However, such powers are not without their problems. To what extent are products made from cells which have been induced to forego their controlled rate of division less safe than they would be were they to have been made from a more restrained cell? Secondly, what are the costs involved in the use of media with specially prepared growth and/or differentiation factors and how are such costs likely to change if these proteinaceous materials are made at maximum efficiency from genetically engineered recombinant cells lines. Finally, we have to recognize that, whereas most material made from animal cells in the past (viruses, interferon and the early antibody-producing technologies), relied upon relatively similar media for cell growth and product formation, the manufacturer of the modern generation of products is more likely to require quite different approaches to the medium formulation for each of these two phases. Also, whereas production operations for the conventional materials lasted between 3 and 8 days, modern and future production cycles are designed to operate for as many months.

The composition of the basal medium and the definition of three different types of medium predicated on the way in which substitutions have been made for the serum component of normal media have been fully described (*29*). However, during the last three years much progress has been made in discovering and characterizing the cellular and, normally, retroviral genes whose products are deeply involved in the control of cell proliferation and differentiation. In addition, a variety of materials can be used to enhance or inhibit such genes or their products. As the effects of these materials are dependent on:

1. The cell in which they are expressed;
2. The environment of that cell at the time of the expression;
3. The amount of material and its intracellular location;
4. The presence or absence of other such genes or gene products;
5. What must be a large number of materials whose attributes have yet to be discovered;

there is a need for a statement of the area of knowledge as it is presently available, secondly there is a need to determine which areas can best be exploited by the animal cell biotechnologist in the furtherance of his aims. Finally there is a need to understand the nature of the biochemistry of cells under the influence of oncogene products so that, when such cells are used to generate products, the safety of both the process and of the product can be assured.

2. THE SALIENT BACKGROUND

2.1. The Objectives

The requirement to provide for a situation such that a bioprocess involving animal cells needs the application of, in the first instance, a medium which will promote cell growth and then, later, a second medium which can maintain cells in a non-growing yet product-generating state, for extended periods of time, is the major challenge in medium formulation. This challenge is rendered more acute as growing cells express fewer of the characteristics of the differentiated state which is most appropriate for product generation. However, whereas each of the two media should conform to such standards as:

low cost;
reproducibility and reliability in formulation and use;
widely applicable for different cell lines/product areas;
formulation from readily available components;
simple methods for formulation and sterilization;

acceptability to regulatory agencies;
the ability for diverse control strategies for nutrients, pH, etc., to be implemented;
the achievement of high yields of cells and product per unit of medium used;

in addition, media required for cell production should:

stimulate cell division by the provision of cell growth factors and removal of growth inhibitors;
enable the cells to grow to high cell densities (in excess of 10^8 cells ml^{-1});
retain those characteristics of the cells required for the production phase;

in contrast, the production medium is most suitably formulated if it:

discourages further cell biomass production and cell division;
promotes the specific kind of differentiation which yields most product;
is effective without proteinaceous supplements or, if protein is required, then with low levels of a definable species of protein.

2.2. The Conventional Position

To meet the objectives of the previous section, media for animal cell bioprocesses have normally been formulated from a mixture based on inorganic salts (including some known and some unknown trace elements), carbohydrates (*36*), amino aids (both "essential" and some selected non-essential), vitamins, an organic or more commonly inorganic salt for pH buffering, a colorimetric indicator of pH, and either one or a small number of complex undefined materials. It is in the variations on these latter materials that the art and the science of medium formulation has been developed.

Whereas such materials as trypticase soy broth and lactalbumin hydrolysate (*47*) are commonly used for medium formulations for BHK suspension cell production for foot-and-mouth disease vaccine other complex materials such as casein, yeast extract, tryptose phosphate broth (*34*), bactopeptone and yeastolate (*26*) have also been added to formulae. Such materials have been used either rarely as serum substitutes (*26*) or, more commonly, as additives to serum-containing medium. In view of the central position of serum in cell culture media, it is worthwhile to review its role and the problems it presents.

2.2.1. The Multiple Roles of Serum

It is as well to appreciate the overall composition of serum as presented in Table I (*13*) as it is to understand that only a few tens of the 1000 or so

TABLE I Plasma Composition (Adapted from (*13*))

	g/100 g of protein*
Albumin	50–65
α_1 antitrypsin	2–4
α_2 macroglobulin	1–4
Transferrin	3–65
α_1 Lipoprotein	4–8
β_1 Lipoprotein	4–14
IgA	1–3
IgM	0.5–2
IgG	13–22

*This is a breakdown of the 60–85 g of protein per litre of plasma.

components have been identified and their properties determined. A summary of the functions of serum includes the provision of:

1. *Factors which promote cell division*—growth factors found in serum include epidermal GF, insulin-like GF's I & II, somatomedin A & C. Macrophage-granulocyte colony stimulating factors, platelet-derived GF, fibroblast GF, endothelial cell GF).
2. *Attachment factors*—fibronectin, fetuin, hydrocortisone, laminin collagen, chondronectin serum spreading factor epiboline (*2*).
3. *Antitrypsin proteins*—Inhibit the action of trypsin commonly used in releasing cells from their substratum.
4. *Trace elements*—Fe, Ca, Mo, Zn, V, Se, Cr, Co, Mn—and the means to enable them to pass through the plasma membrane (transferrin).
5. *Protective action against physical damage*—Albumin seems to be responsible for protection against shear damage and the effect of the air–liquid interface of bubbles, although materials which are difficult to separate from the albumin could also be implicated.
6. *Materials for membrane biosynthesis*—Linoleic acid, cholesterol.

However, while serum is a pluri-potent reagent, its use presents a number of important problems.

2.2.2. Problems Resulting from the Serum Requirement

The main difficulties experienced by users of serum-containing media can be summarized.

1. The variability of the raw material, which necessitates extensive pre-testing and the storage of large quantities at −20°C as tactical and strategic reserves.
2. The high cost of those sera which contain less antibody protein and highest growth promoting potencies (foetal calf serum or newborn calf serum).
3. The protein and lipid content of the serum presents a difficult raw material to a downstream processing operation based on chromatographic procedures.
4. The filter sterilization of serum cannot be relied upon to remove mycoplasma and/or viruses. UV treatment of sera and extensive pre-testing reduce such problems.
5. In some cases serum proteins prevent cell attachment to substrata, yet the attachment factors in the serum will change the nature of the cell–substratum interaction and in so doing affect the cell's physiology.
6. Although some serum preparations can be cytotoxic to the cells under study, an adequate pretesting procedure of different serum batches should obviate this problem.
7. There is little specific control over which particular type of cell is favoured for growth.
8. The large number of components in the serum preclude a detailed scientific investigation of cellular nutrition and physiology.
9. There are often periodic scarcities of specialized sera which lead to both high prices and restricted supply.
10. The serum is unstable as incorporated into pre-formulated medium to manageable proportions.

2.2.3. Resolution of Many of the Serum-associated Problems

As indicated in the preceding section, many of the problems associated with the serum can be eliminated or obviated by a thorough pre-screening and pre-testing programme. Also, a further set of problems is side-stepped by the use of serum for the cell growth medium and virtually eliminating the serum from the production phase of the culture. The provision of adequate stores prevents the problems of scarcity and it is possible, under some circumstances, to treat the serum with polyethylene glycol (8% w/v) to remove contaminating immunoglobulins (*3*) which should improve the separation of monoclonal antibodies, although other problems of high-protein, high-lipid feed materials would still be present. Materials such as pluronic F68 and other foam-stabilizing compounds will stand in for the cell-protective properties of whole serum (*20*).

Notwithstanding the above techniques for decreasing or eliminating the uses of serum, it is becoming increasingly important to be able to design culture media which enable the biotechnologist to deliberately control cell behaviour in the extracorporeal environment of a culture system. To achieve this, there is a need for a wide range of specific chemicals (some of which may be genes) and to be able to apply such materials to arrive at predetermined capabilities. Such a mode of operation would be rendered ineffective were serum to remain a component of the system, as this would establish the cells in a mode which would be beyond the control of the technologist, thus rendering subsequent efforts at modulating growth, differentiation and productivity either difficult or not possible.

2.3. The Present Position

Some 44 basal media fortified with a relatively small number of additional and defined materials have been described for the growth and maintenance of animal cells in culture (*33*). Most such media require insulin and transferrin in addition to other more commonly used materials such as epidermal growth factor, selenite, hydrocortisone or dexamethasone and bovine serum albumin. Such media can be as expensive as using the most expensive serum (Table II) and suffer from the disadvantage of a lack of general utility. It is also clear that while the addition of a growth factor may effect one or two multiplications of a cell population (*10*), it does not have the same effect as a direct and deliberate modification of the cell or the establishment of the cell in an environment which affects the physiology of the cell so as to promote more continuous cell division.

TABLE II The Approximate Cost of Medium Supplementation (Selected)

	For 1 litre of medium £ Sterling	
Foetal calf serum % 10%	10	
Adult bovine serum % 10%	1	
Insulin 5 $\mu g\ ml^{-1}$ } 5.3		12.1 (Insulin to Linoleic acid)
Transferrin 5 $\mu g\ ml^{-1}$ }		
Albumin 1.2 $mg\ ml^{-1}$		
Linoleic acid 5 $\mu g\ ml^{-1}$		
Nu-Serum % 10%		6
HB101, 102, 103		
KC 2000		
Epidermal growth factor, 10 $ng\ ml^{-1}$		0.5
Hydrocortisone/dexamethazone, 1 μM		4

To be able to grow cells and to obtain from them particular products, it will become increasingly necessary to use the mass of information which has been generated by the study of cancer cells and the viral and cellular factors which control growth and differentiation. This information base is burgeoning following the discovery of oncogenes in cells (*7*) by Bishop and Varmus in 1975 and the elucidation of some of the more common pathways whereby external chemicals act as signals to initiate profound changes in the cell. A summary of that knowledge base is presented below.

3. GROWTH FACTORS AND ONCOGENES

There is a synergistic interaction between progress in the knowledge base and progress in the technology base. Additional knowledge opens up new approaches to the resolution of practical problems and the solution of those problems provides us with opportunities to obtain more knowledge. This situation aptly describes the present position with regard to the exploitation of a new knowledge base on growth factors, oncogenes, and gene expression in relation to the need to solve the practical problems resulting from a requirement to grow a wide variety of cell types and obtain materials of value from those grown cells. As previous sections (2.2.1; 2.3) have related, growth factors are part of the repertoire of materials from which media can be formulated. However, we do not make use of the new knowledge which has recently been obtained about the growth factor receptors and their mechanisms of operation, nor do we, as yet, seek to interfere with the signal-transmission process whereby an external agent changes the transcriptional activity of a gene or the translational properties of the transcribed mRNA. Were we able to modulate the way in which this network of interacting elements operates, we would be able to usher in a new era of processes and products based on the use of animal cells in culture. The purpose, then, of this section is to set out a brief summary of the knowledge which has accrued during the last five years in a way which will provoke experiments and technological developments to augment our existing manipulative skills.

3.1. The System Format

Agents external to the cell must perforce make an initial contact with the cell at the cell's periphery. Some agents are lipophilic (e.g. steroid hormones) and experience little difficulty in penetrating the plasma membrane bilayer and penetrating into the cytosol, or even the nuclearsol where, in concert with specific receptor molecules, they exert their effects on gene expression. However, most proteinaceous molecules do not penetrate the plasma membrane but interact with molecules embedded in that membrane. Consequent upon such interactions, a chain of events is entrained which transduces and transforms this initial contacting, firstly across the plasma membrane; then, via the agency of components which are free to move in the cytosol, the effect is transferred to nuclear materials, which in turn cause changes in gene expression. While this simplified format pertains, the connections between the various stages are not equally well understood. Events at the plasma membrane are beginning to be appreciated in some detail and depth; the events which occur in the cytosol and nuclearsol are less well worked out; events at the level of the expression of the genes is understood at an intermediate level.

3.2. The System Components (see Fig. 1 for summary)

There is a series of naturally occurring growth factors and hormones, most of which have been found in the serum (Section 2.2.1), which interact with the cell via receptor molecules whose structure is such that they uniquely respond to specific chemical stimuli. That is, they may well respond to a series of related, or even chemically unrelated molecules, yet they clearly do not respond to all molecules. Such receptors may, on contact with their stimulant or agonist, either be provoked into acting as protein-phosphorylating enzymes in their own right or may activate a secondary system, generally known as a 'G'-protein complex. This, in turn, activates (or inhibits) an enzyme which generates mobile small molecules which are capable of activating a second, generally protein-phosphorylating, enzyme. From the activated growth factor receptors and from the hormone-activated receptors, or from systems which are stimulated by the self production of G-protein-like materials, the focal product in each case appears to be phosphorylated proteins; the phosphate group is confined to tyrosine residues or to serine or threonine residues. The phosphorylated proteins affect the physiology of the cell markedly and indeed can be responsible for its "transformation" (see Section 4) and/or differentiation. As many as ten such proteins are so phosphorylated, but, while a small number of them have been characterized, many more remain to be explored. Thus, our present state of knowledge enables us to recognize the extreme importance of these reactions while operating in relative ignorance of their full implications (*42*).

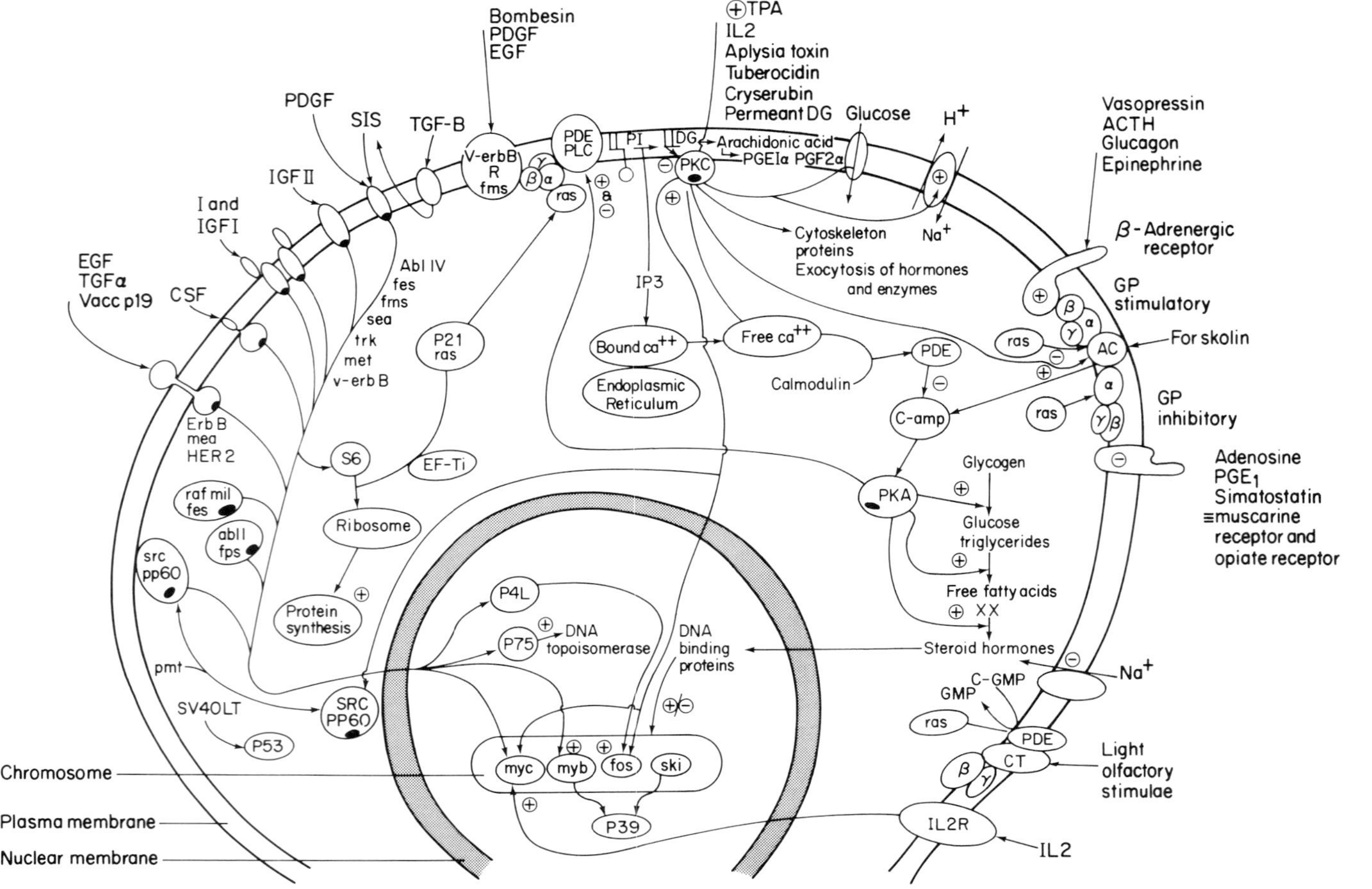

Fig. 1. Schematic outline of growth factor interactions.

3.2.1. The Components and the Oncogenes

Oncogenes are those genes whose products can be implicated in the control of the growth and multiplication of cells (*onkos*—Greek, mass or bulk). There are two forms of such genes: those called protooncogenes, which reside within the chromatin of all "normal" animal cells; and genes which can be almost homologous (except in as few as 1–3 bases) to cellular genes but which are truncated. They have significant areas of homology with the cellular genes but which are transported between cells by retroviruses. The former genes are called c-onc genes while the latter are called v-onc genes (*7*).

The c-onc genes such as C-*erbB* which codes for the receptor for the epidermal growth factor, c-*src*, and others are active during normal embryo development (*24, 40*). This enables embryonic cells to express division times of as little as 20–30 min (*11*). Mutant forms or improper expression of such genes, e.g. c-*ras* and c-*myc* have recently been shown to be involved in as many as 10–20% of tumours isolated from human malignancies (*12, 43*). In the case of the epidermal growth factor receptor, the normal c-*erbB* gene codes for a protein tyrosine kinase enzyme, which is active when stimulated by an extracellular ligand, the epidermal growth factor. However, the viral counterpart of this gene v-*erbB* lacks the agonist recognition part of the receptor and is active in the absence of ligand (*42, 6*). Similarly the c-*ras* proteins can exist in two forms, one of which is normally active, and alternative forms, which may differ by as little as a single amino acid, have over 10 times the activity (*32*). Other products of oncogenes, such as the *src* product pp60, can be associated with the plasma membrane by its acylation with either palmytic or myristic acid, yet may also be associated with the nucleus (*40, 4*) as indeed can the other protein tyrosine kinase c-*abl*. In addition to the membrane-bound protein tyrosine kinase enzymes, a number of similar enzymes are present in

---

Fig. 1

AC	Adenylate cyclase	PDE	Phosphodiesterase
ACTH	Adrenocorticotrophic hormone	PDGF	Platelet-derived growth factor
CSF	Colony-stimulating factor	PGE	Prostaglandin
CT	Cellular transducin	PI	Phosphatidyl inositol
DG	Diacylglycerol	PKA	Protein kinase A
EF	Elongation factor	PKC	Protein kinase C
EGF	Epidermal growth factor	PLC	Phospholipase C
GP	G-protein	PMT	Polyoma virus middle T antigen
I	Insulin	SV40LT	Simian virus 40 large T antigen
IGF	Insulin-like growth factor	TPA	12-*o*-tetradecanoyl-phorbol-1,3-acetate
IL	Interleukin	TGF	Transforming growth factor
IP3	Inositol triphosphate	VACC	Vaccinia virus protein

the cytoplasm. Two sets can be identified (*abl* I and *fps*) while a second set (*raf*, *mil* and *fos*) phosphorylates serine or threonine residues of proteins. Neither of these two sets of molecules has a clear position in the cascade of events which results in cellular reproduction.

Some 30 oncogenes have been discovered in association with the retroviruses (*46*) and most of those have been found to have cellular counterparts. For the most part, the molecular weights of their products, and often the amino acid sequence of these materials, have been determined, as have their putative locations in the cell. What remains to be achieved is a clear understanding of the relationship between these genes, the proteins they express, and the way in which cell growth control is affected by such materials.

3.2.2. The Growth Factor Receptors

About six receptors, different yet related by having domains of structural homology, have been identified for growth factor interactions (*39*). Each receptor traverses the plasma membrane and on the cytosolic side expresses the ability to phosphorylate the tyrosine residues of a variety of protein substrates when stimulated by the agonist growth factor (*22*). The extracellular portion varies, so that unique receptors can be identified for (i) an epidermal growth factor (EGF) group which includes transforming GF-α and the growth-promoting protein from the vaccinia virus (p19); (ii) insulin and the insulin-like GF-I; (iii) insulin-like GF-II; (iv) platelet derived GF; (v) a receptor which responds to colony stimulating factor (CSF) (*19*).

Other receptors are also available. These enable such growth stimulators as the interleukins and transforming growth factor-β to interact with the cells. Neither of these growth factor receptors is directly involved in protein phosphorylation; the former can activate the enzyme protein kinase C (see below) or promote the expression of the nuclear oncogenes c-*myc* and c-*myb* (*21*); the latter stimulates the production of the gene which produces platelet-derived growth factor. Hormones such as vasopressin epinephrine, glucagon, and ACTH also activate receptor molecules and the consequences of this action is to promote an enzyme cascade resulting in the enhanced activity of a protein-phosphorylating enzyme, protein kinase A. Other hormones and chemicals such as somatostatin, prostaglandin E1 and adenosine working via a different receptor are able to inhibit the aforementioned cascade (*9*). Another set of receptor molecules has been discovered which are sensitive to light or which respond to olfactory stimulating materials (*44*).

While the receptors considered so far require a ligand or agonist of biological origin for their activation, chemicals may be synthesized which can by-pass the receptor–ligand complex and which can stimulate directly the next immediate or subsequent links in the chain of interactions which marks the response of a cell to an external stimulus.

3.2.3. *The Consequences of Growth Factor Receptor Activation*

There is little doubt that, following appropriate receptor stimulation, a number of proteins are found to be newly phosphorylated (*23*). It is also held that, once activated, the receptors may cluster together and be endocytosed. In such an event, the protein tyrosine kinase enzyme would be moved into the body of the cell; this would engender more widespread effects than those which could be effected solely at the cell surface. Two effects, however, seem to predominate. The first is that a protein (p42) becomes phosphorylated; although the consequences of this have not been demonstrated, it is conjectured that it could be part of a chain of events which causes the *fos* oncogenes to be transcribed. It has also been reported that a p75 protein becomes phosphorylated and in turn influences a DNA toposomerase with consequences for cell growth (*18*); or it may be involved in the G-protein system (*44*)—see below. This may operate in normal cells, yet in cells which have viral protein tyrosine kinase coding genes, p42 cannot be detected. Alternatively, it has been observed that one of the protein components of the ribosome, S6, becomes phosphorylated when cells are stimulated by growth factors. This change could lead to increases in the rate of protein synthesis and so promote cell division (*45*). It can also be conjectured that as the enzyme becomes activated at the site of the plasma membrane, the cellular proteins prevalent in this area (those associated with the cytoskeleton, in particular) could become phosphorylated and thereby wholly change the physiology of the cell. An alternative speculation involves modifying the glucose transport system, with similar consequences. Much work remains to be done to elucidate the way in which a growth factor causes growth via a protein tyrosine kinase enzymic activity.

3.3. The Protein Kinase C System *(49)*

Protein kinase C is an enzyme which phosphorylates serine and threonine residues. It is a relatively short-lived protein whose half-life is measured in minutes (*27*). This protein occupies a pivotal role in the regulation of the response of the cell to external stimuli. The enzyme is activated following a series of reactions during which an amplification of the initial signal is effected. The reactions involve the activation of a receptor by an agonist, followed by the activation of a second and ubiquitous group of proteins called the G-proteins which in turn stimulate a membrane-bound enzyme to hydrolyse the local phosphatidyl inositol molecules with the production of diacyl glycerol and inositol-1,4,5-triphosphate. The former material in conjunction with the free calcium ions liberated by the latter activates the protein kinase C enzyme with profound and far-reaching consequences.

3.3.1. Receptors Involved in the Protein Kinase C Activation

When cells are exposed to platelet-derived growth factor, or bombesin, an increase in the turnover of the inositol phospholipids of the plasma membrane is observed (*25*). Also, when cells express the truncated v-*erbB* gene or the receptor which responds to colony-stimulating factor (c-*fms*) a similar enhanced turnover results. Such observations indicate that in addition to the protein tyrosine kinase activity which the activated receptor proteins are capable of engendering, they are also capable of participating in the series of reactions which leads to the activation of protein kinase C.

3.3.2. The G-protein Amplification

Activation of the G-protein cluster by a receptor–agonist complex results in the liberation from the group of three proteins (α, β, γ) the portion designated α being activated by having a molecule of GTP in place of a GDP molecule attached to its substance. This exchange of GTP for GDP occurs following the interaction with an activated receptor complex. In this activated and mobile form the α-GTP protein is capable of activating a number of membrane-associated enzyme molecules before it becomes subjected to degradation and down-regulation by an internally catalysed self-hydrolysis which convert it to its α-GDP form. In this state it rejoins the complex which has been left behind and reverts to its quiescent form (*44*).

The expression of the cellular oncogene c-*ras* and the viral oncogene v-*ras* results in the production of a p21 protein which is also capable of binding GTP but which differs from the α-protein–GTP complex in that it is less susceptible to hydrolysis by the phosphatase enzymes. This protein is thought to be able to substitute for the requirement for the activated receptor stimulation of the G-protein system and causes the activation of the membrane-bound enzymes directly. Clearly, the degree of this activation has been shown to be determined by (i) the degree of expression of the c-*ras* gene and (ii) the quality of the c-*ras* product, as mutants of this gene with a single base change yield a p21 protein whose GTP ligand is 10 times more long-lived. The effect of the *ras* protein can be eliminated by the intracellular injection of monoclonal antibodies to the p21 protein. This *ras*-based excitation system by-passes the requirement for the activated receptor stimulation which enables the cell to control its own rate of growth when suitably provoked.

Not only are the *ras* genes deeply implicated in growth control, but different *ras* genes are transcribed in different cell types (*32*). It has also been shown that 90% of the tumours initiated by exogenous carcinogens such as dimethylbenzanthracene (DMBA) and promoted by 12-*o*-tetradecanoylphorbol-13 acetate (TPA) have a mutated *ras* gene associated with them (*43*). It would appear that, as the *ras* gene stands out in its susceptibility to chemically induced point mutations, it would be an ideal candidate for a target for the deliberate manipulation of cell physiology for practical ends as opposed to developing a further understanding of the courses of tumorigenesis.

3.3.3. The Membrane Enzyme: Phospholipase C

The system which begins with the growth factor activation of a receptor and leads to the activation of protein kinase C activation works via an enzyme called phospholipase C (PLC) (or phosphodiesterase or polyphosphoinositide phosphodiesterase). This enzyme hydrolyses a phospholipid called phosphotidylinositol-4,5-biphosphate (PIP_2) to two molecules, diacyl glycerol (DG) and inositol-1,4,5-tris-phosphate (IP_3) (*44, 5*). The former can be further hydrolysed to arachidonic acid (which itself can be converted to two potent agonists promoting cell growth, namely prostaglandins E_1 and F_2 (PGE_1 and PGF_2)) and stearic acid. The mechanism of enzyme activation has not been determined, but it becomes switched on for a relatively short period of time (seconds). Furthermore, it would appear that numerous such enzyme molecules are so activated. Of the resultant molecules, the IP_3 diffuses into the cytosol, whereas the DG exerts its effect by becoming attached to the enzyme protein kinase C.

The consequences of the release of IP_3 from the plasma membrane reverberate throughout the cell. This molecule causes calcium ions bound in an inactive form within the vesicles of the endoplasmic reticulum to be released as free calcium ions into the cytosol. This increase in free calcium ion concentration enables calcium ions to become attached to the protein kinase C enzyme, whose full activation is effected by the additional association of DG molecules. The liberated calcium is also sequestered by a ubiquitous cytoplasmic protein, calmodulin, which, when thus activated is capable of activating the phosphodiesterase enzyme which can convert cyclic AMP to AMP.

3.3.4. Protein Kinase C

Much attention has been devoted to this enzyme since its discovery in 1977 and its implication in the phosphatidyl inositol system in 1984 (*49*). It is a short-lived enzyme associated with the inner surface of the plasma membrane which catalyses the addition of phosphate groups to the serine or threonine residues of susceptible proteins. Its role in such phosphorylations encapsulates a number of cell types (blood cells, endocrine cells, exocrine systems, nervous, muscle and other systems), and when active it provokes such cell types and systems to release various products held in storage vesicles into the extracellular medium (*27*). Since 1983, over 60 papers have been published describing such stimulated excretions. Clearly, as many of the materials excreted are hormones, there will be extensive secondary effects on the cell which liberates materials which could well re-stimulate it to further excretory activity. This self-stimulation is limited, as the PKC also phosphorylate the growth factor receptors, the result of which is to decrease their affinities for the exogenous ligand so that, when so phosphorylated, they resist activation. Additionally, the enzyme phosphorylates proteins which function as receptors, membrane proteins such as the glucose transporter, the Na^+/H^+ antiport the sodium channel APTase. In so doing the effects could be to increase transfer of glucose into the cell for a more active metabolic disposition, an export of hydrogen ions resulting in an increase in the alkalinity of the cytoplasm—a feature which is held to promote DNA synthesis. Further, PKC phosphorylates the proteins of the cytoskeleton. One of the modifications associated with changes in the cell consequent upon adopting a more rapid rate of cell division is a realignment of the cytoskeleton. Were these changes to be induced by PKC, it would then be responsible for initiating the chain of events leading to more rapid growth rates. Such a situation is supported by observations on the phosphorylation of vinculin, which is held to bind the microfilbrils to transmembrane proteins involved in attachment. Were such associations broken, the intensity of the attachment of a cell to a surface would be weakened, thus enabling a less trammelled existence. The increased mobility of plasma membrane-localized molecules would also be promoted by a dissolution of the structures imposed on the cell by the network of intercalating fibres and tubules which make up the cytoskeleton.

There are also a growing number of reports on the effect of PKC on a wide range of enzymes, initiation factors, and other proteins implicated at critical points of the cell's metabolism (*27*).

What makes these 'effector' observations of great importance to the animal cell biotechnologist is that the enzyme is susceptible to the modulation of its activity by exogenous agents. Such agents are generally known as tumour promoters and range in composition from TPA, aplysiatoxin, tuberocidin, cryserubin and molecules which resemble DG but which are capable of permeating the plasma membrane (*41*). Interleukin-2 is also capable of stimulating PKC. The use of such materials during the cell growth phase could produce some interesting effects.

3.4. The Adenylate Cyclase System

An equivalent system to the PKC system of Section 3.2 above can be found in the adenylate cyclase system. The activation of this system leads either to the promotion or decrease of a different protein kinase, protein kinase A

(PKA). The components of the system are much the same as the PKC system, in consisting of at least two groups of receptors whose activation stimulates the formation of a G-protein GTP moiety. However, in this system there exists two such G-proteins, one of which is activated by activated receptors for such hormones as vasopressin, epinephrine, ACTH and glucagon and whose activation leads to an increase in the activity of a membrane associated adenylate cyclase enzyme. The second component of this system derives from a second set of activated receptor molecules which, when they stimulate the particular G-protein system which is in association with them, causes an activation of that system such that the resulting interaction with the adenylate cyclase enzyme leads to a decrease in the level of activity of that enzyme. Hormones whose activated receptors result in such a decrease of enzymic activity are adenosine, PGE, and somatostatin. Other inhibitory systems are those associated with the muscarine and opiate receptor–ligand complexes.

The product of the positive activation of adenylate cyclase (AC) is the production of the mobile molecule cyclic AMP (c-AMP). This compound causes the dissociation of a tetrameric protein which is the inactive form of PKA. Once this dissociation has been achieved, the phosphorylation of the serine group of a variety of metabolically critical enzymes effects major changes in the flow of materials through the pathways of intermediary metabolism. Such changes result from an inhibition of glycogen biosynthesis and an increase in the production of glucose from that material. Further changes include the stimulation of the enzymes which hydrolyse triglycerides to free fatty acids; a second route whereby cellular activity can be stimulated; and, thirdly, PKA promotes the biosynthesis of a number of steroid hormones which can in their turn traverse the nuclear membrane and in association with soluble steroid binding proteins convert those latter proteins to complexes which can bind to DNA and so effect a change in the transcription of the genetic material.

This concatenation of reactions is stayed by the hydrolysis of c-AMP to AMP by a calmodium–calcium complex stimulated phosphodiesterase. From a biotechnologist's view, there are fewer opportunities to deliberately modulate this cascade of reactions. Except for forskolin (*28*), there are few exogenous chemicals which can directly stimulate or inhibit AC.

3.5. Other Receptor–Effector Systems

Proteins equivalent to those found in the G-protein system can also be involved in the cell's response to light and chemicals which cause an olfactory sensation (*44*). Such proteins, known as transducin in vertebrate photoreceptors, have analogies in some invertebrates.

It would seem, therefore, that the systems described above for signal transduction and amplification prevail in organisms as diverse as yeast cells (*16*) and man, and in both cases they are involved in the control of cell proliferation.

3.6. Events in the Nucleus

Four oncogenes can code for proteins which are active in the control of gene expression in the nucleus. They are *myc, myb, fos,* and *ski*. The role of such genes in neoplastic transformation has been reviewed recently (*30*) and their involvement in growth control has been summarized (*8*). When serum-starved cells are provided with serum, the mRNA derived by the transcription of c-*fos* increases 50-fold in 30 minutes, whereas that of c-*myc* increases 20-fold 1 hour after stimulation and declines over a subsequent 18-hour period. This effect can be obtained by substituting purified growth factors (EGF, FGF, IL-2 or PDGF) for the serum. However, the induction of the expression of these two genes does not *per se* lead to an increase in cell proliferation. Yet, once expressed, the cells achieve an increased level of "competence", so that were there a further stimulus to growth within a 16-hour period, growth would ensue.

In addition to growth factor stimulation of the expression of such genes, TPA and hence PKC are also capable of eliciting enhanced transcription of c-*myc* (*15*).

The proteinaceous products of c-*myc* and c-*myb* expression remain in the nucleus and both the c-*myc* and c-*myb* proteins have been shown to be capable of binding to DNA (*35*), in which state c-*myc* has been shown to promote gene expression in a test system while a specific function for c-*myb* has not yet been ascribed. The kinetics of the expression of these two oncogenes lead one to believe that they may have different functions.

The role of the *fos* gene product is likewise obscure. While the kinetics of the expression of this gene depend on the origin of the gene (viral or cellular), the cell in which it is expressed, and the levels at which it is present, its necessary involvement in the events resulting in cell transformation and proliferation is not challenged (*25*).

It is held that the product of the c-*myc* gene can be supplanted by either the adenovirus protein E1A or the polyoma virus protein large T antigen (*17, 38*).

The molecular biology of the nuclear oncogenes can be manipulated by the insertion into the cell of such genes using the retrovirus vector systems. Such a system may transpose modified genes or it can cause the enhanced expression of endogenous genes by enhancer or promoter insertion mechanisms. Indirect methods of causing gene expression are via the activated receptor–PKC pathways as described. There do not as yet seem to be any defined molecules such as steroids which evoke specific gene expression. Other proteins, such as p53 (*37*), appear when cells are proliferating rapidly and the effect of their presence is extended when they associate with proteins such as the SV40 large T antigen (*31, 17*); the biotechnological control of this largely uncharted system requires the determination of more of its characteristics and the provision of readily modulatable manipulative capability at key control points.

Whereas most of the studies reported above have been instigated by the need to understand and hence control cancer in humans, their further use to the animal cell biotechnologist requires that the phenomenon of cell transformation, often alluded to above, be further discussed.

4. THE TRANSFORMATION OF ANIMAL CELLS IN CULTURE

The word "transformation" is used to describe a process which has many and different repercussions. For many purposes it is necessary to be precise about the way in which a cell culture is described, for this often determines the extent to which it can be used to manufacture prophylactic or therapeutic biologicals. As it stands, the word means only that the cells are "changed" in some way. Table III lists 19 ways in which such changes may be recognized: there could well be more. The situation is made more complicated by some of the changes being reversible while others seem to be stable. Also, some of the requirements of animal cells for growth factors can be obscured when they are grown in dense cultures with cell concentrations in excess of 10^8 cells ml^{-1} (*14*). In any event, much of the need to be clear about the degree of transformation of a cell culture may be obviated when the regulatory agencies accept the principles propounded by various expert committees (see Petricciani, Chapter 2 of this volume) and either accept or reject products derived from animal cells in culture based on the absolute amount of DNA which is left residual in a dose of the final product. In this sense, materials made from cells which are known to have the potential for oncogenesis, such as the Namalwa cell (an Epstein–Barr-transformed α-interferon-secreting lymphocyte) and a hybridoma secreting a therapeutically useful antibody, have been licensed for use in clinical trials in the U.S.A. recently.

While the above considerations enable products to be licensed, following the necessary and extensive downstream processing, the biotechnologist still needs to be able both to grow cells and to obtain products from such grown cells. In this sense, there is a need to know the *extent* to which a cell is transformed and the relationship between the degree of transformation and

TABLE III Some Properties of "Transformed" Cells

Changed cell shape
Decreased requirement for substratum
Excretion of proteases
Agglutination by plant lectins
Loss of cell markers (MHC)
Surface materials are more mobile
Faster rates of cell growth and division
Decrease in the contact inhibition of motion
Increased glucose consumption
Form foci in adherent cultures
Growth in soft agar from single cells:
—when fortified with growth promoters;
—in the absence of growth promoters
Decreased requirements for growth promoters
Vinculin is phosphorylated
Delocalization of microfibres below plasma membrane
Decreased production of fibronectin, laminen, filamin
Changes in karyology:
—number of chromosomes;
—character of chromosome
Capacity for tumorigenicity in "nude" animals
Immortalization

the generation of secreted or other types (e.g. lytic) products. Indeed, there may well be a need to grow large quantities of cells as rapidly, cheaply and efficiently as possible in the first instance and then to be able to follow this biomass generation phase with an extended production phase in which the cells do not proliferate and in which they comport their metabolism about those metabolic processes which lead to the most efficient biotransformation of the provided raw material to the finished secreted products. This ideal facility may be realizable if the information and understandings of the process of the transformation of cells in culture are incorporated into the operating strategies of the technologist. However, before such a system can be achieved it is necessary to know how the various growth factors and the products of the different oncogenes interact to bring about the transformed state. There is yet some way to go before the complete picture can be filled in, but information is available, albeit in a confused and disorganized form.

4.1. Growth Factors and Oncogenes in the Transformation of Animal Cells in Culture

Most authors agree that the process of transformation is a multistage, multicomponent set of events. This is well recognized in the chemical carcinogenesis area, where tumour initiators have generally to be followed by tumour promoters before there ensues a neoplastic malignant transformation (i.e. unlimited *in vivo* cell growth). Of course, multiple applications of high concentrations of initiators may, under certain circumstances, with particular cells effect the required transformation without the need of the promoters (*1*). This cytoplasmic stage can be based on the presence of excess *ras* gene product or the presence of a *ras* gene product which is mutated so as to increase the life of the active form of the GTP-ligated protein. Alternatively, the *src* gene product, particularly in association with the polyoma middle T antigen, can provide the environment for a fully consummated transformed state.

The partners of the Stage I transformation system can be found amongst the *myc* and *myb* gene products, from the expressions of the p53 protein in conjunction with the SV40 large T antigen, or the involvement, not necessarily in the nucleus, of the adenovirus E1A or polyoma virus middle T antigen proteins. Various combinations of these materials lead to tumorigenesis. This model of transformation is so successful that the analysis of human tumours has shown that some 10–20% of them have evidence of elevated levels of both the *ras* and *myc* gene products.

There are, however, a number of secondary effects which control the emergence of a particular cell type. The effect of PKC on adenylate kinase can be stimulatory or inhibitory, while the effect of PKA on PLC can likewise lead to stimulation or inhibition (*44*). The effect of phosphorylating a number of the receptor molecules can lead to a down-regulation of their activating abilities. Therefore, the control of the consequence of a particular exogenous stimulus is complex and may not always achieve the desired, conjectured, or intended ends. However, there are a number of pointers to the existence of several biomolecules which can prevent the promulgation of the transformed state when all the other factors have been lined up to promote that state. It has been shown, for instance, that when a transformed cell has been engendered in a cell culture of normal cells, the normal cells are capable of suppressing the manifestations of the transformed state (*47*). Were the cells around the transformed cell to be killed selectively, then the transformed cell would proliferate. Similarly, when transformed cells are positioned on top of a feeder layer of normal cells, the evolution of the transformed state is impeded. Also, normal cell-transformed cell hybrids tend to revert to the normal cell phenotype.

The observations described above indicate that, if it is not already so, it will be possible to provoke cells to proliferate extensively on media devoid of costly growth factors and devoid of unwanted contaminating proteins, and then, perhaps, for the production phase, provide the cells with those biomolecules which can "turn-off" the proliferation activity so that the cells can be made to focus on product generation.

4.2. Cell Differentiation

While the discussion of the preceding section has indicated how the control of cell proliferation, and possibly the degree of transformation, may be controlled by deft manipulations of exogenous components, the deliberate induction of the secretion of a particular biomolecule is yet some way ahead. However, some progress in defining the relationship between particular proteins of the colony-stimulating-factor type and the production of mature colonies of differentiated cells has been shown to be possible for some blood-cell lineages (*16*). Also, it is possible to influence differentiation by the use of such materials as TPA, retinoic acid, dibutyl c-AMP, dimethylsulphoxide, and these materials in different combinations, but methods whereby such knowledge may be usefully applied has not yet been transferred to the domain of the manufacturer of biologicals.

5. CONCLUSIONS

This review of oncogenes and growth factors has been designed to provide a framework for further developments in the ability of animal cell biotechnologists to control the growth of cells in their cultures and to derive, from those grown cells, product generated with the maximum efficiency or lowest cost per unit material. The use of the complex and poorly understood whole serum as a component of cell culture is probably in decline. It will not be eliminated entirely, because existing processes which are licensed will not be changed from their existing protocols. Rather, we can expect to see the deliberate manipulation of the cells with those oncogenes which are most suitable for the product line required and which enable efficienct cell proliferation in cheap, easy-to-standardize media. Also, the newly engineered cells (for both growth and product-oriented genes) will be provided with media tailored appropriately for the particular stage of the process which is in hand.

The years of the first half of the 1980s have been marked by an explosion of information and understanding about the nature and role of oncogenes in animal cells. It would well be that, augmented with an equal or larger body of additional knowledge in this area, the second half of the decade may be noted for the productive application of this new understanding.

REFERENCES

1. Balmain, A. (1986). Molecular events associated with tumour initiation, promotion, and progression in mouse skin. *In* "Oncogenes and Growth Control" (P. Kahn and T. Graf, eds.), pp. 327–331. Springer-Verlag, Berlin.

2. Barnes, D. (1984). Attachment factors in cell culture. *In* "Mammalian Cell Culture" (J. P. Mather, ed.), pp. 195–226. Plenum, New York.
3. Barteling, S. J. (1974). Use of a polyethylene glycol-treated serum for production of foot-and-mouth disease virus (FMDV) in growing BHK-suspended cell culture. *Bull. Off. Int. Epiz.* **81** (11–12), 1243–1254.
4. Ben-Neriah, Y., and Baltimore, D. (1986). Normal and transforming *N*-terminal variants of c-*abl*. *In* "Oncogenes and Growth Control" (P. Kahn and T. Graf, eds.), pp. 106–114. Springer-Verlag, Berlin.
5. Berridge, M. J. (1986). Inositol lipids and cell proliferation. *In* "Oncogenes and Growth Control" (P. Kahn and T. Graf, eds.), pp. 147–153. Springer-Verlag, Berlin.
6. Beug, H., Hayman, M. J., and Vennstrom, B. (1986). Mutational analysis of v-*erbB* oncogene function. *In* "Oncogenes and Growth Control" (P. Kahn and T. Graf, eds.), pp. 85–92. Springer-Verlag, Berlin.
7. Bishop, J. M. (1983). Cellular oncogenes and retroviruses. *Annu. Rev. Biochem.* **52**, 301–354.
8. Bravo, R., and Muller, R. (1986). Involvement of proto-oncogenes in growth control: The induction of c-*fos* and c-*myc* by growth factors. *In* "Oncogenes and Growth Control" (P. Kahn and T. Graf, eds.), pp. 253–258. Springer-Verlag, Berlin.
9. Darnell, J., Lodish, H., and Baltimore, D. (1986). "Molecular Cell Biology," p. 683. Scientific American Books.
10. Darnell, J., Lodish, H., and Baltimore, D. (1986). "Molecular Cell Biology," p. 1044. Scientific American Books.
11. Darnell, J., Lodish, H., and Baltimore, D. (1986). "Molecular Cell Biology," p. 1039. Scientific American Books.
12. Darnell, J., Lodish, H., and Baltimore, D. (1986). "Molecular Cell Biology," p. 1066. Scientific American Books.
13. *Documenta Geigy* (1970). 7th edn. Adapted from table of plasma proteins. Scientific Table, Ciba-Geigy, Basle, 1970.
14. Eagle, H. (1965). Metabolic controls in cultured mammalian cells. *Science* **148**, 42–51.
15. Fahrlander, P. D., and Marcu, K. B. (1986). Regulation of c-*myc* expression in normal and transformed mammalian cells. *In*"Oncogenes and Growth Control" (P. Kahn and T. Graf, eds.), pp. 264–270. Springer-Verlag, Berlin.
16. Fasano, O. (1986). RAS genes and growth control in the yeast *Saccharomyces cerevisiae*. *In* "Oncogenes and Growth Control" (P. Kahn and T. Graf, eds.), pp. 200–208. Springer-Verlag, Berlin.
17. Fried, M., and Prives, C. (1986). The biology of simian virus 40 and polyomavirus. *In* "Cancer Cells 4: DNA Tumor Viruses," pp. 1–16. Cold Spring Harbor Laboratory, New York.
18. Goldberg, A. R., Wang, T. W., and Tse-Dinh, Y-C. (1985). Properties of the major species of tyrosine protein kinase in rat liver: Effects on DNA topoisomerase activity. *In* "Cancer Cells 3: Growth Factors and Transformation," pp. 369–379. Cold Spring Harbor Laboratory, New York.
19. Gough, N. M. (1986). The granulocyte–macrophage colony-stimulating factors. *In* "Oncogenes and Growth Control" (P. Kahn and T. Graf, eds.), pp. 35–42. Springer-Verlag, Berlin.
20. Handa, A. (1986). Gas–liquid interfacial effects on the growth of hybridomas and other suspended mammalian cells. Ph.D Thesis, Birmingham University.
21. Hatakeyama, M., Minamoto, S., Mori, H., and Taniguchi, T. (1986). Structure and function of the human interleukin-2 receptor. *In* "Oncogenes and Growth Control" (P. Kahn and T. Graf, eds.), pp. 128–134. Springer-Verlag, Berlin.
22. Hunter, T., and Cooper, J. A. (1985). Protein–tyrosine kinases. *Annu. Rev. Biochem.* **54**, 897–930.

23. Hunter, T., Alexander, C. B., and Cooper, J. A. (1985). Protein phosphorylation and growth control. *In* "Growth Factors in Biology and Medicine," Ciba Foundation Symposium 116, pp. 188–198. Pitman, London.
24. Jakobovits, A. Y. A. (1986). The expression of growth factors and growth factor receptors during mouse embryogenesis. *In* "Oncogenes and Growth Control" (P. Kahn and T. Graf, eds.), pp. 9–17. Springer-Verlag, Berlin.
25. Jenuwein, T., and Muller, R. (1986). The *fos* oncogene and transformation, *In* "Oncogenes and Growth Control" (P. Kahn and T. Graf, eds.), pp. 278–283. Springer-Verlag, Berlin.
26. Keay, L. (1976). Autoclavable low-cost serum-free cell culture media: The growth of established cell lines and production of viruses. *Biotechnol. Bioeng.*, **18**, 363–382.
27. Kikkawa, R., and Mishizuka, Y. (1986). The role of protein kinase in transmembrane signalling. *Annu. Rev. Cell Biol.* **2**, 149–178.
28. Kliner, S. L., and Carlsen, R. C. (1984). Forskolin activation of adenylate cyclase *in vivo* stimulate nerve regeneration. *Nature (London)* **307**, 405.
29. Lambert, K. J., and Birch, J. R. (1985). Cell growth media. *In* "Animal Cell Biotechnology," Vol. 1 (R. E. Spier and J. B. Griffiths, eds.), pp. 85–122. Academic Press, London.
30. Land, H. (1986). Oncogenes cooperate, but how? *In* "Oncogenes and Growth Control" (P. Kahn and T. Graf, eds.), pp. 304–311. Springer-Verlag, Berlin.
31. Lane, D. P., and Gannon, J. (1986). Monoclonal antibody analysis of the SV40 large T antigen–p53 complex. *In* "Cancer Cells 4: DNA Tumor Viruses," pp. 387–393. Cold Spring Harbor Laboratory, New York.
32. Marshall, C. J. (1986). The *ras* gene family. *In* "Oncogenes and Growth Control" (P. Kahn and T. Graf, eds.), pp. 192–199. Springer-Verlag, Berlin.
33. Maurer, H. R. (1986). Towards a chemically-defined, serum-free media for mammalian cell culture. *In* "Animal Cell Culture—A Practical Approach" (R. I. Freshney, ed.), pp. 13–32. I.R.Z. Press, Oxford.
34. Mizrahi, A., and Avihoo, A. (1976). A simple and cheap medium for BHK cells in suspended cultures. *J. Biol. Stand.* **4**, 51–56.
35. Moelling, K. (1986). Properties of the *myc* and *myb* gene products. *In* "Oncogenes and Growth Control" (P. Kahn and T. Graf, eds.), pp. 271–277. Springer-Verlag, Berlin.
36. Morgan, M. J., and Falk, P. (1986). The utilisation of carbohydrate by animal cells: An approach to their biochemical genetics. *In* "Carbohydrate Metabolism in Cultured Cells" (M. J. Morgan, ed.). Plenum, New York.
37. Oren, M. (1986). P53: Molecular properties and biological activities. *In* "Oncogenes and Growth Control" (P. Kahn and T. Graf, eds.), pp. 284–289. Springer-Verlag, Berlin.
38. Pettersson, V., and Roberts, R. J. (1986). Adenovirus gene expression and replication: A historical review. *In* "Cancer Cells 4: DNA Tumor Viruses," pp. 37–57. Cold Spring Harbor Laboratory, New York.
39. Pfeffer, S., and Ullrich, A. (1986). Structural relationships between growth factor precursors and cell surface receptors. *In* "Oncogenes and Growth Control" (P. Kahn and T. Graf, eds.), pp. 70–76. Springer-Verlag, Berlin.
40. Rohrschneider, H. R. (1986). Tissue-specific expression and possible functions of pp60$^{c\text{-}src}$. *In* "Oncogenes and Growth Control" (P. Kahn and T. Graf, eds.), pp. 27–31. Springer-Verlag, Berlin.
41. Rosner, M. R., McCaffrey, P. G., Friedman, B., and Faulker, J. G. (1985). Modulation of growth factor action by tumour promoters and C-kinase. *In* "Cancer Cells 3: Growth Factors and Transformation," pp. 347–351. Cold Spring Harbor Laboratory, New York.
42. Schlessinger, J. (1986). Regulation of cell growth by the EFG receptor. *In* "Oncogenes and Growth Control" (P. Kahn and T. Graf, eds.), pp. 77–84. Springer-Verlag, Berlin.
43. Schwab, M. (1986). Amplification of proto-oncogenes and tumor progression. *In* "On-

cogenes and Growth Control" (P. Kahn and T. Graf, eds.), pp. 332–339. Springer-Verlag, Berlin.
44. Stryer, L., and Bourne, H. R. (1986). G proteins: A family of signal transducers. *Annu. Rev. Cell Biol.* **2**, 391–419.
45. Thomas, G. (1986). Epidermal growth-factor mediation of S6 phosphorylation during the mitogenic response: A novel S6 kinase. *In* "Oncogenes and Growth Control" (P. Kahn and T. Graf, eds.), pp. 177–183. Springer-Verlag, Berlin.
46. Washley, G. D., and May, J. W. (1970). "Animal Cell Culture Methods." Blackwell Scientific Publications, Oxford.
47. Wyke, J. A., and Green, R. (1986). Suppression of the neoplastic phenotype. *In* "Oncogenes and Growth Control" (P. Kahn and T. Graf, eds.), pp. 340–345. Springer-Verlag, Berlin.
48. Wagner, E. F., and Muller, R. (1986). A role for proto-oncogenes in differentiation? *In* "Oncogenes and Growth Control" (P. Kahn and T. Graf, eds.), pp. 18–26. Springer-Verlag, Berlin.
49. Nishizuka, Y. (1984). The role of protein kinase C in cell surface signal transduction and tumour promotion. *Nature (London)* **308**, 693–698.

4

Immobilized Hybridomas: Oxygen Diffusion

A. D. MURDIN[1*], N. F. KIRKBY[2], R. WILSON[1*], and R. E. SPIER[1]
[1]Department of Microbiology, University of Surrey, Guildford, U.K.

[2]Department of Chemical and Process Engineering, University of Surrey, Guildford, U.K.

* Present address: see list of contributors, page **xiii**.

ANIMAL CELL BIOTECHNOLOGY VOL. 3
ISBN 0-12-657553 3

TERMS

c concentration (μg ml^{-1})
c_m concentration in surrounding medium (μg ml^{-1})
D diffusion coefficient (cm^2 s^{-1})
R respiration rate (μg ml^{-1} s^{-1})
r distance (cm)
r_c radius of a capillary (cm)
r_0 distance at which the oxygen concentration falls to zero (cm)
r_s radius of a sphere (cm)

1. INTRODUCTION

The advent of hybridoma culture on a large scale has given considerable impetus to research concerning the immobilization of anchorage-independent animal cells. The use of immobilized cells solves a number of problems associated with hybridoma culture. For instance, immobilized cells are protected from a potentially stressful environment and may be cultured at high concentrations for extended periods of time. However, there are also problems associated with immobilization. One problem of major importance is that of mass transfer within aggregates of immobilized cells. This chapter considers some of the factors determining mass transfer within cell aggregates and presents mathematical models describing oxygen distribution in flat plates, cylinders, and spheres containing immobilized cells.

2. IMMOBILIZATION

Many methods have been employed for the immobilization of animal cells. This section is not intended to provide anything other than a brief description of immobilization methods currently employed for the culture of hybridomas. For further background information on this subject the reader is directed to recent reviews by Scott (*35*), Lambe and Walker (*17*), Katinger and Scheirer (*14*), and Griffiths (*8*).

Many hybridomas can be cultured using conventional fermenter technology, such as stirred-tank (*31*) or airlift fermenters (*3*). However, for a number of reasons, the conventional technology is less than ideal. Hybridomas are not robust cells; they require a stable, stress-free physical and chemical environment to achieve optimum growth and productivity (*6*). Those culture systems in which the hybridomas are maintained as a suspension of single cells in a growth medium achieve relatively low concentrations of biomass (on the order of 10^6 cells ml^{-1} in batch cultures, 10^7 cells ml^{-1} in perfused suspensions) when compared to ascites cultures (variable, up to ca. 10^8 cells ml^{-1}) or to the growth of animal cells in tissues (ca. 2×10^9 cells ml^{-1}) (*2, 13*). Suspension systems are frequently run on a batch basis, in which the cells produce antibody for only a few days before they die and in which the cells must be removed from the antibody-containing medium as the first step in downstream processing (*3*).

These factors have prompted a considerable body of work upon the immobilization of hybridomas, which is seen as a likely solution to many of the problems. The improvements sought from immobilization are the use of increased cell concentrations within the reactor, the provision of a physically and biochemically protected environment for the cells, increased efficiency in the use of biomass, and simplified downstream processing (the provision of a concentrated, cell-free, product stream). The techniques employed for immobilization can be considered as two classes, those which seek to make use of established reactor technology whilst improving the use of the cells, and those which seek to improve the use of the cells in novel reactor configurations.

Immobilization techniques which seek to make use of a more or less conventional reactor design involve entrapping the cell within some form of particle. This may be a polymeric bead, for instance of agarose (*27*) or alginate (*36*), or a membrane-bounded capsule such as the Damon Biotech system (*31*). Alternatively, this may be a porous particle such as the Verax sponge bead (*12, 40*), porous agarose beads (*28*) or particles of polyester or polyurethane foam (*18, 23*). The immobilized cells are maintained in a protected, biochemically conditioned environment. Beads and porous particles allow a cell-free product stream to be withdrawn from the reactor. The capsules retain antibody, which means that this may only be obtained by recovering and breaking the capsules, but this step does provide a concentrated product for further processing. In all of these cases, the reactors can easily be perfused with a continuous supply of fresh culture medium, extending the productive lifetime of the cells and increasing the cell concentrations obtained in the reactors.

Novel bioreactor designs which do not use particle-entrapped cells achieve immobilization by enclosing the cells within one or more specific compartments. Examples include the many variants of the hollow-fibre bioreactor (*2, 39*), in which cells are enclosed within the shell of a hollow-fibre cartridge, and the membrane bioreactor design of Klement *et al.* (*15*), in which cells are contained in a thin layer between two planar membranes. These systems, too, maintain the cells in a conditioned and protected environment. A cell-free, and in some cases concentrated, product stream can be withdrawn from the reactors and the cells can be perfused with fresh medium to increase cell concentrations and the lifetime of the culture.

Inevitably, such improvements in performance are not obtained without further problems arising. For instance, cells immobilized in beads or capsules have to be produced by conventional means prior to immobilization. The immobilization in these cases is a technically complex step which must be performed under sterile conditions (*27, 31*). The hollow-fibre and membrane bioreactors are complicated pieces of equipment compared to a stirred tank. Furthermore, their design means that scale-up may only be achieved by installing multiple copies of the basic unit, depriving the operator of any benefit from economies of scale. With a few notable exceptions (*2, 33*) there is a distinct lack of data in the literature concerning the scale-up and operation of such reactors for the culture of hybridomas. Much of the published data are available only as abstracts or in manufacturers' releases (see Altshuler *et al.* (*2*) for a summary).

A significant problem for all of these immobilized systems is that of mass transfer within the immobilized cell aggregate (*30*). As a generalization, mass transfer within the aggregate occurs primarily by diffusion (but see Section 3.1.4). The problem of mass transfer is especially acute for oxygen because of its low solubility in water (7.6 $\mu g\ ml^{-1}$ under typical conditions). A simple calculation reveals that 10^6 cells immobilized in a volume of 1 ml will exhaust the oxygen contained in that volume in from 42 min to 6 hr. This assumes respiration rates of 10.88 and 1.28 $\mu g\ 10^6$ cells hr^{-1} respectively, the upper and lower limits reported for animal cells (*37*). Further oxygen must be obtained by diffusion from the surrounding medium. In contrast, glucose is typically present at considerably higher levels in culture medium, sufficient to supply the cells for over 60 hr in the example given. The metabolic activity of hybridomas immobilized in agarose has been observed to decline substantially within 5 hr of immobilization, in a manner consistent with oxygen limitation (Murdin and Al-Rubei, unpublished observations).

3. MATHEMATICAL MODELLING

Since oxygen transfer is likely to be a major limiting factor in the culture of immobilized animal cells, the question arises of how best to predict and monitor its effects. Prediction can be achieved by modelling, and there are many accounts in the literature of the equations describing oxygen transfer (see, e.g., ref. (*21*)). This section presents mathematical models of oxygen diffusion in cylinders, spheres, and flat plates. For a comprehensive account of the mathematics concerning diffusion, see Crank (*5*). A simple method for monitoring the effects of oxygen transfer limitations upon cellular metabolic activity is described in Section 4.

Previous studies of oxygen transfer have been made in a number of systems.

Those most immediately applicable to the present problem include studies of immobilized bacteria (*1*), fungal pellets (*16, 19, 42*), animal tissues (*7*) and solid tumours (*38*), together with the theoretical study of the kinetics of immobilized cells presented by Monbouquette and Ollis (*20*). The work of Tannock (*38*) is particularly useful in the context of immobilized mammalian cells, as it considers diffusion of oxygen and glucose into tissues and the distribution of carbon dioxide and lactic acid within tissues. Monbouquette and Ollis (*20*) treat the subject of immobilization in considerably greater depth than this chapter, and avoid many of the simplifying assumptions made here.

The models presented here describe steady-state oxygen concentration profiles within cell aggregates. Although this chapter is particularly concerned with hybridomas, the models are relevant to any immobilized cells. Oxygen transfer to and within the culture medium surrounding the cell aggregates is not considered here, and oxygen transfer at the interface between the medium and the cell aggregate is dealt with only briefly. These aspects of oxygen transfer have been discussed on many occasions previously (see, e.g., refs. (*21, 30*)).

3.1. Simplifying Assumptions

The modelling of oxygen diffusion within cell aggregates can be greatly simplified by allowing a number of assumptions.

3.1.1. Respiration Rate

It is convenient to assume that cellular respiration rate is independent of oxygen concentration. This is unlikely to be true, though there is some evidence to suggest that the change in respiration rate is small over an appreciable range of oxygen concentration. Thus, in the only published study of its kind, Boraston *et al.* (*4*) found that the respiration rate of a murine hybridoma line decreased by less than 20% as the dissolved oxygen tension was reduced from 100% to 60% of saturation and did not decrease further when the dissolved oxygen tension was reduced to 8% of saturation. The authors correctly observed that this pattern might not hold for other cell lines.

The likely effect of this assumption is to reduce the accuracy of the model at low oxygen concentrations, so that the concentrations of oxygen present deeper within the cell aggregate are higher than predicted. Kobayashi *et al.* (*16*) have presented a model for the diffusion of oxygen within mycelial pellets in which respiration rate is a function of oxygen concentration.

3.1.2. Homogeneity of the Immobilized Cell Aggregate

It is usually a reasonable approximation to assume that the immobilized cell aggregate is homogeneous. However, in some circumstances mass transfer limitations on nutrients other than oxygen may lead to an inhomogeneous distribution of viable cells in immobilized aggregates. Within agarose or alginate beads, for example, where cells have little mobility within the gel matrix, higher growth rates in nutrient-rich volumes may result in relatively high cell concentrations near the bead surface. Radovich (*30*) cites several examples of inhomogeneous growth of immobilized yeasts and bacteria, and Monbouquette and Ollis (*20*) have presented a model of the growth kinetics of cells immobilized in porous matrices which describes this phenomenon.

Significant inhomogeneity may affect not only cell growth and respiration rates but also the solubility and diffusivity of oxygen and other molecules (*11, 34*).

3.1.3. Uniform Fluid Properties Within the Cell Aggregate and the Culture Medium

The models assume that there are no significant differences between the diffusivity and solubility of oxygen in the culture medium and in the cell aggregate.

The data available in the literature support the conclusion of Monbouquette and Ollis (*20*) that diffusivity is a system-specific parameter that varies from system to system, from time to time, and from place to place within a system at any given time. Thus, reported values for the diffusion coefficient of oxygen in various microbial aggregates range from 1.6% to 95% of the value in water (*1, 10, 25, 29, 32, 42*). However, reported values for diffusion coefficients of oxygen and glucose in systems closely analogous to those used for immobilizing hybridomas generally lie in the range 70–100% of the value in water (*1, 9, 26, 32*), though there are exceptions (*10*). This is similar to the diffusion coefficient of oxygen in fermentation media resembling animal cell culture media, which has been measured as 70–80% of the value in water (*11*).

The solubility of oxygen will also be a system-specific variable. Schumpe and Quicker (*34*) have reviewed the subject of gas solubilities in microbial culture media and cultures. It is possible to predict the solubility of gases in media containing only low-molecular-weight solutes, but the methods fail in the presence of significant quantities of complex biological molecules such as proteins. However, the changes in solubility are relatively small under conditions similar to those in animal cell cultures and for the purposes of this chapter the solubility of oxygen can reasonably be approximated by the value in water.

3.1.4. Diffusion

The models presented in this chapter assume that oxygen transfer within a cell aggregate occurs by diffusion alone. This is true if the extracellular fluid in the cell aggregate is stagnant.

There is unlikely to be significant flow within agarose or alginate beads, or in systems in which the cell concentration approaches that found in tissues (ca. 2×10^9 ml^{-1}). There may be flow through open channels in porous beads or foam particles, but this would depend on the cell concentrations present, on the nature of the channels and upon the pressure gradients within the bioreactor. The flow forced through hollow fibre cartridges generates substantial pressure gradients within the extra-capillary space which are capable of driving a flow of medium within and through this space (*39*).

The models presented here consider only mass transfer due to diffusion, and thus significant bulk flow within the cell aggregates will compromise the solution obtained. If the flow within a system is known, it is possible to account for its effect in the mass balance equations given here and to derive models describing mass transfer by flow and diffusion combined.

3.1.5. Oxygen Transfer at the Surface of the Aggregate

The extent of any diffusion barrier between the cell aggregate and the surrounding medium is difficult to assess. The simplest assumption to make is that there is no obstacle to oxygen transfer at this point and that the oxygen concentration at the surface of the aggregate is the same as the concentration in the surrounding medium. The answer then obtained describes the highest achievable oxygen concentrations within the aggregate.

In reality, there may be a stagnant boundary layer at the medium–aggregate interface, which could form a barrier. The use of a membrane to separate cells from medium may present a further barrier. The significance of either of these would have to be determined for each system examined.

3.1.6. Steady-State Conditions

The analysis can be further simplified by considering only the situation in which the system is at steady state with respect to time (i.e. that situation in which oxygen concentration within the system does not change with time). This is most likely to hold true in continuous culture, which is the desired condition for immobilized cell systems.

3.2. Diffusion into a Flat Slab

Consider a thin layer within the slab of unit area, thickness δr at a distance

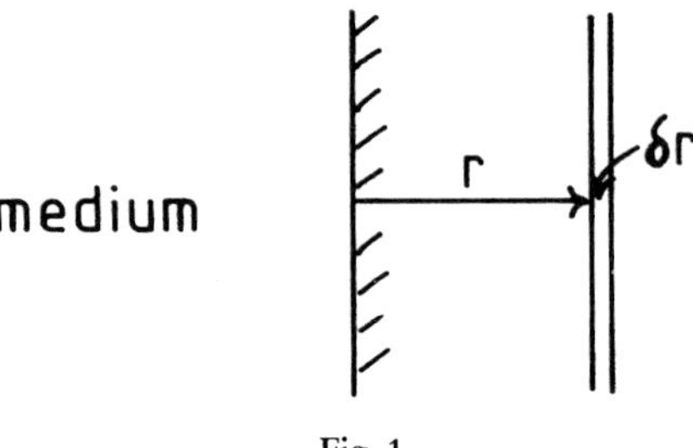

Fig. 1.

r from the interface with the medium (Fig. 1). At steady state the mass balance equation for this layer is:

OXYGEN IN = OXYGEN OUT + OXYGEN USED

$$-D\left[\frac{\mathrm{d}c}{\mathrm{d}r}\right]_r = -D\left[\frac{\mathrm{d}c}{\mathrm{d}r}\right]_{r+\delta r} + R\,\delta r, \tag{1}$$

which may be rearranged to give

$$R = D\left[\frac{[\mathrm{d}c/\mathrm{d}r]_{r+\delta r} - [\mathrm{d}c/\mathrm{d}r]_r}{\delta r}\right], \tag{2}$$

i.e.

$$R = D\left[\frac{\mathrm{d}^2c}{\mathrm{d}r^2}\right]. \tag{3}$$

The general solution to equation (3) is

$$c = ar^2 + br + e, \tag{4}$$

where a, b and e are constants.

Therefore,

$$\frac{\mathrm{d}c}{\mathrm{d}r} = 2ar + b, \tag{5}$$

$$\frac{\mathrm{d}^2c}{\mathrm{d}r^2} = 2a. \tag{6}$$

Solving for a, b and e,

$$a = \frac{R}{2D} \qquad \text{from (3).}$$

Using the boundary conditions

$$c = 0, \qquad \frac{dc}{dr} = 0 \qquad \text{at } r = r_0,$$

$$b = -\frac{Rr_0}{D} \qquad \text{from (5)}$$

and

$$e = \frac{Rr_0^2}{2D} \qquad \text{from (4).}$$

Thus, equation (4) becomes

$$c = \frac{Rr^2}{2D} - \frac{Rrr_0}{D} + \frac{Rr_0^2}{2D},$$

i.e.

$$c = \frac{R}{2D}[r_0 - r]^2. \tag{7}$$

From the boundary condition

$$c = c_m \qquad \text{at } r = 0,$$

$$r_0 = \left(\frac{2Dc_m}{R}\right)^{1/2} \qquad \text{from (7).}$$

Thus,

$$c = \frac{R}{2D}\left[\left(\frac{2Dc_m}{R}\right)^{1/2} - r\right]^2. \tag{8}$$

3.3. Diffusion from a Cylindrical Capillary

Consider a thin shell around the capillary of unit length, radius r, and thickness δr (Fig. 2).

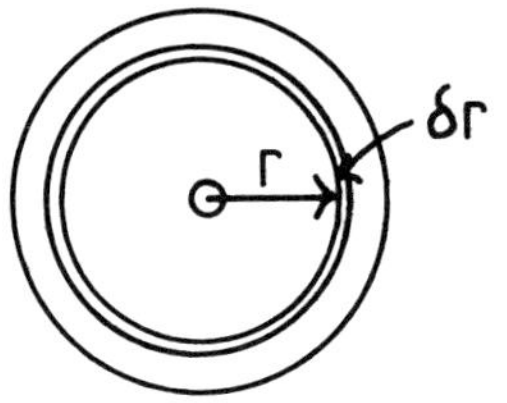

Fig. 2.

At steady state the mass balance equation for this shell is:

OXYGEN IN = OXYGEN OUT + OXYGEN USED

$$-D\,2\pi r\left[\frac{\mathrm{d}c}{\mathrm{d}r}\right]_r = -D\,2\pi(r+\delta r)\left[\frac{\mathrm{d}c}{\mathrm{d}r}\right]_{r+\delta r} + R\,2\pi r\,\delta r. \tag{9}$$

Which may be rearranged to give

$$R = D\left[\frac{[\mathrm{d}c/\mathrm{d}r]_{r+\delta r} - [\mathrm{d}c/\mathrm{d}r]_r}{\delta r}\right] + \frac{D}{r}\left[\frac{\mathrm{d}c}{\mathrm{d}r}\right]_{r+\delta r}, \tag{10}$$

i.e.

$$R = D\left[\frac{\mathrm{d}^2c}{\mathrm{d}r^2} + \frac{1}{r}\frac{\mathrm{d}c}{\mathrm{d}r}\right]. \tag{11}$$

The general solution to equation (11) is

$$c = ar^2 + b\log r + e, \tag{12}$$

where a, b, and e are constants.

Therefore,

$$\frac{\mathrm{d}c}{\mathrm{d}r} = 2ar + \frac{b}{r}, \tag{13}$$

$$\frac{\mathrm{d}^2c}{\mathrm{d}r^2} = 2a + \frac{b}{r^2}. \tag{14}$$

Solving for a, b, and e,

$$a = \frac{R}{4D} \qquad \text{from (11)}.$$

Using the boundary conditions

$$c = c_m \qquad \text{at } r = r_c \text{ (radius of the capillary)},$$

$$c = 0 \qquad \text{and} \qquad \frac{\mathrm{d}c}{\mathrm{d}r} = 0 \qquad \text{at } r = r_0,$$

$$b = -\frac{Rr_0^2}{2D} \qquad \text{from (13)},$$

and

$$e = c_m - \frac{Rr_c^2}{4D} + \frac{Rr_0^2}{2D}\log r_c \qquad \text{from (12)}.$$

Thus, equation (12) becomes

$$c = c_m + \frac{R}{4D}\left[r^2 - r_c^2 - 2r_0^2 \log \frac{r}{r_c}\right]. \tag{15}$$

An approximate value for r_0 may be obtained in a number of ways. Consider the case of a cylinder oxygenated by diffusion from the external surface (analogous to the case for a sphere described below) rather than from an internal capillary. The radius of such a cylinder for which c is just 0 at the centre (i.e. radius $= r_0$) is

$$\text{radius} = \left(\frac{4Dc_m}{R}\right)^{1/2}. \tag{16}$$

Knowing the radius, the surface-area-to-volume ratio of the cylinder can be calculated. Assuming that this ratio represents the maximum oxygenated volume-to-surface-area ratio that can be achieved in the system, it is possible to calculate the volume and thus the thickness of the shell that can be oxygenated by a capillary of a given surface area. Then r_0 is equal to that thickness plus the radius of the capillary.

A better estimate of r_0 may be obtained by substituting the boundary conditions $c = 0$ at $r = r_0$ into equation (15), then using Newton's method or a Taylor series expansion to obtain a numerical approximation to r_0.

3.4. Diffusion into a Sphere

Consider a thin shell within the sphere of radius r and thickness δr (Fig. 3).

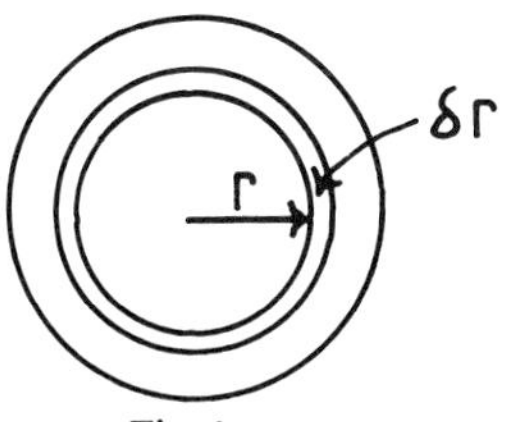

Fig. 3.

At steady state the mass balance equation for this shell is:

OXYGEN IN = OXYGEN OUT + OXYGEN USED

$$D\,4\pi(r + \delta r)^2\left[\frac{dc}{dr}\right]_{r+\delta r} = D\,4\pi r^2\left[\frac{dc}{dr}\right]_r + R\,4\pi r^2\,\delta r. \tag{17}$$

Neglecting terms in δr^2, this may be rearranged to give

$$R = D\left[\frac{[\mathrm{d}c/\mathrm{d}r]_{r+\delta r} - [\mathrm{d}c/\mathrm{d}r]_r}{\delta r}\right] + \frac{2D}{r}\left[\frac{\mathrm{d}c}{\mathrm{d}r}\right]_{r+\delta r}, \tag{18}$$

i.e.

$$R = D\left[\frac{\mathrm{d}^2 c}{\mathrm{d}r^2} + \frac{2}{r}\frac{\mathrm{d}c}{\mathrm{d}r}\right]. \tag{19}$$

The general solution to equation (19) is

$$c = ar^2 + b + \frac{e}{r}, \tag{20}$$

where a, b, and e are constants.

Therefore,

$$\frac{\mathrm{d}c}{\mathrm{d}r} = 2ar - \frac{e}{r^2}, \tag{21}$$

$$\frac{\mathrm{d}^2 c}{\mathrm{d}r^2} = 2a + \frac{2e}{r^3}. \tag{22}$$

Solving for a, b, and e,

$$a = \frac{R}{6D} \qquad \text{from (19).}$$

For a sphere which is aerobic throughout its volume, the boundary conditions are

$$c = c_m \qquad \text{at } r = r_s \text{ (radius of the sphere),}$$

$$\frac{\mathrm{d}c}{\mathrm{d}r} = 0 \qquad \text{at } r = 0.$$

Thus,

$$e = 0$$

and

$$b = c_m - \frac{Rr_s^2}{6D} \qquad \text{from (20).}$$

Thus equation (20) becomes

$$c = c_m - \frac{R}{6D}[r_s^2 - r^2]. \tag{23}$$

For a sphere which contains an anaerobic volume at its centre, the boundary conditions change to

$$c = c_m \qquad \text{at } r = r_s,$$

$$c = 0 \qquad \text{and} \qquad \frac{dc}{dr} = 0 \qquad \text{at } r = r_s - r_0.$$

Solving for b and e by substitution as above,

$$e = \frac{R(r_s - r_0)^3}{3D} \qquad \text{from (21),}$$

and

$$b = -\frac{R(r_s - r_0)^2}{2D} \qquad \text{from (20).}$$

Thus,

$$c = \frac{Rr^2}{6D} - \frac{R(r_s - r_0)^2}{2D} + \frac{R(r_s - r_0)^3}{3Dr}. \tag{24}$$

An approximate value for r_0 may be obtained in a number of ways. The radius of a sphere in which c is just 0 at the centre (i.e. $r_s = r_0$) is

$$\text{radius} = \left(\frac{6Dc_m}{R}\right)^{1/2} \qquad \text{from (23).}$$

Knowing the radius, the surface-area-to-volume ratio of the sphere can be calculated. Assuming that this ratio represents the maximum oxygenated volume-to-surface-area ratio that can be achieved in the system, it is possible to calculate the volume and thus the thickness (r_0) of the shell that can be oxygenated within a sphere of given radius (and thus known surface area).

A better estimate of r_0 may also be obtained by substituting the boundary conditions $c = 0$ at $r = (r_s - r_0)$ into equation (24), then using Newton's method or a Taylor series expansion to obtain a numerical approximation to r_0.

3.5. Calculated Oxygen Concentration Profiles

Figures 4 to 7 show calculated oxygen concentration profiles within spheres under various conditions.

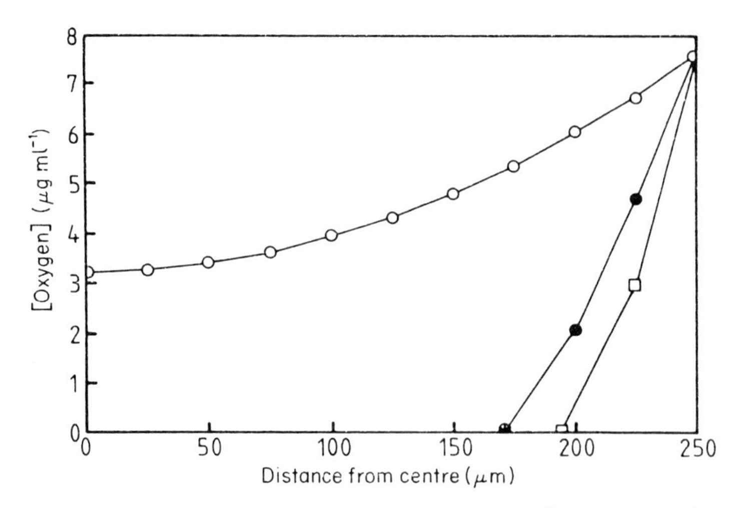

Fig. 4. The effect of cellular respiration rate upon oxygen diffusion into a sphere of radius 250 μm containing 2×10^9 cells ml^{-1}. Oxygen concentration profiles were calculated using equations (23) and (24) for cellular respiration of 0.56 ng oxygen (10^6 cells)$^{-1}$ s^{-1} (open circles), 1.94 ng oxygen (10^6 cells)$^{-1}$ s^{-1} (solid circles), and 3.06 ng oxygen (10^6 cells)$^{-1}$ s^{-1} (open squares). The surrounding medium was assumed to be saturated with oxygen (7.6 μg ml^{-1}) and the diffusion coefficient of oxygen was assumed to be 2.65×10^{-5} cm^2 s^{-1}.

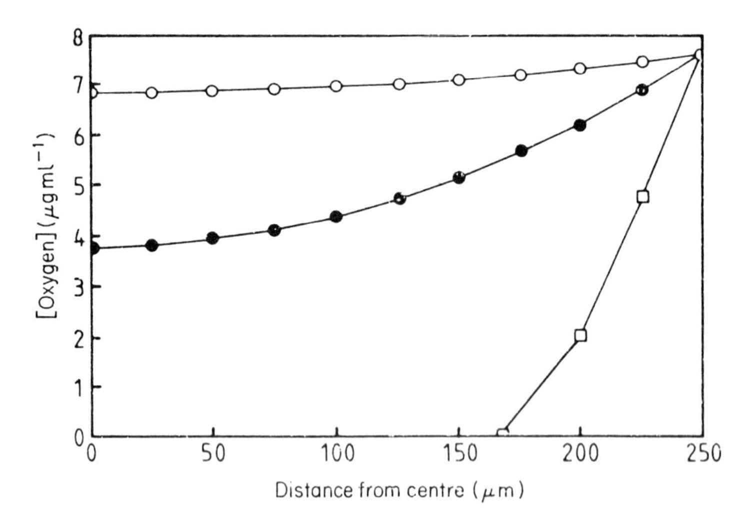

Fig. 5. The effect of cell concentration upon the diffusion of oxygen into a sphere of radius 250 μm. Oxygen concentration profiles were calculated using equations (23) and (24) for cell concentrations of 10^8 cells ml^{-1} (open circles), 5×10^8 cells ml^{-1} (solid circles), and 2×10^9 cells ml^{-1} (open squares), with a cellular respiration rate of 1.94 ng oxygen (10^6 cells)$^{-1}$ s^{-1}. The surrounding medium was assumed to be saturated with oxygen (7.6 μg ml^{-1}) and the diffusion coefficient of oxygen was assumed to be 2.65×10^{-5} cm^2 s^{-1}.

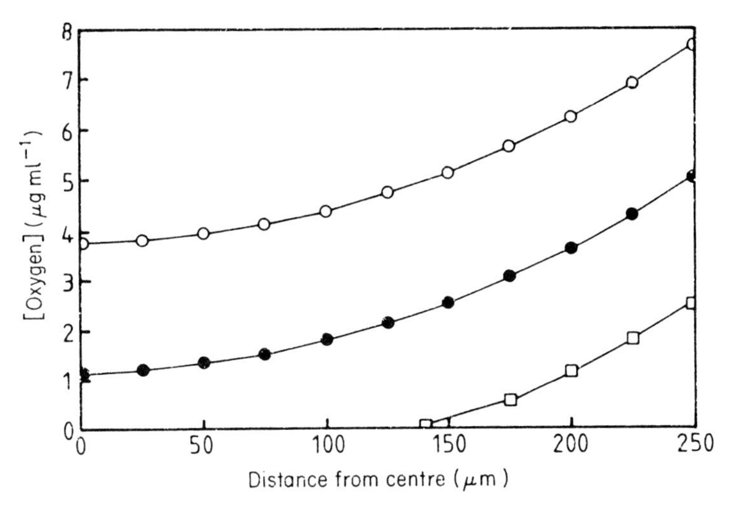

Fig. 6. The effect of oxygen concentration in the surrounding medium upon oxygen diffusion into a sphere of radius 250 μm containing 5×10^8 cells ml^{-1}. Oxygen concentration profiles within the sphere were calculated using equations (23) and (24) for external oxygen concentrations of 7.6 μg ml^{-1} (open circles), 5 μg ml^{-1} (solid circles), and 2.5 μg ml^{-1} (open squares). Cellular respiration rate was taken to be 1.94 ng oxygen (10^6 cells)$^{-1}$ s^{-1} and the diffusion coefficient of oxygen was assumed to be 2.65×10^{-5} cm^2 s^{-1}.

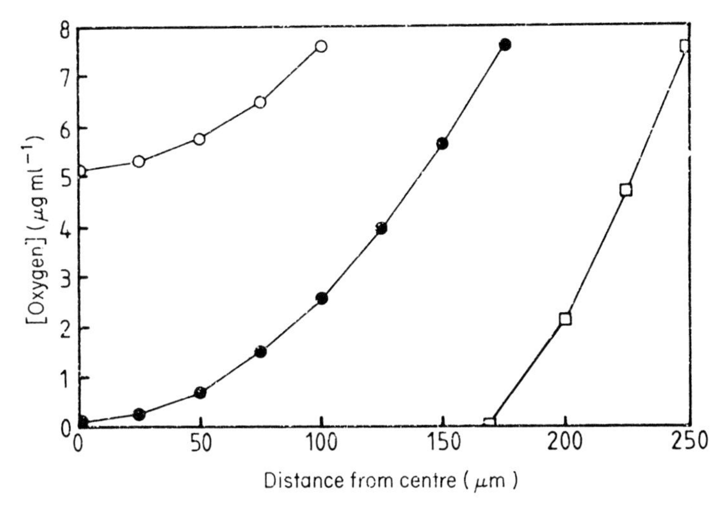

Fig. 7. The effect of size on oxygen diffusion into spheres. Oxygen concentration profiles within spheres of radius 250 μm (open squares), 175 μm (solid circles), and 100 μm (open circles) were calculated using equations (23) and (24). The cell concentration within the spheres was taken as 2×10^9 cells ml^{-1}, cellular respiration rate as 1.94 ng oxygen (10^6 cells)$^{-1}$ s^{-1}. The surrounding medium was assumed to be saturated with oxygen (7.6 μg ml^{-1}) and the diffusion coefficient of oxygen was assumed to be 2.65×10^{-5} cm^2 s^{-1}.

4. MEASUREMENT OF THE METABOLISM OF IMMOBILIZED CELLS

Direct monitoring of oxygen concentrations within immobilized cell aggregates is technically difficult and this chapter does not consider the problem. However, a measurement of the metabolic activity of immobilized cells may be made *in situ* by staining the cells with 3-(4,5-dimethylthiazol-2-yl)-2,5-diphenyl tetrazolium bromide (MTT) (*41*) and subsequently analysing photographs of sectioned, stained immobilized cell preparations (*24*). MTT is soluble in water, forming a pale yellow solution. The reduced form is blue and insoluble in water. Reduced MTT is deposited as blue crystals within actively metabolizing cells (*22, 41*). MTT is specifically reduced by the action of mitochondrial dehydrogenases and so provides a measure of the extent of cellular metabolic activity (*22*).

Figure 8 shows results obtained using this method. Cells immobilized at 5×10^6 ml^{-1} in 1% agarose were incubated under RPMI 1640 plus 1% foetal calf serum in an atmosphere of 95% air, 5% CO_2. After 42 hr 500 μg ml^{-1} of MTT was added to the supernatant medium, and the incubation was continued overnight. The agarose slab was then recovered and sectioned vertically (*41*). Oxidized, yellow, MTT had diffused throughout the slab. Blue, reduced, MTT had been deposited close to the surface of the slab. The sections were photographed and the photographs were analysed to determine the brightness as a function of depth below the surface. Brightness was inversely related to the amount of reduced MTT present. The brightness levels were used as an index of cellular metabolism.

Despite the numerous assumptions made in deriving the mathematical model, there is a good fit between the measured metabolic activity of the immobilized hybridomas and the calculated steady-state oxygen concentrations in this instance. It is possible to describe the measured metabolic activity in terms of the calculated oxygen concentrations (i.e. as a function r_0) (see Fig. 5).

5. CONCLUDING REMARKS

The models presented here provide a useful guide to the design of bioreactors intended for culturing immobilized hybridomas. For instance, using the models it is possible to determine the dimensions of an immobilized cell aggregate within which the oxygen concentration does not fall below a desired value and consequently to set limits on the useful size of spherical beads or upon the separation of fibres in a hollow-fibre cartridge.

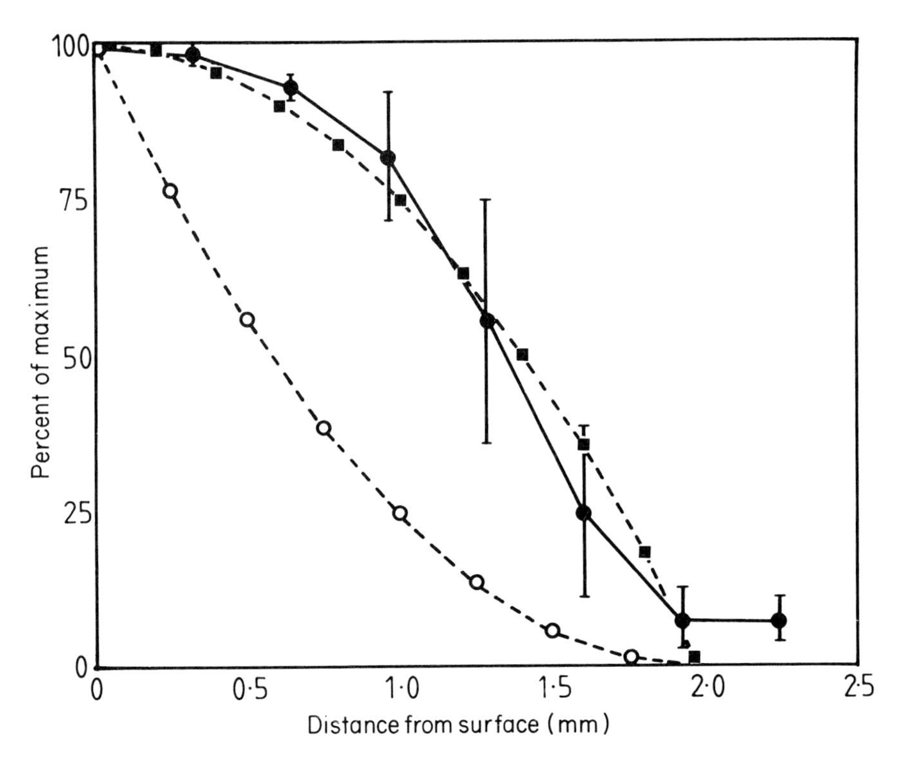

Fig. 8. Comparison of the metabolic activity of cells of a hybridoma line immobilized in an agarose slab (solid circles) with the calculated oxygen concentration profile within the slab (open circles). Both sets of data are expressed as percentages of the values at the surface of the slab. Metabolic activity was assessed by measuring MTT reduction as described in the text. The results shown are the means of four determinations. The error bars show standard deviations. Oxygen concentration profiles were calculated using equation (8), assuming that oxygen concentration in the surrounding medium was 7.22 μg ml^{-1} (i.e. saturated with respect to the gas phase), that the cellular respiration rate was 1.94 ng oxygen (10^6 cells)$^{-1}$ s^{-1} and that the diffusion coefficient of oxygen was 2.65 $\times$ 10^{-5} cm^2 s^{-1}. The measured metabolic activity is described by the relationship

$$M = \left[1 - \frac{r^2}{r_0^2}\right] \times 100\%,$$

where M is the metabolic activity expressed as a percentage of maximum. Calculated values of M are shown as solid squares.

The use of the cytometabolism diffusion test of Wilson and Spier (*41*) in conjunction with photographic image analysis provides a simple but powerful means for assessing the metabolism of immobilized cells. The combination of this technique with the mathematical models (which may be adapted to molecules other than oxygen) provides the basis of a method for studying the effect of various factors on the metabolism of immobilized cells *in situ* and, ultimately, for predicting the performance of cells immobilized in a defined situation.

REFERENCES

1. Adlercreutz, P. (1986). Oxygen supply to immobilized cells. 5. Theoretical calculations and experimental data for the oxidation of glycerol by immobilized gluconobacter oxydans cells with oxygen or p-benzoquinone as the electron acceptor. *Biotechnol. Bioeng.* **28**, 223–232.
2. Altshuler, G. L., Dziewulski, D. M., Sowek, J. A., and Belfort, G. (1987). Continuous hybridoma growth and monoclonal antibody production in hollow fiber reactors-separators. *Biotechnol. Bioeng.* **28**, 646–658.
3. Birch, J. R., Thompson, P. W., Boraston, R., Oliver, S., and Lambert, K. (1987). *In* "Animal and Plant Cells: Process Possibilities" (C. Webb and F. Mavituna, eds.), pp. 162–171. Ellis Horwood Ltd, Chichester.
4. Boraston, R., Thompson, P. W., Garland, S., and Birch, J. R. (1984). Growth and oxygen requirements of antibody producing mouse hybridoma cells in culture. *Dev. Biol. Stand.* **55**, 103–111.
5. Crank, J. (1975). "The Mathematics of Diffusion", 2nd edn. Clarendon Press, Oxford.
6. Emery, A. N., Lavery, M., Williams, B., and Handa, A. (1987). Large scale hybridoma culture. *In* "Animal and Plant Cells: Process Possibilities" (C. Webb and F. Mavituna, eds.), pp. 137–146. Ellis Horwood Ltd, Chichester.
7. Forster, R. E. (1963). Factors affecting the rate of exchange of O_2 between blood and tissues. *In* "Oxygen in the Animal Organism" (F. Dickens and E. Neil, eds.), pp. 393–409. Int. Symp. IUB/IUBS **31**.
8. Griffiths, B. (1986). Scaling up of animal cell cultures. *In* "Animal Cell Culture: A Practical Approach" (R. I. Freshney, ed.), pp. 33–69. IRL Press Ltd, Oxford.
9. Hannoun, B. J. M., and Stephanopoulos, G. (1986). Diffusion coefficients of glucose and ethanol in cell-free and cell-occupied calcium alginate membranes. *Biotechnol. Bioeng.* **28**, 829–835.
10. Hiemstra, H., Dijkhuizen, L., and Harder, W. (1983). Diffusion of oxygen in alginate gels related to the kinetics of methanol oxidation by immobilized *Hansenula polymorpha* cells. *Eur. J. Appl. Microbiol. Technol.* **18**, 189–196.
11. Ho, C. S., Ju, L.-K., and Ho, C.-T. (1986). Measuring oxygen diffusion coefficients with polarographic oxygen electrodes II. Fermentation media. *Biotechnol. Bioeng.* **28**, 1086–1092.
12. Karkare, S. B., Phillips, P. G., Burke, D. H., and Dean, R. C., Jr. (1985). Continuous production of monoclonal antibodies by chemostatic and immobilized hybridoma culture. *In* "Large-Scale Mammalian Cell Culture" (J. Feder and W. R. Tolbert, eds.), pp. 127–149. Academic Press, Orlando.
13. Katinger, H. (1987). Principles of animal cell fermentation. *Dev. Biol. Stand.* **66**, 195–209.
14. Katinger, H., and Scheirer, W. (1985). Mass cultivation and production of animal cells. *In*

"Animal Cell Biotechnology" (R. E. Spier and J. B. Griffiths, eds.), Vol. 1, pp. 167–193. Academic Press, London.
15. Klement, G., Scheirer, W., and Katinger, H. W. D. (1987). Construction of a large scale membrane reactor system with different compartments for cells, medium and product. *Dev. Biol. Stand.* **66**, 221–226.
16. Kobayashi, T., van Dedem, G., and Moo-Young, M. (1973). Oxygen transfer into mycelial pellets. *Biotechnol. Bioeng.* **15**, 27–45.
17. Lambe, C. A., and Walker, A. G. (1987). Reactor requirements for animal cells. *In* "Animal and Plant Cells: Process Possibilities" (C. Webb and F. Mavituna, eds.), pp. 116–124. Ellis Horwood Ltd, Chichester.
18. Lazar, A., Reuveny, S., Mizrahi, A., Avtalion, M., Whiteside, J. P., and Spier, R. E. (1987). Production of biologicals by animal cells immobilized on a polyurethane foam matrix. Paper presented at the ESACT-OHOLO joint meeting "Modern Approaches to Animal Cell Technology", Tiberias, Israel.
19. Metz, B., and Kossen, N. W. F. (1977). The growth of mold in the form of pellets—a literature review. *Biotechnol. Bioeng.* **19**, 781–799.
20. Monbouquette, H. G., and Ollis, D. F. (1986). A structured model for immobilized cell kinetics. *Ann. N.Y. Acad. Sci.* **469**, 230–244.
21. Moo-Young, M., and Blanch, H. W. (1981). Design of biochemical reactors—mass transfer criteria for simple and complex systems. *Adv. Biochem. Eng.* **19**, 1–69.
22. Mosmann, T. (1983). Rapid colorimetric assay for cellular growth and survival: application to proliferation and cytotoxicity assays. *J. Immunol. Methods* **65**, 55–63.
23. Murdin, A. D., Thorpe, J. S., and Spier, R. E. (1987). Immobilisation of hybridomas in packed bed bioreactors. Paper presented at the ESACT-OHOLO joint meeting "Modern Approaches to Animal Cell Technology", Tiberias, Israel.
24. Murdin, A. D., Wilson, R., Kirkby, N., and Spier, R. E. (1987). Examination of a simple model for the diffusion of oxygen into dense masses of animal cells. Paper presented at the ESACT-OHOLO joint meeting "Modern Approaches to Animal Cell Technology", Tiberias, Israel.
25. Ngian, K. F., and Lin, S. H. (1976). Diffusion coefficient of oxygen in microbial aggregates. *Biotechnol. Bioeng.* **18**, 1623–1627.
26. Nguyen, A.-L., and Luong, J. H. T. (1986). Diffusion in k-carrageenan gel beads. *Biotechnol. Bioeng.* **28**, 1261–1267.
27. Nilsson, K., Scheirer, W., Merton, O. W., Ostberg, L., Liehl, E., Katinger, H. W. D., and Mosbach, K. (1983). Entrapment of animal cells for the production of monoclonal antibodies and other biomolecules. *Nature (London)* **302**, 629–630.
28. Nilsson, K., Buzsaky, F., and Mosbach, K. (1986). Growth of anchorage-dependent cells on macroporous microcarriers. *Biotechnology* **4**, 989–990.
29. Onuma, M., Omura, T., Umita, T., and Aizawa, J. (1985). Diffusion coefficient and its dependency on some biochemical factors. *Biotechnol. Bioeng.* **27**, 1533–1539.
30. Radovich, J. M. (1985). Mass transfer effects in fermentations using immobilized whole cells. *Enzyme Microb. Technol.* **7**, 2–10.
31. Rupp, R. G. (1985). Use of cellular microencapsulation in large-scale production of monoclonal antibodies. *In* "Large-Scale Mammalian Cell Culture" (J. Feder and W. R. Tolbert, eds.), pp. 19–38. Academic Press, Orlando.
32. Sato, K., and Toda, K. (1983). Oxygen uptake rate of immobilized growing *Candida lipolytica*. *J. Ferment. Technol.* **61**, 239–245.
33. Schönherr, O. T., van Gelder, P. T. J. A., van Hees, P. J., van Os, A. M. J. M., and Roelofs, H. W. M. (1987). A hollow fibre dialysis system for the *in vitro* production of monoclonal antibodies replacing *in vivo* production in mice. *Dev. Biol. Stand.* **66**, 211–220.

34. Schumpe, A., and Quicker, G. (1982). Gas solubilities in microbial culture media. *Adv. Biochem. Eng.* **24**, 1–38.
35. Scott, C. D. (1987). Immobilized cells: a review of recent literature. *Enzyme Microb. Technol.* **9**, 66–73.
36. Sinacore, M. S. (1984). Gel entrapment: applications in production of biologicals and mass culturing of animal cells. *Karyon Technology News* **1**. Karyon Technology Inc., 333 Providence Highway, Norwood, MA 02062, USA. Cited by Lambe and Walker (*17*).
37. Spier, R. E., and Griffiths, B. (1984). An examination of the data and concepts germane to the oxygenation of cultured animal cells. *Dev. Biol. Stand.* **55**, 81–92.
38. Tannock, I. F. (1968). The relation between cell proliferation and the vascular system in a transplanted mouse mammary tumour. *Brit. J. Cancer* **22**, 258–273.
39. Tharakan, J. P., and Chau, P. C. (1986). Operation and pressure distribution of immobilized cell hollow fiber bioreactors. *Biotechnol. Bioeng.* **28**, 1064–1071.
40. Verax Corporation (1986). Continuous cell culture with fluidized sponge beads for large-scale production of medical proteins. Verax TM-184A. Verax Corporation, Lebanon, NH 03766, USA.
41. Wilson, R., and Spier, R. E. (1987). A cytometabolism diffusion test (CMDT) for the determination of animal cell bioreactor parameters. *Enzyme Microb. Technol.* (accepted for publication).
42. Yano, T., Kodama, T., and Yamada, K. (1961). Fundamental studies on the aerobic fermentation. Part VIII. Oxygen transfer within a mold pellet. *Agr. Biol. Chem.* **25**, 580–584.

5

Sensors for the Control of Mammalian Cell Processes

OTTO-WILHELM MERTEN
Institut Pasteur,
Technologie Cellulaire,
Paris, France

ANIMAL CELL BIOTECHNOLOGY VOL. 3
ISBN 0-12-657553 3

1. INTRODUCTION

The rapid development of the techniques of cell culture using different fermentation systems and process modes has required a sensing technology for the monitoring of physiological and biochemical parameters of the fermentation process. Today cell technology is used mainly to produce viral vaccines and other biologicals, like monoclonal antibodies or interferon, or products from genetically engineered cells. Although the cell lines used for production are adapted to the *in vitro* systems, these systems are far from being optimal. The main reasons are that little is known about the physiology of the cell lines used, and also that each cell line has its own individuality. One therefore tries to imitate the *in vivo* situation without knowing exactly what is necessary and what is not. This situation can be improved by using on-line detection systems for the biochemical and physiological parameters for researching the cell physiology and for process optimization and scale-up. Monitoring of fermentation parameters is interesting for the control of the day-to-day maintenance of optimal conditions, product quality, and downstream processing.

Unfortunately, some cell lines of interest are genetically unstable (see, e.g., (*1*)) when they are cultivated in large scale. This lack of stability is a considerable inconvenience when these cell lines are used in a production process. An on-line detection system can be employed for the surveillance of the maintenance of a technically improved cell strain under selective conditions. During the selection of a production clone, biosensors can be used to identify the features requiring strain engineering and to predict and verify the effects of strain engineering. Finally, the development of mathematical models can be done only when enough information about a bioprocess is available. To achieve that and for testing these models, on-line detection systems are necessary. The availability of computers and microprocessors provides us with the possibility of multiparametric monitoring. The data generated can be used for a simple feedback control loop, or, when enough knowledge about the cell physiology is available, they can be used for the development of more sophisticated control circuits. But so far it is only possible to measure the external environment of a cell (except for ATP, DNA, and NADH content of cells). We are only able to influence the cells' internal processes via the alteration, adaptation, or regulation of external parameters. With this regulation, stimuli are transmitted to the intracellular compartments of the cells. Hence, only indirect control of the cells and the culture is possible and a lot of additional research has to be done to reach the state of more direct intracellular control/regulation of the cell physiology.

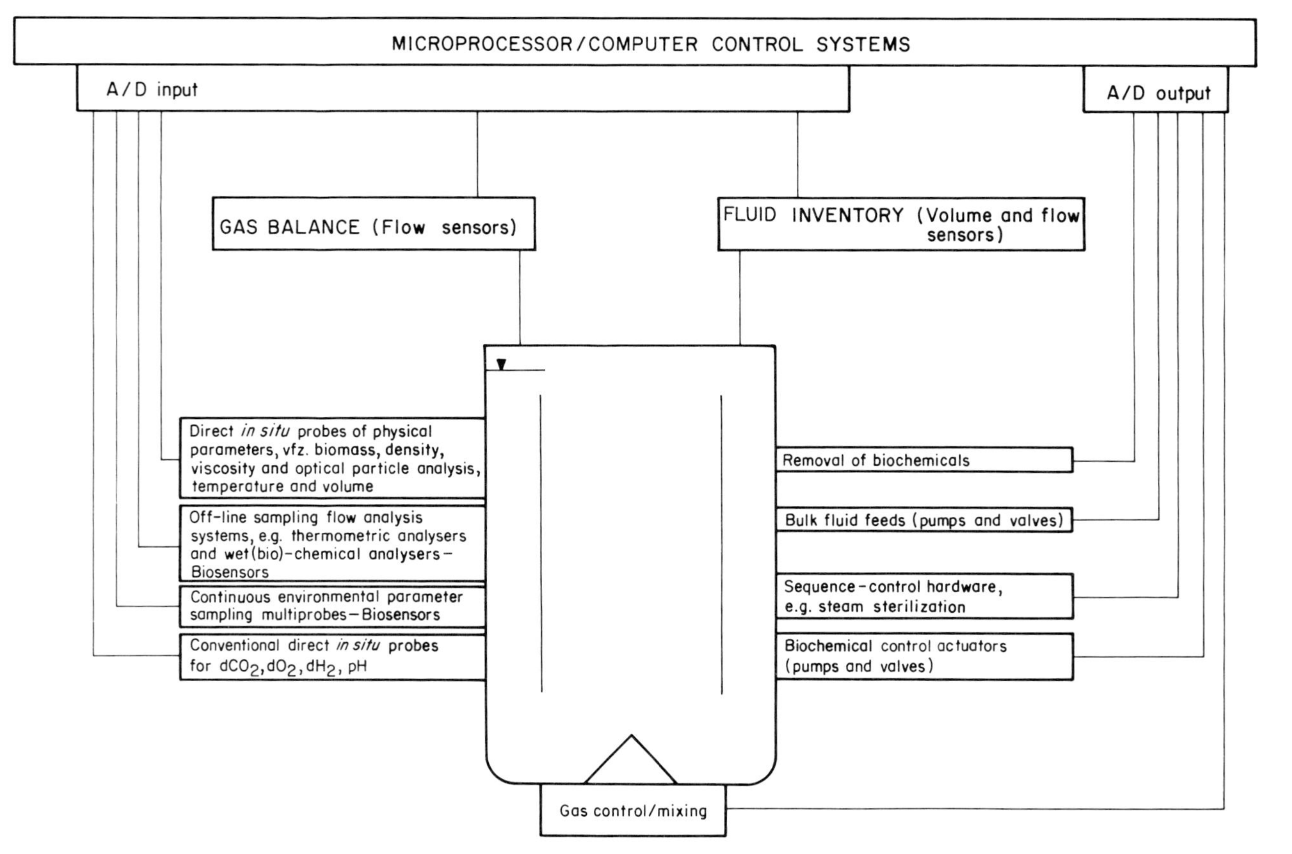

Fig. 1. Applications of sensing systems to direct digital fermenter control system.

To meet all these purposes, many different sensors are necessary. Unfortunately, only few sensing devices are available at present. Classical detection systems include electrodes for monitoring pH, pO_2, pCO_2, and redox potential. Additionally, physical parameters, such as volume, weight, temperature, viscosity, optical particle analysis, and to some extent biomass determination can be sensed in an on-line mode. Off-gas analysis, using mass spectrometry, can be used for the calculation of respiratory and other gas balances (for review, see Lloyd *et al.* (*2*)). But most of the biochemical and physiological parameters have still to be determined in an off-line mode. In Fig. 1 the applications of such biochemical and physical sensors for the control of a fermentation process are shown. This chapter deals with the modern development of on-line sensing methods for the control of fermentation processes, with special reference to mammalian cell culture. Sensing systems will be described in brief, their advantages and disadvantages will be shown, and some considerations concerning the problem of the complex cell culture medium will be given. Chemical detection systems, mentioned above, will not be described in detail. As only the application of biosensors for animal cell fermentation is reviewed, sensing devices based on the use of immobilized microorganisms owing to the potential danger of contamination of the culture are not included. For reviews, see Suzuki and Karube (*3*) and Karube and Suzuki (*4*).

2. BIOCHEMICAL AND PHYSIOLOGICAL PARAMETERS

The cell culture medium has on one hand to provide the cells with the necessary nutrients and growth factors and on the other has a function as a waste repository (Fig. 2). The cells are very sensitive to the concentrations of the substance present. Below a certain concentration, the nutrients will diffuse outwards from the cell. This is of particular importance with respect to certain growth factors produced by the cells themselves under certain conditions. For instance, this might be the reason why high-cell-density cultures (more than 10^7 cells ml^{-1}) can grow without addition of serum to the growth medium in many cases (*6, 7*).

The fact that the cell culture media should imitate the *in vivo* environment of cells to some extent provides us with the problem that these media are serum-supplemented in most cases, to provide the source of hormones and growth factors in low-density cell cultures (up to 3×10^6 cells ml^{-1}). Not only because of the relatively high costs of serum but mainly because the serum is not a chemically defined reagent, and is therefore non-standardizable, a lot of effort was expended in replacing the serum by defined compounds. However, these serum-free media still contain many different substances

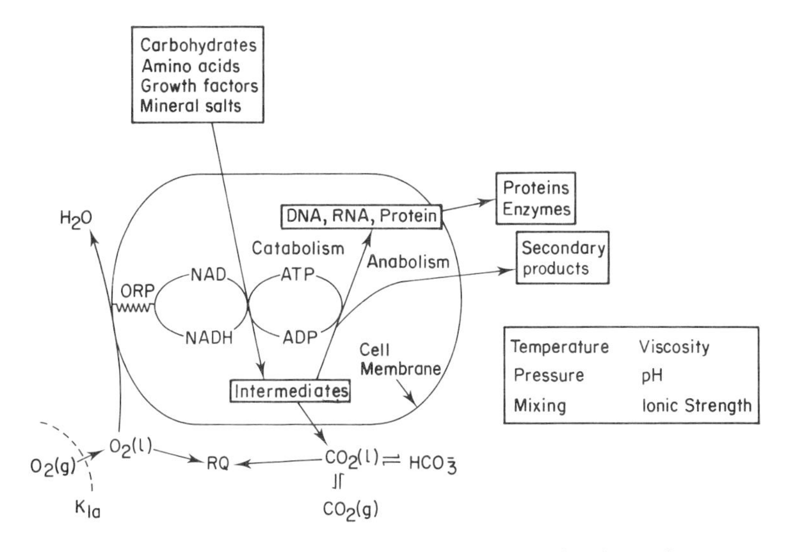

Fig. 2. Generalized model of animal cell metabolism (taken from ref. (*5*)).

which are more or less necessary for cell propagation. Table I presents some supplements for serum-free media used for the culture of hybridoma cell lines. It is evident that besides the basal medium formulation many different additives are used which are more or less essential (already mentioned above).

The control of a probable limiting substance (for growth or production) is necessary, especially in high-density culture systems. These media compositions are sufficient in most instances of low-density culture systems, but may be inadequately composed for high-cell-density culture systems. The

TABLE I Supplements for Serum-free Cultures of Hybridomas[a]

Basal medium	Supplement
MEM	Insulin
RPMI-1640	Transferrin[b]
Ham's F12 : DMEM = 1 : 1	Ethanolamine
RPMI-1640 : DMEM : Ham's F12 = 2 : 1 : 1	Testosterone, hydrocortisone, cholesterol, bovine serum albumin (BSA), casein
IMDM	Linoleic acid, oleic acid-BSA (fatty-acid-free), soy-bean lipid, human low-density lipoprotein
	Vitamins
	Trace elements

[a] References: (*8*)–(*15*).
[b] Transferrin is supplied to all formulations.

development of new culture system configurations providing the possibility of high cell densities shows that cells in a high-cell-density environment have other nutritional demands and that they change their metabolism and physiology. Two examples are given. (i) Serum-free hybridoma cultures require the presence of transferrin in a low-density culture system. Using high-density culture systems, like hollow-fibres, transferrin is an essential medium compound only at the start of a new culture, but it can be absent during the continuation of the culture (*6, 7*). (ii) A second effect of high-density culturing systems is the elevated specific production of biologicals (e.g. 2–10 times better production of monoclonal antibodies by hybridomas (*7, 16*), and 11 times better production of hCG by JEG-7 (*17*).

Some examples concerning medium components are given in the following. Medium components, such as the carbon source used, which is mainly glucose, or some amino acids, like glutamine, are too concentrated and can therefore cause problems. Too high a glucose concentration causes the production of high amounts of lactic acid (Crabtree effect) which causes an acidification of the medium and an unnecessary waste of the energy source. Cells cultivated in a medium with too high a concentration of glutamine (also a carbon source), produce high amounts of NH_4^+, which is toxic for the cultured cells (*18, 19*). A feedback control system for these compounds would provide a method for adding both carbon sources to the culture according to the demand of the cells. The waste products would be minimized. Hu *et al.* (*20*) presented a feed-forward regulation model system for hybridomas for keeping the concentrations of glucose and glutamine lower than usual, but without the use of sensors for a feedback control. The influence on these carbon sources is not easy to understand because replacing glucose by other sugars, such as galactose, maltose, or fructose, would provide the same influence on the cell growth as would glucose, but accompanied by a lower production of lactate (*18*). The reduction of the sugar concentration is in general only possible when the glutamine concentration is elevated in parallel, which causes elevated production of NH_4^+.

The presence of amino acids is necessary, but the question remains which amino acid has to be present and at what concentration. A lot of work has been done in this field, with the resulting finding that different cell lines have different demands and produce different amino acids. In addition, these demands also depend on the cultivation system used and the physiological state of the cells. Roberts *et al.* (*21*) presented some results for the MOPC-31C mouse plasmacytoma cell line in a stationary batch culture. It consumed glutamine, isoleucine, methionine, and valine, and tyrosine and phenylalanine slightly. Aspartic acid, glutamic acid, glycine, proline, and serine were produced. The other amino acids were not influenced. Polastri *et al.* (*22*) published results for virus production using Vero cells on microcarriers in a spinner culture system. During cell growth, glutamine, histidine, arginine, tryptophan, and methionine were consumed, glutamine during the virus production as well. Alanine and serine were produced during cell growth and virus production. The concentrations of the other amino acids were influenced slightly. The last example is given for hybridomas: generally, glutamine is consumed and alanine is produced (*23, 24*). For instance, one mouse–mouse hybridoma consumed leucine, serine (totally), isoleucine, methionine, arginine (totally), glutamine (totally), and phenylalanine, tyrosine, and valine slightly in a static batch culture. In parallel, they produced alanine, asparagine, glycine (slightly), and glutamic acid (during the lag-phase and the beginning of the log-phase). The others were not affected. In comparison, a human–human–mouse hybridoma consumed only glutamine and arginine and produced glutamic acid, alanine, and proline in a static batch culture. The concentrations of the other amino acids decreased slightly during the log-phase and increased at the end of the log-phase and the beginning of the stationary phase (*24*).

These examples imply that the development of control systems is very important for process control and process optimization. But it also becomes evident that control has to be established for each different process and that change of the process system also requires change of the control system, because of the change of the cell physiology. Before these control systems are available, the application of biosensors has to provide enough data on the different culturing systems to allow establishment of useful mathematical models (cf. with Section 1). The following physical, biochemical, and physiological parameters shown in Table II are of general interest irrespective of the culturing system used. Table II is divided into two sections, which present the physical parameters (which sensing devices are autoclavable generally and are not described here), and the biological parameters with subdivision into biomass—cells, nutrients, wastes, and products. The determination of the cell number can be done in a direct way, using the methods mentioned in Table II, or indirectly by the measurement of the ATP or NAD/NADH content of the cells, which varies with the change in the physiology, and the DNA content of the cells, which is constant. These methods are described in Section 4.1. In addition, the accumulation of wastes like alanine can be an interesting parameter for determination of metabolically active cells (*23, 24*); alternatively, the amount of accumulated lactate dehydrogenase can serve as a parameter for the dead cell number. The calculation of RQ (the respiration coefficient) is already a classical microbiological method for the control of cell metabolism (for detail, see Fleischaker *et al.* (*5*)).

In principle, it is necessary because of the lack of information to control all nutrients and wastes produced. If enough knowledge of the cell physiology is available, only the key parameters have to be controlled and regulated.

TABLE II Physical and Biological Control Parameters and Possible Sensing Devices

Parameter	Device
Physical	
Pressure	Manometer
Temperature	pT-100
Volume	Device for measuring weight or the niveau
Weight	Device for measuring weight
Flow of liquids or gases	Flow meter
pH	Electrode
pO_2	Electrode
pCO_2	Electrode
Off-gas analysis (O_2, CO_2, etc.)	Mass spectrometry, infrared analysis
Redox potential	Redox sensor
Biological	
Cells, biomass, cell number	Optical methods
	Fuel cell type electrode
	Piezoelectric membrane systems
	Magneto-inductive monitoring of impedance properties of cells
	Acoustic resonance densitometry
	Electrical counting
ATP	On-line FIA (luminometer)
NAD/NADH	Fluorimeter
DNA	On-line FIA (fluorimeter)
Nutrients	
Carbohydrate source	Enzyme electrodes, ENFETs, FIA, equipped with enzyme reactors and subsequent detector, luminometer, or affinity systems
Amino acids	
Hormones	
Wastes	
Lactic acid	As for nutrients
NH_4^+	
Amino acids	
Enzymes	
Secreted products	
Virus	As for nutrients
Antigen	
Monoclonal antibody	
Enzymes	
Immunomodulators	

Finally, the product is one of the key parameters and the main consideration is to minimize the medium throughput combined with a maximal product output. The quotient of product to medium quantity has to be maximized, therefore.

Some applications of sensors for the determination of biological parameters presented in Table II are described in detail in Section 4.

3. SAMPLING DEVICES, TRANSDUCERS, BIOSENSORS, AND ON-LINE DETECTION SYSTEMS

3.1. Biosensor Transducers

Although this field is already well-reviewed (e.g. *25, 26*) it is necessary to give a brief description of the transducing elements that can be used in biosensor technology. Gronow *et al.* (*27*) gave the following definition of a biosensor: "A biosensor is an analytical tool or system consisting of an immobilized biological material (e.g., an enzyme, antibody, whole cell, organelle, or combination thereof) in intimate contact with a suitable transducer device which will convert the biochemical signal into a quantifiable electrical signal." This signal can be electronically amplified, stored, and subsequently displayed. Table III lists typical transducers that have been exploited. Generally, they can be categorized into six different types: potentiometric, amperometric, optical, calorimetric, conductimetric, and those measuring mass changes. The construction of a sensor is shown in Fig. 3. It should be mentioned that in some cases the enzyme has not to be coupled directly to the transducer (e.g. enzyme reactor with subsequent detector; see Fig. 7(e)).

In the following sections the different transducing devices are described briefly.

3.1.1. Potentiometric Devices

Potentiometric devices operate under equilibrium conditions and measure the accumulation of charge density at the electrode surface brought about by

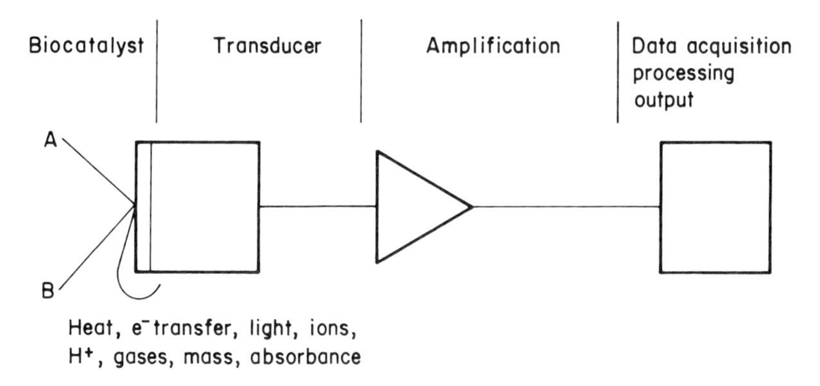

Fig. 3. Schematic of a generalized biosensor. The biocatalyst converts substrate A into product B with a concurrent change in a physicochemical parameter which is converted into an electrical signal by the transducer, amplified and suitably processed and outputted (taken from ref. (*25*)).

TABLE III Biosensor Transducers, Operation Modes, and Applications

Measurement mode	Transducer system	Species detected	Typical application
Potentiometric	Ion-selective electrode (ISE)	H^+, K^+, Na^+, NH_4^+, Ca^{2+}, Li^+, I^-, CN^-	Ions in biological media, enzyme electrodes enzyme immunosensors
	Gas-sensing electrodes	NH_3, CO_2	Gases, enzymes, organelles, cell or tissue electrodes for substrates and inhibitors, enzyme immunoelectrodes
	Field-effect transistor (FET)	H^+, H_2, NH_3	Ions, gases, enzyme substrates and immunological analytes
Amperometric	Enzyme electrodes	O_2, H_2O_2, I_2, NADH, mediators	Enzyme substrates and immunological systems
Optical	Photodiode (optoelectronic, fibre-optic and waveguide devices in conjunction with a light-emitting diode)	Light absorption, fluorescence	pH, enzyme substrates, immunological analytes
	Photomultiplier (in conjunction with fibre-optic)	Light emission, bio- and chemiluminescence	Enzyme substrates
Calorimetric	Thermistor	Heat of reaction	Enzymes, organelles, whole cell or tissue sensors for substrates, products and inhibitors, gases, pollutants, antibiotics, vitamins, etc., immunological analytes
Conductance	Conductimeter	Increase of solution conductance	Enzyme substrates
Mass change	Piezoelectric crystals	Mass absorbed	Volatile gases, vapours, immunological analytes

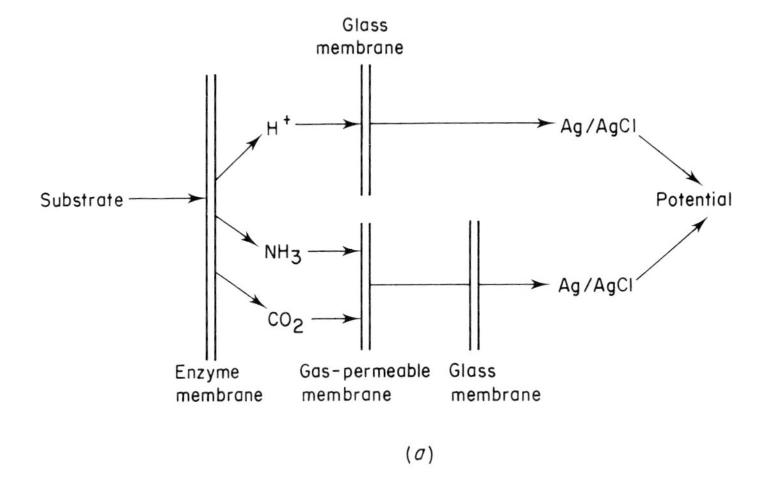

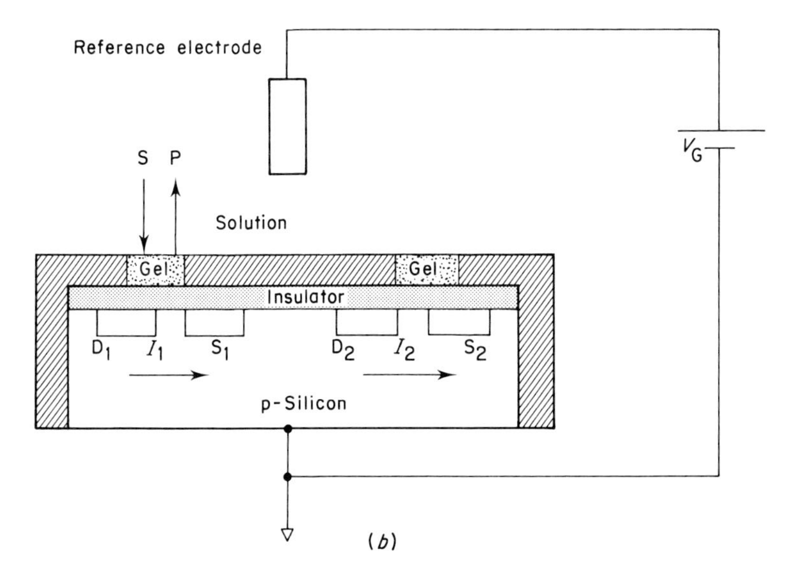

Fig. 4. Potentiometric devices. (*a*) Classical assembly of the electrode types: pH (or ISE) electrode and gas-sensing electrode. (*b*) Schematic diagram of the ENFET chip. D and S are transistor drain and source, respectively. V is the applied gate voltage and I is the drain-to-source current. The enzyme is coupled to gate 1 (taken from ref. (*29*)).

a surface reaction or other selective processes. Three different transducer systems are available (Table III): the ion-selective electrode (ISE), the best known potentiometric device, used for the monitoring of cations and anions; the gas-sensing electrodes for the detection of gases; and the field-effect transistor (FET), used mainly as an ion-selective FET (ISFET) for the detection of ions and gases. The coupling of these devices with an enzyme reaction leads to the potentiometric enzyme electrodes or FETs, in which the immobilized enzyme is coated over the conventional ISE or gas-sensing electrode (shown in Fig. 4(*a*)) or in which the enzyme is immobilized over the surface of an ISFET gate (gate 1 in Fig. 4(*b*)). Many potentiometric-based enzyme sensors have been described, e.g. for the determination of glucose (see Table IX), or lacetate, pyruvate, and amino acids (see Table X), and of other substances, like penicillin (*28*), and for the detection of immunoreactions, like PIMIA (potentiometric ionophore modulation immunoassay) (Table XI, and Fig. 12). The ENFETs are relatively new and were used for the determination of glucose ((*29*), Table IX), penicillin (*30*), urea (*31*), and creatinine (*31*), for instance. Although the ENFETs provide many advantages over the ISEs, such as cheapness, robustness because of their solid state nature, small size, and the possibility of electronic compensation of changes in temperature (*32, 33*), the fact that considerable problems remain in the appropriate engineering of the sensing surface and in the interpretation of the mechanisms of sensor response has prevented widespread use so far of ISFETs and ENFETs.

3.1.2. Amperometric Devices

Amperometric devices are based on the principle that at a certain fixed potential the current intensity (electrode polarization) is proportional only to the surface reaction of the electrode (e.g. in the measurement of O_2). There is a linear function of the concentration of the analyte and the redox reaction. For instance, the indirect detection of the redox reaction via the determination of O_2, of H_2O_2, or directly via electron mediators like meldola blue (*34*), nile blue (*35*), ferrocene (*36*), or via FAD/$FADH_2$ and NAD/NADH at the electrode surface (*37*) was employed for determination of glucose (Table IX), lactate and pyruvate (Table X), etc., by the use of the corresponding immobilized enzyme. Table III shows a short selection and Fig. 5 illustrates the various modes of mediated electron transfer to the electrodes.

3.1.3. Optical Transducing Devices

Optical transducing devices, largely based on fibre-optic technology, vary widely, but are usually sensitive to changes in fluorescence or other properties, such as light transmissivity, that may result from biological processes. Very important advantages are miniaturization, low costs, and disposability. Classical fibre-optic devices were used, for instance, for the determination of pH or for the determination of pH changes caused by an enzyme reaction,

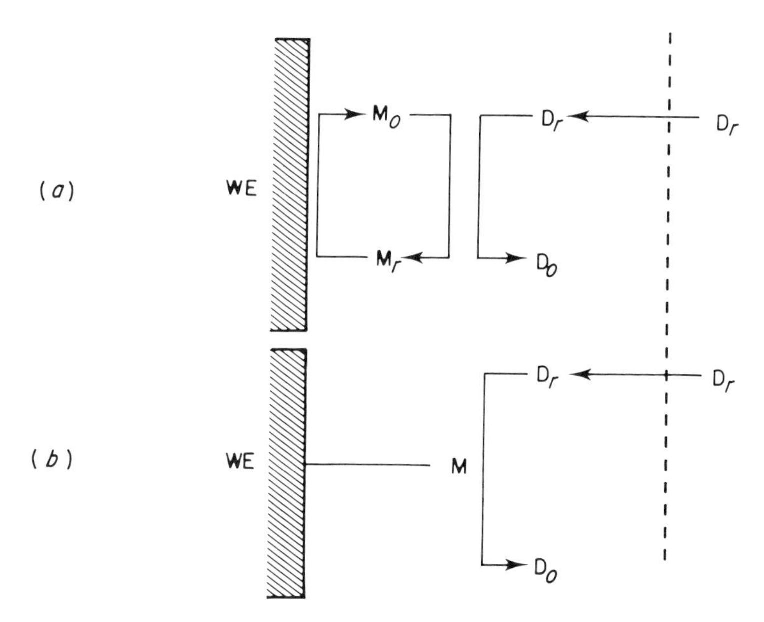

Fig. 5. Various modes of mediated electron transfer to electrode devices: (*a*) via soluble mediator, (*b*) via immobilized mediator. M, mediator; WE, working electrode; D, electroactive material; subscripts *o* and *r* refer to oxidized and reduced forms, respectively (taken from ref. (*26*).

both via the use of a pH-sensitive dye. Lowe *et al.* (*38*) demonstrated these applications for the determination of glucose, urea, and penicillin. Other applications are the glucose affinity sensor, described by Schultz *et al.* (*39*) (shown in Table IX and in Fig. 10) and the detection of antigens or antibodies via surface plasmon resonance (*40*) or at glass–liquid interfaces (Fig. 14) (*41, 42*). A very simple principle is the application of enzyme reactors with subsequent optical devices for the determination of colour-producing reactions (e.g. the Technicon system, Table IX). Details are given in Section 4.

Because of the simplicity of luminescence techniques (the chemically or biologically induced light is observed directly without the need for an excitation source and light-degrading optics) in comparison with the classical optical methods, these will find ever wider application. On-line measurement methods, based on luminescence techniques, were used in the determination of ATP (*43, 44*), NADH (*45*), bile acids (*46*), etc., in which the enzymes are coupled on the inside of nylon tubes.

3.1.4. Calorimetric Devices

Calorimetric devices are based on the measurement of the thermal effects caused by a reaction (mainly enzyme reaction), by virtue of the resistance of the device being sensitive to temperature changes. The construction principle

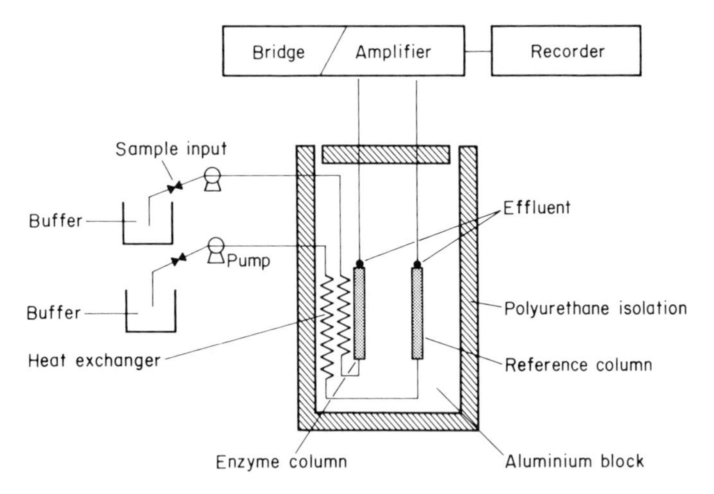

Fig. 6. Enzyme thermistor (taken from ref. (*172*).

is shown in Fig. 6. Generally, the enzymes (enzyme thermistor) or the antigen or antibody (TELISA—thermistor enzyme linked immunosorbent assay) are coupled to a solid support (e.g., enzyme reactor). These devices were used for the determination of glucose (*47*) (Table IX), lactate and pyruvate (*48*) (Table X); and of antibodies or antigens (*49*) (Table XI, and Fig. 11).

3.1.5. Conductometric Devices

Conductometric devices function by the measurement of the time dependence of the conductance change when an AC signal is switched between a pair of reference electrodes (equipped, for example, with a "blank" membrane) and of catalytically active electrodes (e.g., equipped with a membrane-coupled enzyme). Enzymes such as urease produce ionic products (NH_4^+, HCO_3^-) from neutral substrates (urea) and can thus be monitored by the increasing conductance of the solution (*25*).

3.1.6. Piezoelectric Crystal Devices

Piezoelectric devices are based on the measurement of a mass change by a certain reaction. They have been used for the measurement of volatile substances and can be used for the analysis of bioreactor off-gas (*50, 51, 52*). Roederer and Bastiaans (*53*) described a device for the determination of antigens or antibodies in samples. Details are given in Section 4, Table XI, and Fig. 13.

3.1.7. Other Principles

Although gas chromatography and HPLC are not biosensing techniques, they must be mentioned briefly. They can be used instead of a specific detector because of their ability to separate and subsequently detect different classes of substances, such as amino acids, peptides, high-molecular-weight acids, sugars, etc., in an on-line mode. Examples were given by McLaughlin *et al.* (*54*) for gas chromatography and by Rousseau (*55*), McLaughlin *et al.* (*54*), and Dinwoodie and Mehnert (*56*) for HPLC.

3.2. Sensor Applications; Sampling Devices

The application of biosensors for the control of bioreactors is difficult. Generally, the following requirements have to be met.

1. The biosensor should be autoclavable/steam-sterilizable.
2. The biosensor should show a long lifetime.
3. The biosensor should not show drifting phenomena.
4. The possibility of recalibration should be available.
5. The device should show fast response characteristics (not so important in animal cell fermentations, but required for use in FIA).
6. The sensor should work in real-time or nearly real-time, when applied in flow-line systems.
7. The sensitivity of the device should be sufficient, when applied to FIA systems, to cope with the sample dilution.
8. The whole device (sensor, sampling system) should not be too expensive.

The main drawback in the direct application of the biosensor to on-line determination of physiological parameters in cell fermentation is still that these sensors are not autoclavable, because of the inactivation of the biological material coupled to the sensor. Although this is a basic problem, there are many possibilities for the connection of the sensor to the fermenter, as shown in Fig. 7. The direct insertion of a sensor is only possible when it is steam-sterilizable (e.g., pH-probe, pCO_2-probe, or optical sensor, shown Figs 7(*a*)–(*c*). In most cases direct insertion of the sensor into the bioreactor is

---

Fig. 7 (overleaf). Possible connections of sensing devices to the fermenter; all connections are shown without filtration devices. (*a*) Direct *in situ* probe (e.g. pH-probe). (*b*) Direct *in situ* probe with sensing surface protection (e.g. optical sensors). (*c*) *In situ* probe with a membrane protected sensor (e.g. pO_2-probe). (*d*) Flow systems with sensors. (i) *In situ* dialysis system where the sensors are fixed in the downstream. (ii) Semicontinuous sampling by flow injection analysis principles (*171*). (iii) Semicontinuous sampling by flow injection analysis principles with a dialysis unit (*146*). (iv) Continuous flow-line sampling. (*e*) Flow systems with enzyme reactors or other separation/detection systems. (i) System with the use of an enzyme reactor and a detector (e.g. (*34*)). (ii) System with the use of parallel to the sample added enzyme and a detector (merging zone FIA (e.g. (*170*)). (iii) System equipped with a separation system (HPLC, GC) and detector only with a filtration system applicable.

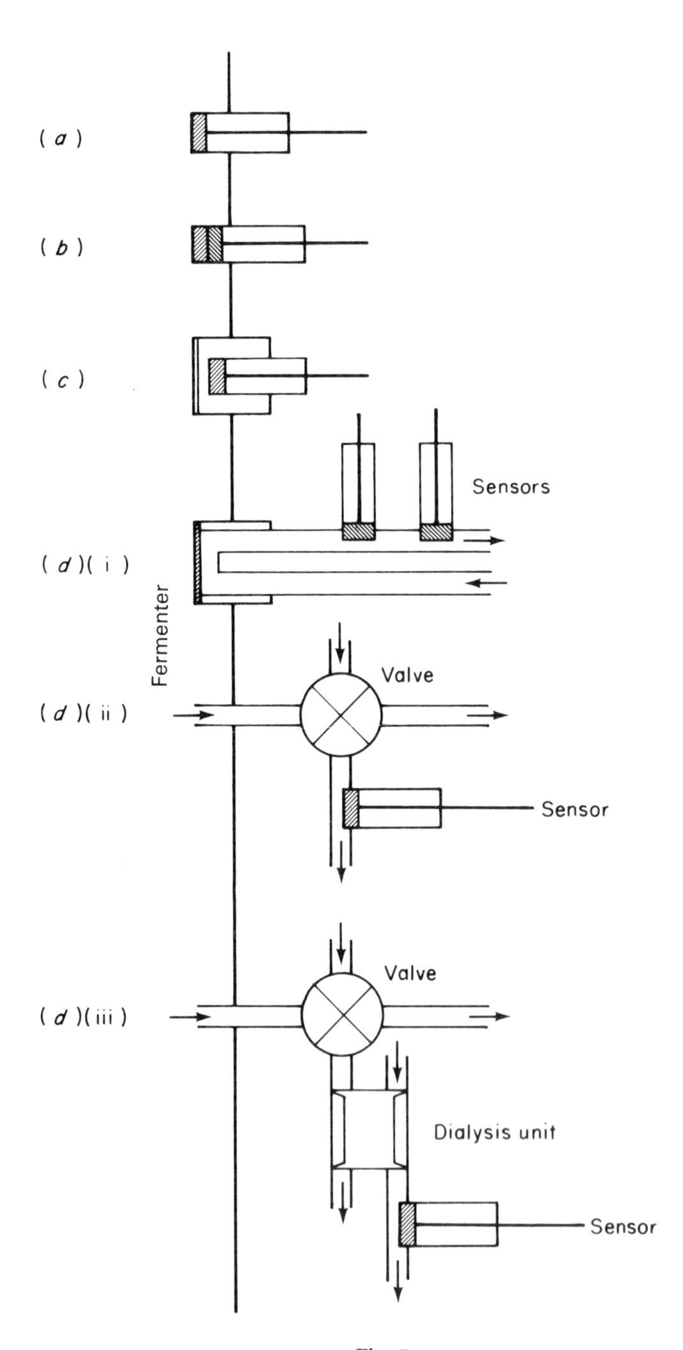
(a)
(b)
(c)
Sensors
(d)(i)
Fermenter
Valve
(d)(ii)
Sensor
Valve
(d)(iii)
Dialysis unit
Sensor

Fig. 7.

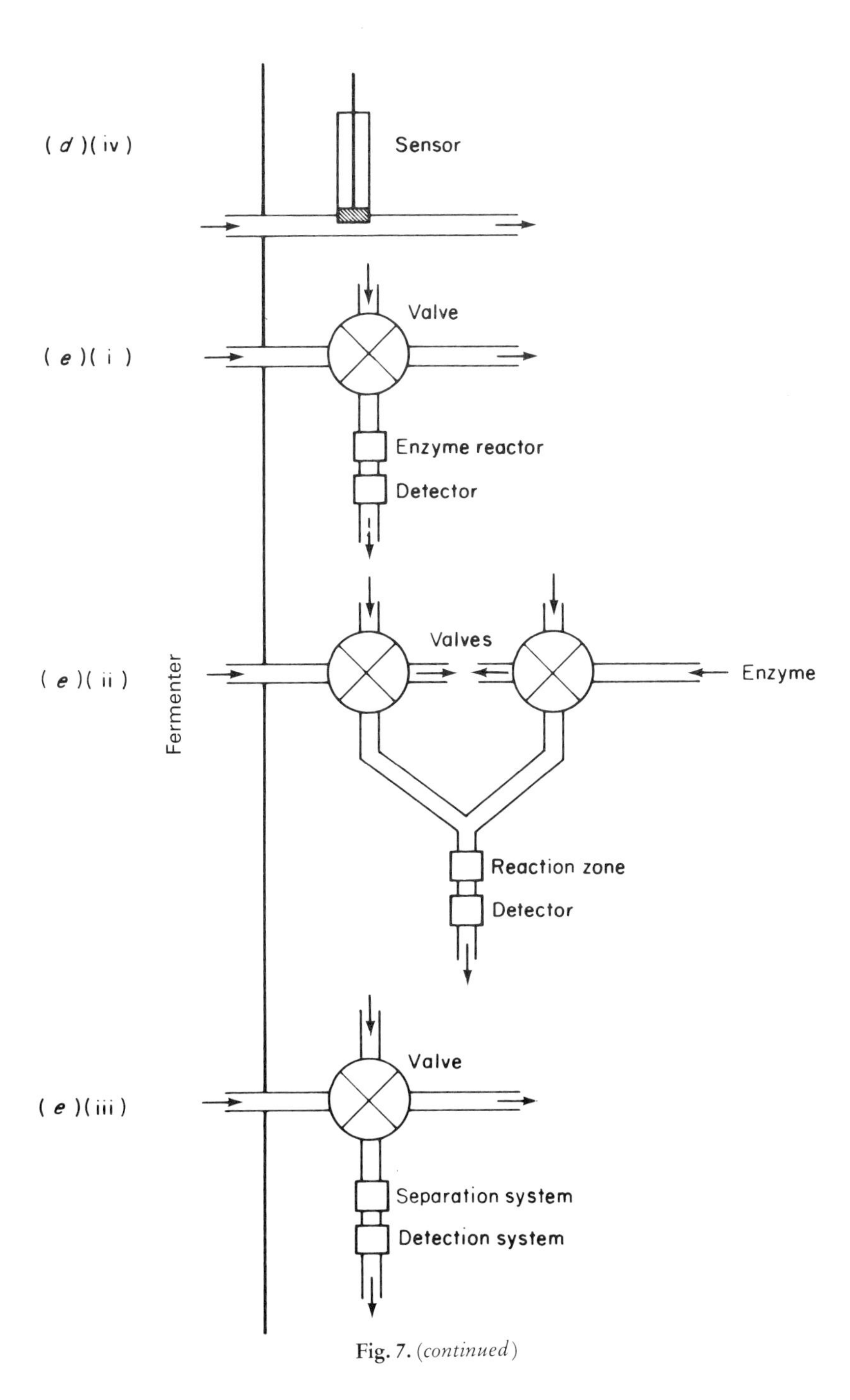

Fig. 7. *(continued)*

not possible. The main reason is the lack of autoclavibility. Another is the necessity for repeated recalibration of the sensor because of drifting problems. A very short lifetime sensor has to be replaced from time to time. The necessity for repeated recalibration and the ability for changing the sensor are already catered for in some sensors which are applied directly *in situ* (e.g., the INGOLD changeable electrode; recalibration of an *in situ* electrode, see e.g. (*57*)). But all three problems together can be only solved by the indirect application of these sensors. The possible configurations are shown in Figs 7(*d*)–(*e*). All principles shown in these figures can be used with or without filtration systems, employing dialysis, ultrafiltration, or microfiltration membranes. Using cell cultivation systems, already equipped with a filtration system, like hollow-fibre bioreactors, sampling devices without additional filters can be employed, as cell debris and other cellular materials are already retained by these culture systems. The overall advantage of filtration devices is the separation of cells from the supernatant, with positive effects on the subsequent biosensor (for details see Section 3.3), the disadvantages are fouling, clogging, and poisoning problems, which are general problems of membrane-based biosensors themselves. Fouling, clogging, and poisoning, which are caused by the development of a proteinaceous film (*58*), by the deposition of cells on the membranes, or by fatty acids, interfere with the sensor directly or indirectly (in the case of a filtration device), causing drift, high noise, longer response time, loss in sensitivity, and finally damage to the filtration device or sensing system. According to Clarke *et al.* (*59*), pre-fouled membranes should show the best performance because of their relative resistance to further fouling, which is an interesting consideration with regard to filter-based sampling devices. Figure 7(*d*) presents a number of configurations for attachment of an enzyme electrode to a fermentation vessel. Figures 7(*e*)(i) and (ii) show two other sensing configurations, which can be equipped with a filtration system or not (this is valid for Fig. 7(*d*) also). In Fig. 7(*e*)(i), the system is based on an enzyme reactor with a subsequent detector, where the working life depends on the stability of the immobilized enzyme ((*34, 35, 47, 48, 60, 61*), Technicon system, Table IX). The other system, a (stopped-) flow-injection-analysis (FIA) system with merging zones, is a flow-line system too, but not a real biosensor (Fig. 7(*e*)(ii)). In this case, soluble enzyme is added to the sample in parallel. For each sample new enzyme is used (*62, 63, 64*). Figure 7(*e*)(iii), finally, presents a configuration which is used for the on-line connection of a gas chromatograph (*54*) or HPLC (*54, 55, 56*) to a fermenter.

It is necessary to mention that all systems employing GC, HPLC or enzyme reactors have to be equipped with a filtration system. The advantages and disadvantages of the different biosensing systems and configurations are presented in Section 3.3. Figures 8(*a*) and 8(*b*) present some possible filtration

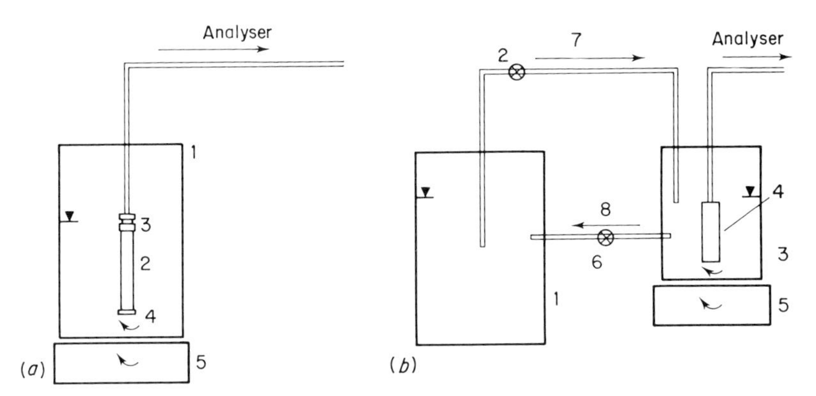

Fig. 8. Possible filtration systems for automatic cell-free sampling. (*a*) Invasive system: (1) fermenter, (2) rotating porcelain filter (1 μm), (3) rotating seal, (4) magnetic bar, (5) magnetic stirrer. (*b*) External system: (1) fermenter, (2, 6) pump, (3) filtration unit, (4) rotating filter (1 μm), (5) magnetic stirrer, (7) cell suspension, (8) cell suspension minus cell-free sample.

devices, whereby two different groups can be defined: the invasive and the external filtration systems. A summary of these devices is outlined in Table IV. The fact that fouling and clogging of the filtration membranes are the main problems has led to special constructions for preventing these drawbacks and for improving their lifetime. The first group, the invasive filtration devices, show the main drawback that they cannot be changed during a fermentation process, once they have clogged. In spite of this limitation, some applications have been published. To prevent, or slow down, the clogging several devices, firstly dialysis units (*65, 66, 67, 68*) and secondly flat-membrane devices (*69*) were placed near the stirrer of the fermenter or they were themselves equipped with small magnetic stirrers (*68*) to induce a continuous flow at the membrane surface. The flat-membrane device was backflushed periodically (*69*). In spite of special construction, the lifetime did not exceed 7–14 days in the case of the dialysis units, used in refs (*65*) and (*68*). The dialysis units provide another advantage—that the sample is taken already diluted, which is a frequent requirement for the working range of the subsequent sensor. This is done mainly by the proper selection of the dialysis membrane, its area, and the dialysis fluid flow rate.

Another solution was published by Feder and Tolbert (*75*) and Tolbert *et al.* (*76*) who used a cylindrical rotating (200–300 rpm) porcelain filter (pore diameter 1 μm) for perfusion culture systems. This system would be quite useful for application as a cell-free sampling device, especially because of the relative long working time.

These invasive filtration systems, although easy to handle, have the draw-

TABLE IV Filtration Systems

Filtration system	Cell recirculation	Sterilization	Lifetime (days)	Agitation	Time to reach 90% steady state (min)	Membrane	See figure	Reference
Invasive systems							8.a	
Dialysis unit	*In situ* cell retention	*In situ*	7	Stirrer of the fermenter	4.7	Dialysis (MW 12 000)[b]	7.d.i	*65*
Dialysis unit combined with enzyme electrode	*In situ* cell retention	*In situ*, afterwards the enzyme electrode is added	Some days	Stirrer of the fermenter	5–10	Dialysis		*66, 67*
Dialysis unit	*In situ* cell retention	*In situ*	7–14	Magnetic stirrer, tangential filtration	?	Ultrafilter (MW 30 000)		*68*
Flat membrane	*In situ* cell retention	*In situ*	?	Stirrer of the fermenter[a]	?	Nylon, 0.2 μm		*69*
External systems							8.b	
Amicon filtration cell	Cell recirculation	Steam	20	Stirred	?	0.2 μm, 20 cm^2	8.b	*70*
Flat-membrane cell	Cell recirculation	Steam	20	Stirred	10	0.2 μm, 45 cm^2	8.b	*71*
Amicon hollow-fibre	Cell recirculation	Not steam-sterilizable, other methods	?	—	Some minutes	10^5 daltons, 3×10^5 daltons	8.b	*54*
Gelman Acroflux capsule	Cell recirculation	Not steam-sterilizable, other methods	?	—	?	0.2 μm, 900 cm^2	8.b	*56*
Millipore Pellicon	Cell recirculation	Steam	?	—	?	0.2 μm, 1 ft^2	8.b	*72*
Radial flow ceramic filter module	Cell recirculation	Steam	?	—	1–2	0.2 μm	8.b	*73*
Continuous filtration device	No	Steam	?	Filter is changed after each sampling	7	Filter		*74*

[a] The membrane is backflushed periodically.
[b] Molecular weight cut-off in daltons.

back of a limited lifetime. Once they have clogged, their use is ended and they can not be changed. Therefore, external filtration systems are recommended, in which medium and cells are recirculated from the bioreactor to the filtration device and back (Table IV, Fig. 8(*b*)). The advantage is the possibility for sterile change of the system, once clogged, without stopping the bioreactor. Three main groups can be identified: (1) filtration systems with agitation and cell recirculation; (2) filtration systems without agitation with cell recirculation (hollow fibre); and (3) filtration systems without agitation, but with continuous change of the filter and cell loss.

1. *External filtration system with agitation*. Filtration devices of the first group were used by Chotani and Constantinides (*70*) and Kroner and Kula (*71*). The cell suspension (microorganisms) was cycled from the fermenter to the filtration cell and back. The space over the filter is stirred magnetically to prevent clogging. The disadvantage of this system is that magnetic stirring can damage the animal cells. Therefore, two other systems, one described by Tolbert *et al.* (*76*) and Tolbert and Feder (*75*) as an external perfusion system for a high-cell-density culture, the other described by Rebsamen *et al.* (*77*) as a rotating filtration system (tangential filtration) for cell separation, should be the better choice: both have been used for cell culture and showed excellent features over long times (the shortest one month (*75*)), and they should be applicable as continuously working sampling systems.

2. The second group represents all hollow-fibre filtration units. Many different systems, presented in Table IV, have been used, but only in connection with microbial fermentations. Their main problems are that they are constructed without regard to the shear sensitivity of animal cells. The relative high flow rate (to prevent clogging, up to 50 ml min^{-1} (*70, 72*)) and the spacers of the hollow fibres cause cell damage. No data on the lifetime prior to clogging are available. One caution for the cell recycling systems must be entered: they can be employed only if special pumps (not peristaltic ones) are used which do not damage the cells. Pumps such as those used by Rebsamen *et al.* (*77*), Tolbert *et al.* (*79*), or Vick Roy *et al.* (*78*) are recommended.

3. The last possibility is a system in which new filters are always used (*74*). In addition, the flow line between the bioreactor and the filtration unit is steam-sterilized after each sampling. The only, but minor, disadvantage is the loss of cells, which might be a problem in small-laboratory bioreactors.

3.3. Problems Arising During the Use of Biosensors; Requirements for the Use of Biosensors

3.3.1. Sampling Devices

Although there exists an excellent review of the development and application of biosensing devices for bioreactor control (*59*) some problems which might arise will be outlined below. Generally, samples can be taken with or without the use of filtration devices. Using sampling devices without filtration units can entail the following drawbacks. Besides the lack of a filtration unit which is used as a sterile barrier to prevent infections of the culture, the removal of cells from their environment will change their physiology. The main changes are the decrease of the temperature and the reduction of the oxygen partial pressure, which might influence the subsequent measurement of parameters which are in direct relation to the cell physiology (such as ATP, the NAD/NADH relation). Short tubings between bioreactor and detection device help prevent these problems. But because medium compounds (nutrients, product, wastes) are to be detected in most cases, sampling devices using filtration units are recommended. There are good reasons for doing so: first, the bioreactor is protected against possible infections; second, the cell loss is reduced in most of the cases; and third, the sample is already filtered, which will increase the lifetime of the subsequent detection system. The problems arising from the fouling and clogging of these filtration devices were mentioned above (Section 3.2).

Summarizing, the sampling devices of a flow-line sensing system have to meet the following requirements.

1. The removal and analysis of the sample has to be representative.
2. The environment (e.g., temperature, pressure) of the sensing system has to be constant and must not be influenced by the sample except by the parameter to be detected.
3. There should not be any contamination, loss or vaporization of the sample.
4. Recalibration and a system providing recalibration is a necessity.
5. Finally, quantitative means of sample dilution has to be provided, when necessary.

Although these requirements are not necessary for all applications, their provision should be possible in an optimal and simple way. The sampling time (time between bioreactor and sensing device) is only problematic when changes of the analyte are expected (see above). Considering the low rate of change of parameters in a low-density animal cell fermentation, a sampling and detection time of some minutes would not cause problems. The situation might be different in high-cell-density fermentation systems, but data with respect to these systems are not so far available.

3.3.2. Flow Systems

If the sensing device is not applied directly to the fermentation vessel, which is the general case, flow-line systems, like flow injection analysis (FIA) or air-segmented flow systems (Technicon) are used. The advantages or disadvantages of these systems are not discussed here; for details see ref. (*80*). A review of the application of FIA in on-line process control was published recently (*81*). One important feature of the FIA has to be mentioned in detail here: by a proper selection of the injection frequency, a nearly complete return to the baseline is achievable between two successive injections. In this way it is very easy to correct for slow fluctuations of the background (*81*). Another advantage of the FIA is the adaptation of the detection range of the sensor by changing the sample size or the flow rate (i.e. variation of the contact time between sample and sensor).

In most cases, immobilized or entrapped enzyme- or (in the case of immunosensors) antigen- and antibody-based sensing devices are used. The main reasons for the preference for use of immobilized enzymes in biosensors are better selectivity, time stability (*82, 83, 84*), and undoubtedly enzyme economy. According to Guilbault the time stability is dependent on various factors (*85*): the type of entrapment (the chemical type provides the best stability), the enzyme content, and purity of the enzyme support (more enzyme per sensor is recommended to provide a longer lifetime), the optimum conditions of the enzyme (optimal working conditions, such as optimal pH range), provide longer lifetime or reduced inactivation), and finally the stability of the base sensor. (In most cases, the base sensors are more stable than the immobilized enzyme, but this factor has to be considered in the use of some shorter-lifetime electrodes, such as liquid membrane electrodes.) Unfortunately, these immobilized enzyme, antibody or antigen systems have some drawbacks: loss of activity caused by leakage of the immobilized material (*82, 83*), autoinactivation of the enzyme (*86*), gradual occlusion of the active surface, and in the case of enzyme reactors carryover of the sample, blocking of the column by particulate matter, and compressibility of column material, which causes increase in the back pressure and eventual decrease of the flow (*82*). By replacing the enzyme reactor with a nylon tube immobilized enzyme (*44, 45, 46*), the drawbacks of enzyme columns are prevented. In addition, a greater operational stability can be obtained. Because the loss of immobilized enzyme is inevitable, and there must be sufficient activity in reserve, large amounts of enzyme must be insolubilized (already mentioned above). Therefore, Ruzicka and Hansen (*62*) have stated that enzyme consumption is more or less the same, both in FIA-systems with merging zones and in systems working with large amounts of immobilized enzyme. The FIA system has the advantage that the same enzyme activity is always available for the reaction.

3.3.3. Sensing Devices

In all cases where a non-specific influence of some medium compounds on the response of the sensor is to be expected, reference systems have to be installed, as by Decristoforo and Knauseder (*87*) for the determination of cephalosporin, or by Danielsson *et al.* (*60*) for the determination of glucose. This approach is valid for all enzyme electrodes, where changes of pO_2, pCO_2, or NH_4^+, etc., are detected, where unspecific heat caused by the dilution of the sample influences the thermistor signals (*87, 88*); where the colour of the sample has to be compensated in the cases of optical sensing systems, and where unspecific adsorption of proteins on the sensor has to be compensated. In addition, the influence of the sample matrix has to be taken into account so that the ideal situation would be that the difference between sample and flow buffer (flow-line system) is the presence or absence of the analyte only. In this case, the influence of the differences between the matrix of the sample and the flow buffer are minimized (for details, see refs (*89, 90*)). Transducer elements (ISEs, etc.) cannot be expected to maintain their calibration for long periods, and large drifts as well as changes in selectivity and working range can be expected. In most cases, ISEs (like the pH electrode) are calibrated only at a single point. Instead, a two-point calibration is recommended, in which the likely ranges of the analyte concentrations should be covered during a fermentation process. This is, for instance, provided by an FIA system. The detection range of biosensors can be adapted by the use of dialysis units or FIA systems (see above).

A common problem in biosensor applications that utilize enzymes, and that use two substrates (e.g. glucose oxidase catalyses the oxidation of glucose with O_2 (the second substrate) to gluconic acid and H_2O_2 (see equation (1)), is the dependence of the O_2 partial pressure of the sample (this applies to many oxidases without the use of electron mediators). Much work has been done to solve this problem. Primarily, it should be noted that the enzyme reaction is diffusion-limited. Special care has to be taken as the limiting step of the reaction is the diffusion of glucose (in the above-mentioned example) and not the diffusion of O_2 (*91*).

Some groups have described solutions in which the oxygen partial pressure was increased. Romette *et al.* (*92*) used a flow system in which the oxygen sensor was covered with a special membrane. The enzyme was immobilized to this membrane which dissolved the oxygen about 20 times better than can pure water. Between every measurement, air was flushed through the sensor to compensate for the oxygen used. Gough *et al.* (*93*) described a special construction based on an oxygen electrode in which the glucose diffuses only from one side to the sensor, whereas the oxygen can diffuse from two sides to the immobilized enzyme. Another solution was mentioned by Danielsson *et al.* (*60*) and Cleland and Enfors (*66, 67*). Both groups used catalase to regenerate half of the used oxygen, with the effect that the glucose oxidase is protected from the deleterious effects of H_2O_2 (*94*).

Potentiometric devices. Both potentiometric transducers (ISE and FET) exhibit non-linear, but logarithmic, dependence on the concentration of the analyte. This means that it is not the peak height but the absolute value at the peak that is of interest. The baseline can be rather indeterminate and often shows large fluctuations. This is specially troublesome in the case when low analyte concentrations have to be detected. Additionally, in cases of a complex sample matrix like that of fermentations, it is not the true concentration but the activity of the analyte that is measured. It is clearly important to convert the free activities into total concentrations. The algorithmic formulation is described clearly in the review by Clarke *et al.* (*90*). Some examples of this situation are described in the review by Merten *et al.* (*89*).

In cases where cross influences are expected, reference detection systems have to be employed. This is valid for all potentiometric-based enzyme electrodes, registering the change of pH, NH_3 (in case of a gas electrode), NH_4^+ (in the case of ISE), or the change of pCO_2 (gas electrode), whereby the reference electrode compensates for the influences of the medium. In addition, the nonactin-based NH_4^+ electrodes show interferences with K^+ (15%) and Na^+ (about 0.13%) (*95*). This electrode can be used only if there is no great change in the K^+ and Na^+ concentration in the sample. The PIMIA (Fig. 12) shows the main disadvantage: that it is necessary to keep the background K^+ concentration constant, because there is a great response to changes in the K^+ concentration.

Amperometric devices. These devices, and also the optical, enthalpimetric, and conductimetric devices, show a linear response to changes in the analyte concentration. Thus the magnitude of the signal can be measured from the baseline, which allows correction of the drift. In most of the cases O_2 depletion or H_2O_2 increase is measured by the appropriate electrode configuration (the electrode is polarized at about −650 mV or at 600–800 mV, respectively). These two classical measuring systems have a lot of disadvantages. The oxygen measuring system depends on the O_2 partial pressure of the system and it is necessary to promote transport of the dissolved oxygen to the enzyme (which was already mentioned above). The determination of the produced H_2O_2 can be disturbed by peroxidases, which are produced in large quantities by mammalian cells (a membrane with a low molecular weight cut-off prevents this disturbance). Another important point for the determination of H_2O_2 is that attention has to be paid to the influence of uric acid, ascorbic acid, tyrosine, iron(II), and other reducing substances (*91*) on the electrode response. This can be easily prevented by the use of an appropriate membrane, preventing the diffusion of these substances to the electrode (e.g. asymmetric membranes were used by Tsuchida and Yoda (*96*) and Tsuchida *et al.* (*97*); membranes with a low molecular weight cut-off were used by Ho and Asouzu (*98*); a cellulase acetate membrane placed between the electrode and the enzyme layer was used by Mullen *et al.* (*99*)). Other electrode configurations provide the following advantages over the H_2O_2 or O_2 electrode: for improving the life-span of the enzyme used, it is necessary to get rid of the H_2O_2, and to improve the accuracy it is necessary to get rid of the pO_2 and H_2O_2 measurement. The direct electron transfer via electron mediators, already mentioned in Section 3.1.2, shows promising results. Using mediators, the following requirements have to be fulfilled: it should accept electrons rapidly from the reduced enzyme; and should exhibit good electro-chemistry at a practical electrode. It should be immobilized easily and be stable in both the reduced and oxidized forms, the reduced form should not react with oxygen. Its redox potential should be low, so that no extraneous compounds can be oxidized; it should be independent of the pH (*91*).

Appelqvist *et al.* (*34*) and Yao *et al.* (*101*), for instance, used an electrode potential of −50 mV, which prevented the oxidation of reducing substances.

The direct transfer of electrons from the prosthetic group of the enzyme ($FADH_2$ of glucose-oxidase (*102, 103*) for instance) to the electrode surface still remains problematic, because the continuous reduction and oxidation of the FAD/$FADH_2$ system is not completely reversible. At present this system is stable for some days only.

The recently described system using PES (phenazine ethosulphate) or DCPIP (2,6-dichlorophenolindophenol) as electron acceptor (*104*) for glucose determination, using a quinoprotein—glucose dehydrogenase—is too unstable.

In general, electrochemical sensors are influenced by media, which might form deposits on the surface in such a way that the nature or rate of electron transfer reactions at the electrode surface will be changed. The possibility of electrical interference has to be kept in mind.

Optical devices. These devices show one especial disadvantage. Tissue culture media that exhibit a tendency to formation of deposits will cause problems because deposits on optical windows will affect the transmission of light, in a classical system, or will change the evanescent wave at the interface between two optical media (*41, 42*: optical detection of the Ag–Ab reaction (antigen–antibody) at the glass–liquid interface based immunoassay). Another obvious disadvantage is the need to minimize or exclude ambient light from the vicinity of such sensors. However, pulsed light systems have largely overcome this problem in many applications. These drawbacks can be prevented by using other detection systems (electrochemical, enthalpimetric, or piezoelectric devices), because these are insensitive to turbidity or colour. But even optical devices, in this case fibre-optic devices, can be used to overcome some of these problems, and provide some advantages over other types of biochemical sensors. Clarke *et al.* (*59*) summarized these advantages as:

1. Absence of electrical interference.
2. No analogous requirement for a reference electrode.
3. Sensing reagents do not always have to be in direct contact with the fibre-optic probe.
4. Straightforward replacement of sensing elements is facilitated.
5. Optical sensing principles can offer significant cost advantages.
6. Sensors can be made more fouling-resistant without influencing their analytical performance (*105*).
7. Optical sensing principles can be envisaged for most analytes, most significantly where other sensing principles do not exist.
8. Construction of multisensor probes.
9. The robustness and ease of sterilization of fibre-optic probes.

Calorimetric devices. These devices are strongly affected by artifactual signals such as heat caused by differences of pH, ionic strength or viscosity between sample and perfusing buffer passing the thermistor and by adsorption and desorption of various compounds, which may take place when the sample passes through the column (*88, 89*). Such interactions between the sample (e.g. proteins) and the matrix may generate heat non-specifically (*88*). The main compensation possibility, published by Mattiasson *et al.* (*88*), is the use of a split-flow enzyme thermistor, whereby two identical microcolumns are employed, one of which contains an immobilized enzyme preparation, the other an inert support material.

Conductometric devices. These may be influenced by the high conductivity of the cell culture medium.

Piezoelectric devices. The main problems, arising from the use of these devices in on-line immunoassays (MGIA—microgravimetric immunoassay, Fig. 13), is the poor sensitivity and the non-specific adsorption of compounds of the medium or supernatant, although reference systems are used (Section 4.3).

4. APPLICATIONS

Although there are many important parameters in animal cell fermentation, only some applications are shown in detail. The determination of biomass, of glucose, and of the product will be given in detail; some other applications are presented in Table X, where only the most interesting are outlined—measurement of lactate, galactose, pyruvate, NH_4^+, and amino acids. The detection of product, in general, and enzymes, in particular, is mentioned briefly.

4.1. Determination of the Biomass

During a fermentation process, the determination of the viable and total cell count is one of the most important measurements; it is done discretely by counting the trypan-blue-stained and unstained cell suspension in a haemocytometer. This method cannot easily be automated. As shown in Table V, chemical or microscopic methods have disadvantages, such as the need for taking samples, the addition of reagents, off-line rather than real-time character, and the discontinuous mode of determination. In comparison, physical methods show advantages in that they operate in real-time, *in situ*, and they are non-destructive.

Only two chemical methods are mentioned here briefly. First, the determination of ATP by bioluminescence assays; second, the determination of double-stranded DNA by mithramycin staining and measurement of the fluorescence. Details are shown in Table VI. It should be noted briefly that the NADH content of the cells can be measured by luminescence measurements (for details, see ref. (*45*)), but the problems arising from sampling, etc., are similar to those for measurement of ATP.

In summary, the determination of the cell ATP content is possible automatically, but the problem of the influence of the physiological state remains (*110*). Despite this drawback, the determination of the ATP content and of the NAD/NADH-relation in the cells (see below) provides information about the physiological state. The optimal production of a product is linked to a certain stage of the cell physiology and the cell growth (e.g. monoclonal antibody production is dependent of the stage of the growth of the culture

TABLE V Comparison of Physical and Chemical Methods for the Determination of the Cell Count

Physical methods	Chemical methods
Non-destructive of the sample	Destructive of the sample
Addition of reagents not necessary	Addition of reagents necessary
Mostly real-time	Not real-time
Continuously working	Discontinuously working
In situ	Sampling necessary

TABLE VI Details of the Measurements of ATP and DNA for Chemical Determination of Biomass

	Determination of ATP	Determination of DNA
Principle	*ATP* + luciferin + O_2 – oxyluciferin + AMP + PP_i + CO_2 + *light*	*DNA* + mithramycin – interaction of DNA with mithramycin – *fluorescence*
Separation cells–medium	Not necesssary	Necessary, if phenol-red-containing media are used
Extraction	Necessary for ATP	Necessary for DNA
Addition of reagents	1. Extraction buffer 2. Luciferin/luciferase 3. Internal standard: option	1. Mithramycin-solution
Range	1×10^5–2×10^6 cells ml^{-1}	1×10^5–1.5×10^6 cells ml^{-1}
Problems	Variation of cellular ATP content with the change of the physiological conditions	Centrifugation is necessary if phenol-red-containing media are used; sonication for cell disruption is necessary
Application	Only possible in well-known standard processes, because of the influence of the physiological state	Possible after the solution of associated problems
References	(*106*) (*107*)	(*108*) (*109*)

(*110*)). Using continuous cultivation (low- and high-cell-density systems) information about the physiological state of the culture is necessary for optimal control. Unfortunately, on-line determination of ATP has been used only in the case of yeast fermentations (*107*), not for control of animal cell fermentations.

The determination of the cell DNA content is an exact method because of the stable content of DNA in cells. The major problem in automation is the necessity for cell separation, because of the influence of the medium, colour and sonication. But both problems should be solved by the use of phenol-red-free media and the use of DNA extracting reagents. The advantage of the systems is the determination of the viable cell count.

Many papers have been published on the use of fluorescence sensors for the determination of intracellular NADH, mainly in microbial fermentations (*111, 112, 113, 114*). Although the NADH content correlates well with the cell density during the lag and exponential growth phase of a batch culture, the intracellular NADH content rises during the stationary phase, indicating changes in the physiology of the microorganisms (*112, 114*). This was found to be valid for animal cells, too (*115*). The advantage is that the NADH

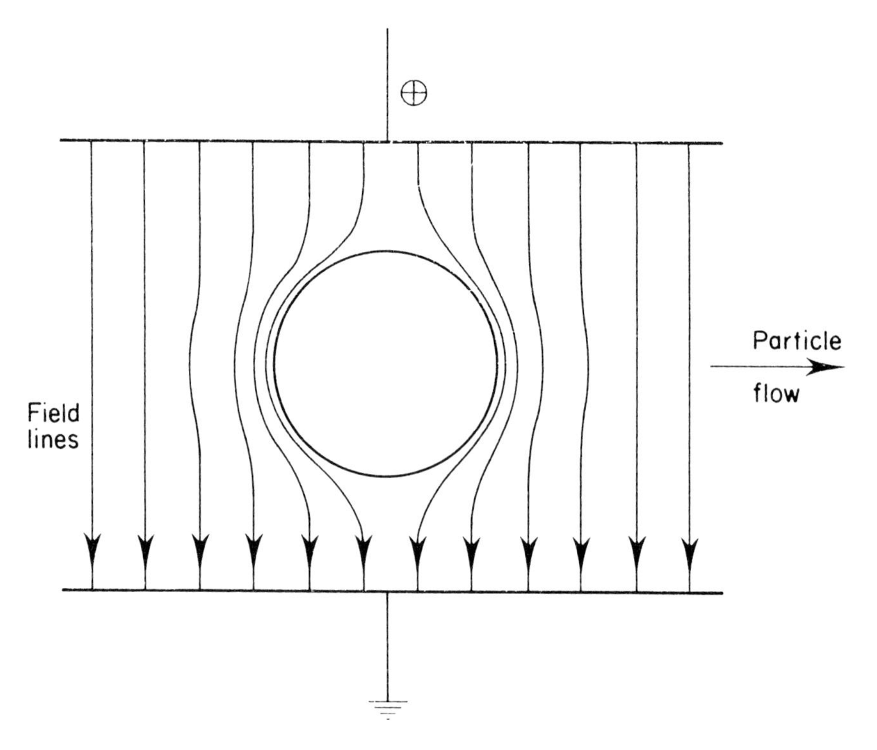

Fig. 9. The principle of an electronic counting device (Coulter counter). *Principle:* The electronic counting technique monitors the effect of cells on an electric field as the cells traverse the field. The cells are suspended in the growth medium which is electronically conductive, and have to flow through a small aperture across which an electric field is applied via a constant current source. When the relatively non-conducting cells pass through the field, the electrical resistance within the aperture increases, giving rise to a transiently increased voltage drop across the aperture. Under certain conditions this magnitude is proportional to the size of the cells. The pulses per time correspond with the cell number per ml. *Advantages:* Differentiation between living and dead cells (*125*), possible counting of adherent cells directly on microcarriers (*126*). It is possible to measure particles in the range 0.4–800 μm (Coulter Electronics Inc). *Disadvantages:* The sample signal has to be much greater than the background signal. Unusually large pulses have to be minimized. The cell concentration has to be in a certain range: high cell concentrations increase the occurrence of coincident counts; low cell concentrations decrease the signal-to-noise ratio. The path of the cells through the field should be uniform. The magnitude of the applied field has to be below a critical value; above this value dielectric breakdown is caused. The on-line application of this system might be disturbed by the time-dependent changes in the medium conductivity. *On-line application:* Yes. *References:* (*119*), (*126*), (*125*). (Taken from ref. (*89*).)

content of the cells reacts immediately to changes of the medium, which was shown by Meyer and Beyeler (*111*) in a continuous culture of yeast. This fast reaction of the cells, which is superior to the classical control via OUR (volumetric oxygen uptake rate) and RQ, allows a faster and therefore more effective control and regulation of the fermentation process (*111*). Leist *et al.* (*115*) used an on-line fluorometer for the control of a fermentation of Bowes 4 melanoma cells. The response correlated with the cell number, but was not caused by the cells. They supposed that the application of a fluorometer probe might be useful, but that the results have to be analysed carefully.

Five other relevant methods are shown, which can be employed in an on-line mode: photometry, with the two principles of nephelometry and turbidimetry; electrochemically based systems; acoustic resonance densitometry (*116*); dielectric (impedimetric) monitoring (*116*); and electrical counting. Details, principles, advantages, disadvantages, and the possibility of detecting the viable or total cell number are shown for these methods in Tables VII and VIII and in Fig. 9. Some details are mentioned briefly here. The main disadvantage of all photometric systems is the fouling of the optical surface. Certain improvements are envisaged through the use of fibre-optic techniques and/or laser optics (*105*; see also Section 3). The electrochemical systems, the fuel cell type electrode, which was published by Matsunaga *et al.* (*121, 122*), and the piezoelectric membrane system, published by Ishimori *et al.* (*123*), suffer the drawbacks that they are too insensitive because their working range lies between 10^6 and 10^8 cells ml^{-1} (for yeasts and bacteria). The system of Matsunaga *et al.* (*121, 122*) has the additional disadvantage of working with dialysis membranes, which increases the response time. The main disadvantage of acoustic densitometry (*116*) is the need for a filtration unit for separation of the cells in order to get a difference measurement between the cell-free medium and the cell suspension in the same medium. For problems linked to filtration systems, see Section 3.

The fuel cell type system, and the two following principles, devices monitoring the impedance properties of cells and electrical counting, measure the living cell, in contrast to the physical methods mentioned above.

Devices for monitoring impedance properties are based on the measurement of the β-dispersion (*116, 124*).

The electrical counting, using a Coulter counter, is presented in Fig. 9. The presence of foreign particles, which is a main drawback in most of its applications, poses few problems in cell culture, because all media are filtered. Cell debris might influence the results slightly.

An excellent review of the estimation of microbial biomass has been published by Harris and Kell (*119*).

TABLE VII Comparison of Photometric Methods for Determination of Biomass (Cell Count)

Principle	Advantages	Disadvantages	On-line	References
Nephelometry				
Straightforward measurements of light scattered by the cells in a suspension give signals which are directly proportional to the cell count	Background automatically 0; linear correlation between the signal and the cell count; detection limit of 1.6×10^5 cells ml^{-1} (*E. coli*)	No discrimination between living and dead cells; temperature control necessary; long use in fermenters is impossible because cells can grow on the optical surface	Yes	(*117*, *118*, *119*)
Infrared nephelometry				
Light: 900 nm	Linear correlation between the signal and the cell count; detection limit of 10^5 cells ml^{-1} (hybridomas)	As for normal nephelometry and additionally the influence of the environmental light has to be constant	Yes	(*120*)
Turbidimetry				
Measurement of the transmitted light or the optical density	Rapid and easy measurements	Gas bubbles may disturb the measurements; the cells can grow on the optical surface; no discrimination between living and dead cells; detection limit of 2.4×10^6 cells ml^{-1} (*E. coli*)	Yes	(*118*)

TABLE VIII Comparison of Electrical Methods for the Determination of Biomass (Cell Count)

Principle	Advantages	Disadvantages	On-line	Cell-count	References
Fuel cell type electrode with dialysis membrane; determination of the electroactivity of the cells without the influence of the electroactive compounds of the medium; two or three electrode systems are used, where one of these systems is covered with a dialysis membrane	No influence of O_2 or phosphate; response time: 3–5 min	Influence of pH and temperature (response increases with increasing pH or temperature); too insensitive. (1) System: Pt-electrode and Ag_2O_2 electrode, which was decomposed by sterilization; detection limit 10^7–10^8 cells ml^{-1} for yeast, 4×10^8–4×10^9 cells ml^{-1} for bacteria. (2) System: two Pt-electrodes, and one saturated calomel electrode; detection limit 10^8–4×10^9 cells ml^{-1} for *B. subtilis*	Yes	Viable	(*121, 122*)
Piezoelectric membrane system; ultrasonic determination of the cell suspension with two piezoelectric membranes; one membrane is charged with a certain voltage; the output voltage of the other membrane increases with the increasing cell density	?	Influence of the following parameters: frequency of the applied voltage, pH, buffering capacity, temperature, density of the medium, adiabatic compressibility; may damage the cells (depends on the input frequency); detection limit 10^6–10^{10} cells ml^{-1} for yeast 10^6–10^8 cells ml^{-1} for bacteria	Yes	Total	(*123*)

(*continued*)

TABLE VIII (*continued*)

Principle	Advantages	Disadvantages	On-line	Cell-count	References
Magneto-inductive monitoring of impedance properties: any electrical charge in motion produces a magnetic field; an alternating potential difference applied to the primary coil induces a sinusoidal current in any conductive medium placed within the flow cell; the current within the flow cell produces an alternating magnetic field, which produces an alternating potential difference in the secondary (sensing) coil. The potential difference in the second coil depends on the conductivity of the cell contents at any applied frequency	The α-dispersion (80–500 kHz) correlates directly with the cell mass (for bacteria), the β-dispersion (0.5–100 MHz) is a result of the bulk impedance network of the cell membrane, largely, which can be well related to the measurement of the viability because the progressive destruction of the cells is an early indicator of the death of the cells. Non-invasive, as a flow cell, sterilizable, robust	?	Yes	Viable	(*116, 124*)
Acoustic resonance densitometry: the amplifier, electromagnets and sample test cell constitute a closed oscillatory circuit whose frequency of oscillation depends on the mass of the flow cell; any change in the density of the contents of the test cell causes a change in the resonant frequency of the system	Non-invasive, as a flow cell, sterilizable	A tangential filtration unit (or other systems) is necessary for determination using this system, because the medium alone and the cell suspension in the medium have to be tested, to compensate for the influence of the medium	Yes	Total	(*116, 124*)

Note: The corresponding figures can be found in the references cited or in ref. (*89*).

4.2. Determination of the Glucose Concentration

The most important carbohydrate source for animal cell culture is glucose. Although other sugars are sometimes used, we shall focus in this section on the different principles for determination of glucose. These principles are also applicable to the determination of other medium compounds (see Section 4.3, Table X). Details of glucose determination are listed in Table IX. Most of the published biosensors are based on the catalysis of the enzyme glucose oxidase (EC 1.1.3.4) which is a flavoprotein. It catalyses the following reaction:

$$\text{Glucose} + O_2 + H_2O \rightarrow \text{gluconic acid } (H^+) + H_2O_2 + dT. \qquad (1)$$

For the determination of glucose, the following substrates or products of the action of the enzyme can be measured. It is possible to determine, first, the decrease of the oxygen concentration of the sample, second, the production of protons, and third, the production of H_2O_2 (equation (2)), all with specific electrodes.

$$H_2O_2 \rightarrow 2H^+ + O_2 + 2e^-. \qquad (2)$$

Another method for the determination of glucose is based on the measurement of the heat produced, using an enzyme thermistor (*47*). For this application, the enzyme catalysis of glucose oxidase (equation (1)) is coupled with the action of catalase (EC 1.11.1.6), shown in equation (3); the catalase has the effect of increasing the heat produced and of producing oxygen which can be used by the glucose oxidase again; and it has the advantage of destroying the H_2O_2 which can be deleterious to the enzyme (*66, 67, 82, 94*):

$$H_2O_2 \rightarrow H_2O + \tfrac{1}{2}O_2 + dT. \qquad (3)$$

The hydrogen peroxide can be determined with other methods, too, as it is shown in equation (4). This reaction, the oxidation of I^- to iodine by H_2O_2, is catalysed by peroxidase (EC 1.11.1.7) or a molybdene (VI) catalyst. Here, the decreasing concentration of I^- is measured by means of a I^--selective electrode (*127, 128, 129*):

$$H_2O_2 + 2I^- + 2H^+ \rightarrow I_2 + 2H_2O. \qquad (4)$$

Yao *et al.* (*101*) used a peroxidase electrode to determine the H_2O_2 produced and measured the hexacyanoferrate(III) amperometrically (equations (5) and (6)), in combination with reaction (1).

$$H_2O_2 + 2Fe(CN)_6^{4-} + 2H^+ \rightarrow 2H_2O + 2Fe(CN)_6^{3-}; \qquad (5)$$

$$Fe(CN)_6^{3-} + e^- \rightarrow Fe(CN)_6^{4-}. \qquad (6)$$

TABLE IX **Comparison of Estimation Methods for Glucose with Respect to the Connection**

Principle (equation)	Determined parameter	Dependence on pH	°C	dO_2	Influences on the measuring device
O_2 electrode (1)	Decreasing O_2	?	+5%/+°C	Yes	pO_2
O_2 electrode (1)	Decreasing O_2				pO_2, but stabilized
O_2 electrode (1,3)	Decreasing O_2	?	?	Stabilized	pO_2, but stabilized
O_2 electrode (1,3)	Decreasing O_2	pH 7.0	?	Stabilized: O_2 production by the electrode	pO_2, but stabilized
pH electrode (1)	Acidification	Strong buffer required	?	O_2 saturation required	pH, buffer capacity
pH electrode (1)	Acidification	?			pH, buffer capacity
pH electrode (1)	Acidification	?	+2.3%/+°C	Above 25 mmHg	pH, buffer capacity
pH electrode (1)	Acidification	?	?	More than 0.2 m*M* O_2	pH, buffer capacity
pH dye, fibre-optic (1)	Acidification	?	?		pH, buffer capacity
H_2O_2 electrode (1,2)	Increasing H_2O_2	No	Temperated	No	Ascorbic, uric, amino acids prevented by low membrane cut-off
H_2O_2 electrode (1,2)	Increasing H_2O_2	4–7 optimum	55°C optimum	?	Asymmetric membrane prevents influences from ascorbic acid, glutathione
H_2O_2 electrode (1,2)	Increasing H_2O_2	?	?	Minimal 5.5 kPa	?
H_2O_2 electrode (1,2)	Increasing H_2O_2	No	Temperated	No	Ascorbic, uric, amino acids prevented by low membrane cut-off
H_2O_2 electrode (1,2)	Increasing H_2O_2	?	?	?	?
H_2O_2 electrode (1,2)	Increasing H_2O_2	4.5 optimum	?	?	?
H_2O_2 electrode (1,2)	Increasing H_2O_2	?	?	?	?

Autoclavable	Connection to the fermenter (Fig.)	Drift	Range in g l^{-1} (linear)	Lifetime	Response time (min)	References
No	Dialysis or flow system (7(*c*), 7(*d*))	?	Up to 4	?	?	(*138, 139*)
No	Dialysis or flow system (7(*c*), 7(*d*))	?	0–4	?	Less than 1	(*92*)
No	Dialysis or flow system (7(*c*), 7(*d*))	?	0–3	?	?	(*140*)
(No)	Dialysis system (7 (*d*)(i))	Stable	1–50 dependent on the buffer flow	Days	5–10	(*66, 67*)
No	Dialysis or flow system (7(*c*), 7(*d*))	?	0.174–17.4	20 d[c]	?	(*28*)
Yes, the enzyme is applied later	Dialysis or flow system (7(*c*), 7(*d*))	?				(*141*)
No	Dialysis or flow system (7(*c*), 7(*d*))	?	0–5	14 d[c]	2–5	(*142, 143*)
No	Dialysis or flow system (7(*c*), 7(*d*))	?	?	?	?	(*29*)
No	Dialysis or flow system (7(*c*), 7(*d*))	?	0–12.6	14 d[c]	?	(*38*)
No	Dialysis or flow system (7(*c*), 7(*d*))	Relatively stable	0.5–4	m[a]	Less than 1	(*135*), YSI[b]
No	Dialysis or flow system (7(*c*), 7(*d*))	Retains 55% activity after 1000 tests	0–10	28 d[c]	1/6	(*96*)
No	Dialysis or flow system (7(*c*), 7(*d*))	?	Up to 6.96	Some days	5–10	(*144*)
No	FIA-system (1 ml min^{-1}) (7(*d*)(i)–(ii))	8%/45 days	0–0.8	m[a]	0.75	(*98*)
No	Dialysis or flow system (7(*c*), 7(*d*))	1 mV/day	0.84–2.34	3 m[a]	?	(*145*)
No	FIA-system (2 ml min^{-1}) (7(*d*), 7(*b*))	?	10^{-4}–10^{-1}	10 m[a]	0.75	(*146*)
No	FIA-system (0.5 ml min^{-1}) (7(*d*)(ii)–(iii))	?	0.02–17	Some weeks	0.75	(*147*)

(continued)

TABLE IX (*continued*)

Principle (equation)	Determined parameter	pH	Dependence on °C	dO_2	Influences on the measuring device
Thermistor (enzyme reactor) (1, 3)	Increasing H_2O_2, increasing temperature	No	Temperated	No	Solvation heat, preventable by having the same matrix for sample and flow buffer
Thermistor (enzyme reactor) (1, 3)	Increasing H_2O_2, increasing temperature	No	Temperated	No	Solvation heat, preventable by having the same matrix for sample and flow buffer
I^- electrode (1, 4)	Increasing H_2O_2, decreasing I^-	?	?	?	?
H_2O_2 electrode (1, 5, 6)	Increasing H_2O_2, increasing hexacyano-ferrate(III)	5.5 optimum	?	?	Ascorbic acid, preventable by electrode potential (−50 mV vs. Ag/AgCl)
H_2O_2 electrode (1, 7)	Increasing H_2O_2	7.4 (6–8)	<2%/°C (15–30°C)	Above 30 mmHg	pO_2 of the sample; cellulose acetate membrane prevents influence of reducing substances
Potentiometric (1, 8)	Change of the ratio of quinone to hydroquinone	5.5 optimum	?	?	Ascorbic acid, compensated by reference electrode
Mediator electrode (9–11)	Increasing ferrocene	No	+4%/+°C	No	Ascorbic acid, 0.13 m*M*, +4%
Mediator electrode (12–14)	Increasing $FADH_2$	?	?	No	?
Optical fibre (15, 16)	Increasing fluorescence	pH 7: −5%/−0.1	+4%/+°C	No	?
Photometer (340 nm), soluble enzyme is added (17)	Increasing NADH	?	Temperated	No	Sample colour
Mediator electrode (enzyme reactor) (17–21)	Increasing NADH	pH 6	?	No	Electrode potential: −50 mV vs. SCE[f]

Autoclavable	Connection to the fermenter (Fig.)	Drift	Range in g l^{-1} (linear)	Lifetime	Response time (min)	References
No	FIA-system (1 ml min^{-1}) with reference system (7(*d*)(ii)–(iii))	?	0.002–0.16	m[a] ?	?	(*60*)
No	FIA-system (1 ml min^{-1}) (7(*d*)(ii)–(iii))	?	0.7–35	?	?	(*47*)
No	Flow system (7(*d*))	?	0.02–2	More than 14 d[c]	1–2	(*127, 128*)
No	FIA-system (1.5 ml min^{-1}) (7(*d*)(ii)–(iii))	10%/2 months (per[d]) 10%/6 months (GO[e])	0.05–10	m[a]	0.6	(*101*)
No	Dialysis or flow system (7(*c*), 7(*d*))	15–30% increase after 15 samples	0–9 (0–6.3 after 15 samples)	1 m[a] at 4°C	0.5–1.5	(*99*)
No	Flow system with dialysis unit (1.23 ml min^{-1}) (7(*d*)(ii))	?	0.007–1.8	?	Seconds	(*130*)
No	Dialysis or flow system (7(*c*), 7(*d*))	?	0.2–5.2	?	1–1.5	(*36*)
No	Dialysis or flow system (7(*c*), 7(*d*))	?	0.002–1.2	2 weeks	6	(*102*)
No	Dialysis or flow system (7(*c*), 7(*d*))	15%/15 d[c]	0.5–4	?	5–10	(*39*), (*131*)
No	FIA-system	0	0–7		0.25	(*63*)
	FIA-dialysis system (2 ml min^{-1}) (7(*d*)(ii)–(iii))	0	0–3.5		0.25	(*63*)
No	FIA-system (1 ml min^{-1}) (7(*d*)(ii)–(iii))	?	0–0.02	?	0.86	(*34*)

(*continued*)

TABLE IX (*continued*)

Principle (equation)	Determined parameter	pH	Dependence on °C	dO_2	Influences on the measuring device
Mediator electrode (22)	Increasing reduced electron acceptor (PES[g] DCPIP[h])	pH 8.7	?	No	Ascorbic acid, urate glutathione compensated via response time
Mediator electrode (22)	Increasing reduced electron acceptor (PES[g])	pH 8.0–8.5 optimum	43°C max.	No	No influence of other sugars, high pO_2 provokes accelerated inactivation
Immobilized glucose-salt-coated wire	Potential change, or polarographic	?	?	No	No ?
Photometer (340 nm) (enzyme reactor) (23, 24)	Increasing NADH	?	?	No	Sample colour

Note: ENFET, enzyme-field-effect transistor; FIA, flow-injection-analysis; *a*, month(s); *b*, Yellow Springs Instrument Co., Inc., Glucose Analyzer Model 27; *c*, day(s); *d*, peroxidase; *e*, glucose oxidase; *f*, saturated calomel electrode; *g*, phenazine ethosulfate; *h*, 2,6-dichlorophenolindophenol; *i*, hour(s).

Unfortunately, in all these electrode configurations, H_2O_2 is involved, which can be deleterious to the enzyme used, because of accelerated inactivation (*100*). As a result, sensor configurations, preventing the production of H_2O_2 were established. Gorton and Bhatti (*130*) employed a flow system in which the glucose oxidase was immobilized in an enzyme reactor. The overall reaction is shown in equation (7). In the presence of glucose, the reaction will alter the original ratio of quinone/hydroquinone, which is measured potentiometrically.

$$\text{Glucose} + \text{Q} + \text{H}_2\text{O} \rightarrow \text{gluconic acid} + \text{QH}_2. \quad (7)$$
$$(\text{Q} = 1,4\text{-benzoquinone, QH}_2 = \text{hydroquinone})$$

Interference from serum components, like ascorbic acid, is compensated by a reference electrode system. The hydroquinone produced in reaction (7) can be detected electrochemically, according to equation (8) (*100*):

$$\text{hydroquinone} \rightarrow \text{benzoquinone} + 2\text{H}^+ + 2\text{e}^-. \quad (8)$$

Autoclavable	Connection to the fermenter (Fig.)	Drift	Range in $g\,l^{-1}$ (linear)	Lifetime	Response time (min)	References
No	Dialysis or flow system (*7*(*c*), *7*(*d*))	PES: $-0.5\,mM$ min^{-1}	0–0.14	2 h[i]	3–5	(*104*)
		DCPIP: stable	0–0.18	24 h[i]	0.5 (1st order differen-tial)	
No	Dialysis or flow system (*7*(*c*), *7*(*d*))	20%/13.5 h[i] at 30°C	0.9–2.7 (non-linear)	Some h[i]	Less than 1.5	(*132*)
?	Dialysis or flow system (*7*(*c*), *7*(*d*)) direct?	No	0.4–2	?	2–4	(*133*)
No	Flow system with dialysis unit (*7*(*d*)(ii))	?	?	?	?	Technicon

Because these indirect measurements of glucose are often inaccurate, it is interesting to look for direct measuring systems. Using the same enzyme system, the configurations described below are possible. Other electron acceptors than oxygen are then used and direct electron transfer from the enzyme to the electrode (via mediators, is possible (see Section 3). Cass *et al.* (*36*), for instance, described a suitable ferrocene-mediated electrode with the following reactions (equations (9)–(11)):

$$\text{Glucose} + \text{GOD(ox)} \rightarrow \text{gluconolactone} + \text{GOD(red)}; \quad (9)$$

$$\text{GOD(red)} + 2\text{Fecp}_2\text{R}^+ \rightarrow \text{GOD(ox)} + 2\text{Fecp}_2\text{R} + 2\text{H}^+; \quad (10)$$

$$2\text{Fecp}_2\text{R} \rightarrow 2\text{Fecp}_2\text{R}^+ + 2e^-. \quad (11)$$

$$(\text{Fecp}_2\text{R}^+ = \text{ferricinium ion, Fecp}_2\text{R} = \text{ferrocene})$$

Some papers describe the possibility of direct electron-transfer from the $FADH_2$, the prosthetic group of the glucose oxidase, to the electrode (*102, 103*). The principle is shown in equations (12), (13), and (14):

$$\text{Glucose} + \text{FAD} \rightarrow \text{gluconolactone} + \text{FADH}_2; \quad (12)$$
$$\text{FADH}_2 + \text{O}_2 \rightarrow \text{FAD} + \text{H}_2\text{O}_2; \quad (13)$$
$$\text{H}_2\text{O}_2 \rightarrow 2\text{H}^+ + 2\text{e}^- + \text{O}_2. \quad (14)$$

There still remains the major problem that the continuous reduction and oxidation of the FAD/$FADH_2$ system is not completely reversible. The system is only stable for some days.

The last direct method is the principle of the affinity sensor, which is shown in Fig. 10 (*39, 131*). The principle is that the free glucose of the sample diffuses through the dialysis fibre into the sensing element. Inside, the glucose has to compete with the FITC-dextran for the binding to ConA. At equilibrium, the level of free fluorescein-conjugate in the hollow fibre lumen is measured via the optical fibre and is correlated to the concentration of the glucose:

$$\text{Glucose} + \text{ConA}_{\text{immobilized}} \underset{+\text{glucose}}{\rightleftharpoons} \text{ConA}_{\text{immobilized}} \ldots \text{glucose}; \quad (15)$$
$$\text{FITC-dextran} + \text{ConA}_{\text{immobilized}} \underset{+\text{glucose}}{\rightleftharpoons} \text{ConA}_{\text{immobilized}} \ldots \text{FITC-dextran}. \quad (16)$$

Up to now, three other possible enzyme systems for the determination of glucose have been published: an often-used enzyme, glucose dehydrogenase (EC 1.1.1.47), catalyses reaction (17):

$$\text{Glucose} + \text{NAD}^+ + \text{H}_2\text{O} \rightarrow \text{gluconic acid} (\text{H}^+) + \text{NADH} + \text{H}^+. \quad (17)$$

Hansen *et al.* (*63*) used this system for the flow injection analysis of glucose and photometrically determined the NADH produced. Appelqvist *et al.* (*34*) determined the NADH amperometrically via a mediator (meldola blue, 7-dimethylaminobenzophenoxazinium salt (MB)) (equations (17)–(21)). This also was employed in a FIA system.

$$\text{NADH} + \text{MB}^+ \rightarrow \text{NADHMB}^+; \quad (18)$$
$$\text{NADHMB}^+ \rightarrow \text{NAD}^+ + \text{MBH}; \quad (19)$$
$$\text{MBH} \rightarrow \text{MB}^+ + \text{e}^- + \text{H}^+; \quad (20)$$
$$\text{NADH} \rightarrow \text{NAD}^+ + 2\text{e}^- + \text{H}^+. \quad (21)$$

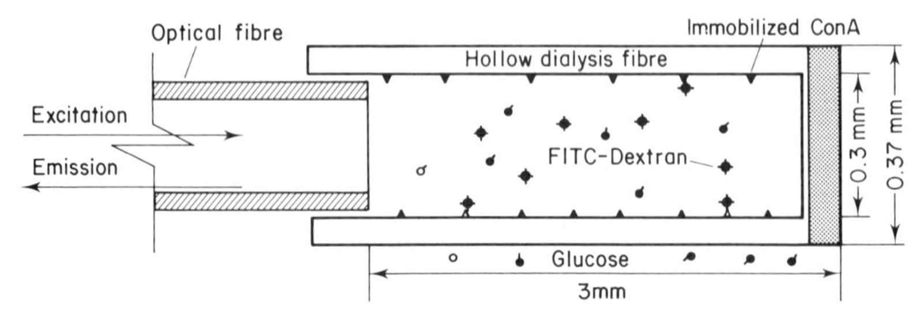

Fig. 10. Schematic diagram of affinity sensor transducer element (taken from ref. (*39*)).

Recently, the application of another glucose dehydrogenase (EC 1.1.99.17) was described (*104*). In comparison with glucose dehydrogenase (EC 1.1.1.47), this dehydrogenase is a quinoprotein enzyme with pyrroloquinolin quinone as prosthetic group. It does not depend on the oxygen partial pressure, and does not use NAD or NADP. Different electron acceptors can be used instead. The general catalytic reaction is as shown in equation (22):

$$\text{Glucose} + \text{acceptor} \rightarrow \text{gluconolactone} + \text{reduced acceptor} \quad (22)$$

Mullen *et al.* (*104*) employed PES (phenazine ethosulphate) and DCPIP (2,6-dichlorophenolindophenol) as electron acceptors. Using PES had the advantage of a fast response (3–5 min), but had the disadvantage that the system was very unstable. DCPIP as an acceptor was more stable (up to some days), but showed the feature of a long response time, up to 30 min. Monitoring the first-order differential of the response reduced the time to 30 s. The response of ascorbate, urate, and glutathione could be separated from the response of glucose because the response of the three substances was fast and unstable, whereas that of glucose was slow and stable. D'Costa *et al.* (*132*) utilized the same enzyme, but used ferrocene as electron acceptor, as was done for the glucose oxidase by Cass *et al.* (*36*). The electrode is O_2-independent, shows a non-linear response to the glucose concentration (5–15 mM l^{-1}), is not influenced by other sugars, and shows higher catalytic activity than glucose oxidase. The stability of the electrode depends strongly on the O_2 partial pressure of the environment and increases with a lower pO_2. This enzyme electrode might be a useful electrode for future applications.

The last electrode system, working without enzymes(!), was published recently by El Degheidy *et al.* (*133*). This system is based on a coated wire, consisting of a membrane-immobilized glucose salt. The different glucose concentrations are measured potentiometrically or polarographically. The advantage of this system is the lack of those problems that arise with the use of enzyme systems. The electrode is O_2-insensitive and shows a linear response to glucose concentrations between 0.4 and 2 g l^{-1}. Although no detailed results on stability and lifetime were given, the authors mentioned that leaching of any substance from the electrode was not detectable and that the response to different glucose concentrations was reliable and reproducible over an extended period.

Finally, the Technicon system, a flow system, should be mentioned. The two immobilized enzymes hexokinase and glucose-6-phosphate dehydrogenase were used (*84, 134*). The reagents ATP and NAD have to be added to the flow buffer. The catalysed reactions are as given in equations (23) and (24):

$$\text{Glucose} + \text{ATP} \rightarrow \text{glucose-6-phosphate} + \text{ADP}; \quad (23)$$

$$\text{Glucose-6-phosphate} + H_2O + NAD^+ \rightarrow \text{6-phosphogluconic acid} + NADH + H^+. \quad (24)$$

The NADH produced is measured spectrophotometrically. The disadvantage of this system is the expense of the reagents and the possible disturbance due to the multistep nature of the reaction. The influences of different parameters on the measuring devices and the dependencies of the methods are presented in Table IX as far as availability of data permits (see Section 3).

It is evident that the system based on the H_2O_2-determination (equations (1, 5, 6)) (*101, 135*) is the most suitable for bioreactor control, so far, because it shows the longest half-life, an acceptable drift, and a short response time, especially important for FIA systems. These systems are not influenced by the pH (physiological), the temperature (controlled) and the dissolved O_2, because the buffers are O_2-saturated, and in the case of the system described by Clark (*135*) oxygen is produced by the determination of H_2O_2. All other detection systems have to be improved for lifetime, drift, and response time.

TABLE X Determination Methods Medium Compounds

Analyte, enzyme	Principle	Determined parameter	pH	Dependence on °C	dO_2
Galactose					
Galactose oxidase (EC 1.1.3.9)	Mediator electrode (enzyme reactor)	Increasing H_2O_2, reduction of mediator[a]	7.0	Ambient	Yes
Lactate					
Lactate oxidase (EC 1.1.3.2)	O_2 electrode[i]	Decreasing O_2	5.5–7.5	?	Yes
Lactate oxidase (EC 1.13.12.4)	O_2 electrode[i]	Decreasing O_2	6.0	+10%/+3°C	Buffer contains enough O_2
3 enzymes[c]	Thermistor (enzyme reactor)	Increasing temperature	7.0	Temperated	O_2 saturated buffer
Lactate dehydrogenase (EC 1.1.1.27)	Mediator electrode (enzyme reactor)	Increasing NADH			
Lactate dehydrogenase (EC 1.1.1.27)	Photometer (340 nm), soluble enzyme is added	Increasing NADH			

In general, the detection range is of only secondary importance, because it can be adapted to the required range using an FIA system (see also Section 3).

4.3. Determination of Other Medium Compounds

Sensors for the determination of other medium compounds, such as galactose, lactate, pyruvate, NH_4^+, and amino acids are not discussed in detail, since the detection principles are often similar to those already mentioned in Section 4.2. For more information, the publications of Guilbault (*85*, *136*), Aston and Turner (*137*), Karube and Suzuki (*4*), and Merten *et al.* (*89*) are recommended. Nevertheless, some applications are shown in detail in Table X which seem to be the most suitable ones for the application in animal cell cultivation, with respect to lifetime, influences from other compounds, or detection range. As has been mentioned, the detection range is variable, when dialysis or FIA systems are used.

Influences of other medium compounds	Connection to the fermenter (Fig.)	Drift	Range $mM\,l^{-1}$ (linear)	Lifetime	References
pO_2, H_2O_2, dihydroxyacetone of the sample	FIA-system ($1\,ml\,min^{-1}$) (7(*d*)(ii)–(iii))	0.3%/d	0.002–60	More than 1 m[b]	(*148*)
pO_2 of the sample	Dialysis or flow system (7(*c*), 7(*d*))	?	0.02–0.2	More than 1 m[b]	(*149*)
pO_2 of the sample	FIA-system ($1\,ml\,min^{-1}$) (7(*d*)(ii)–(iii))	No	0.02–0.2	More than 1 m[b]	(*150*)
Solvation heat, preventable by reference system or having no differences in sample and buffer matrix	FIA-system ($1\,ml\,min^{-1}$) (7(*d*)(ii)–(iii))	?	0.025–1 (WONA[d]) 10^{-5}–10^{-1} (WNL[e])	?	(*48*)
?	FIA-system ($0.5\,ml\,min^{-1}$) buffer contains NAD (7(*d*)(ii)–(iii))	?	10^{-2}–1	300 days	(*35*)
Sample colour	FIA-system ($2\,ml\,min^{-1}$) (7(*d*)(ii)–(iii))	—	0.2–2		(*64*)

(continued)

TABLE X (*continued*)

Analyte, enzyme	Principle	Determined parameter	pH	Dependence on °C	dO_2
Lactate dehydrogenase (EC 1.1.1.27)	Pt-electrode[i]	Increasing $Fe(CN)_6^{4-}$	7.4	+3.5%/+°C	No
Pyruvate					
3 enzymes[c]	Thermistor (enzyme reactor)	Increasing temperature	7.0	Temperated	O_2 saturated buffer
Pyruvate oxidase (EC 1.2.3.3)	O_2 electrode[i]	Decreasing O_2	7.05–7.35	?	Yes
NH_3, NH_4					
	NH_3 gas electrode	pH has to be more than 10: $NH_4^+ + OH^- \rightarrow NH_3 + H_2O$			No
(Nonactin based)	Ion-selective electrode	NH_4^+			
Alanine					
Alanine dehydrogenase (EC 1.4.1.1)	Mediator electrode (enzyme reactor)	Increasing NADH	?	?	No
Asparagine					
Asparaginase (EC 3.5.1.1)	NH_4^+ electrode[i]	Increasing NH_4^+	7.5–8.7 optimum	?	—
Glutamate					
Glutamate dehydrogenase (EC 1.4.1.3)	mediator electrode (enzyme reactor)	increasing NADH	?	?	—
Leucine					
Leucine dehydrogenase (EC 1.4.1.9)	mediator electrode (enzyme reactor)	increasing NADH	?	?	—
Lysine					
Lysine-α-oxidase (EC ?)	O_2 electrode[i]	Decreasing O_2	8.7	25°C	Air wash between each test
Lysine decarboxylase (EC 4.1.1.18)	CO_2 electrode[i]	Increasing CO_2	6.0	?	—

Influences of other medium compounds	Connection to the fermenter (Fig.)	Drift	Range $mM\,l^{-1}$ (linear)	Lifetime	References
?	Dialysis or flow system (7(*c*), 7(*d*))	?	0–1.5	6 w[f]	(*151*)
Solvation heat preventable by reference system or having no differences in sample and buffer matrix	FIA-system ($1\ ml\ min^{-1}$) buffer contains NAD (7(*d*)(ii)–(iii))	?	10^{-5}–10^{-1}	?	(*48*)
pO_2 of the sample	Dialysis or flow system buffer[g]	10 d stable	0.06–0.8	?	(*152*)
	Dialysis or flow system (7(*c*), 7(*d*))	Stable	0.005–10	?	com.[h]
K^+: 15%, Na^+: 0.13%,	Flow system (7.3)	Yes	1–100	Some weeks	(*95, 153*)
?	FIA-system ($0.5\ ml\ min^{-1}$); buffer contains NAD (7(*d*)(ii)–(iii))	?	0.1–1	120 d	(*35*)
As for NH_4^+ electrode, NH_4^+ of the medium, compensated by reference system	Dialysis or flow system (7(*c*), 7(*d*))	?		2–4 w[f]	(*154*)
?	FIA-system ($0.5\ ml\ min^{-1}$) buffer contains NAD (7(*d*)(ii)–(iii))	?	0.1–10	146 d	(*35*)
?	FIA-system ($0.5\ ml\ min^{-1}$) buffer contains NAD (7(*d*)(ii)–(iii))	?	0.1–1	More than 150 d	(*35*)
Arginine, phenylalanine L-ornithin	Dialysis or flow system (7(*c*), 7(*d*))	?	0.2–4	3000 times	(*155*)
CO_2 of the medium, compensated by reference system	Dialysis or flow system (7(*c*), 7(*d*))	No	0.05–100	60 d	(*156*)

(*continued*)

TABLE X (*Continued*)

Analyte, enzyme	Principle	Determined parameter	pH	Dependence on °C	dO_2
Methionine					
Methionine-lyase (EC 4.4.1.11)	NH_3 gas electrode[i]	Increasing NH_3	8.7	?	—
Tyrosine					
Tyrosine carboxylase (EC 4.1.1.25)	CO_2 electrode[i]	Increasing CO_2	4.8	23°C: 8 min for 2 m*M*, 37° 5 min for 2 m*M*	—

Notes: *a*, hexacyanoferrate(III) as mediator;
b, month(s);
c, Enzymes: lactate oxidase (EC 1.1.3.2), catalase (EC 1.11.1.6), lactate dehydrogenase (EC 1.1.1.27); all are immobilized on an enzyme reactor;
d, without NADH in flow buffer;
e, with NADH in flow buffer;
f, week(s);
g, buffer contains: phosphate (0.5 m*M*), thiamine pyrophosphate (0.06 m*M*);
h, commercial;
i, enzyme is immobilized onto the electrode;
j, buffer contains: pyrophosphate (0.05 *M*), p-5′-p (10^{-4} *M*).

4.4. Determination of the Product Concentration in Animal Cell Culture Supernatants

One of the most important parameters in technical cell culture is the production yield of the product. This chapter will present only detection methods for secreted products. If the product of interest is not secreted, the cells have to be sampled via a flow-line system, lysed, and the product has to be extracted (as for the determination of ATP).

Only discrete sampling has so far been possible, done once or only a few times a day. But often it is of great interest to follow the production kinetics in an on-line mode. In this case, semi-continuous sampling or on-line determination is important. To achieve that, the principles outlined below are important. They should normally be reversible systems; the transducer should work without the addition of reagents; and they should be very quick. Unfortunately, most of the systems do not have these features. Only one system exists, optical-fibre-based fluorescence immunosensor, which works completely without reagents. The principle, the so-called affinity sensor, has already been described in Section 4.2 for the determination of glucose (Fig. 10) (*39*). Instead of the ConA, specific polyclonal or, better, monoclonal antibodies with certain affinities are used. The analyte has to compete with an

Influences of other medium compounds	Connection to the fermenter (Fig.)	Drift	Range $mM\,l^{-1}$ (linear)	Lifetime	References
NH_4^+ of the medium has to be compensated by reference system	Dialysis or flow system (7(*c*), 7(*d*)) buffer[j]	Stable	0.01–10	3 m[b]	(*157*)
CO_2 of the medium has to be compensated by reference systems	Dialysis or flow system (7(*c*), 7(*d*))	-1.4×10^{-2} $kPa\,h^{-1}$	0.04–2.6	90 d	(*158*)

FITC-conjugated–analyte analogue. Changes in content of the analyte in the supernatant change the proportion of bound-to-free FITC-conjugate, which is measured. This sensor is the only one which can be used in a complete on-line mode in a flow system (Fig. 7(*d*)).

All other systems which are shown below require the addition of reagents. They can be constructed as competitive, direct, or two-site sensors. The competitive and the direct type have the advantage of requiring only the addition of reagents, maximally once, the two-site type twice. All types require additional washing steps, incubation steps, and elution steps for preparation or reuse. Therefore, all systems are based on the flow injection principle. The immobilized antibodies must have a certain affinity for the analyte. This affinity must be high enough to achieve a sufficient sensitivity, but it has to be in such a range that the elution of the analyte is possible without damaging the immobilized antibody. This is a very important consideration for the lifetime of the systems mentioned.

The following systems are dealt with in detail: TELISA, enzyme sensors working with oxygen electrodes; PIMIA; MGIA; and a system based on the measurement of the changing of the evanescent wave by an antigen–antibody reaction. Details are given in Table XI. Mattiasson *et al.* (*49*) and Borrebaeck *et al.* (*159*) described the application of a thermistor for the immunological

TABLE XI Comparison of Immunosensors and One FIA-detection System

Principle	Competitive direct, two-site mode	Unspecific adsorption	Analyte	Detection limit	Working cycles	Half-life time	References
Optical fibre based fluorescence immunoassay	Competitive[a]	?	?	?	?	?	(*165*, based on *39*)
TELISA (Fig. 11)	Competitive[b]	Yes	HSA	10^{-13} M l^{-1}, VC: 1%	100	?	(*159*)
Electrode-based enzyme immunoassays (equations 25–28)	Competitive[b]	?	HCG	2×10^{-2}–10^{2}, IU ml^{-1}, VC: 5%	1, then change of membrane	—	(*160*)
	Competitive[b]	No	Theophylline	1–5 $\times 10^{-6}$ M l^{-1}, VC: 5.8%	Maximal 10	60% loss/ 6 cycles	(*162*)
	Two-site[b]	?	HSA	10^{-8}–10^{-6} g ml^{-1}, VC: 10%	Several	?	(*163*)
	Two-site[b]	?	HB_sAg	0.1–100 μg l^{-1}, SD: 7–12%	Many	20% loss/ 6 months	(*161*)
PIMIA (Fig. 12)	Competitive, direct[b]	No, but influence of K^+	Antibody to digoxin	μg ml^{-1}	?	Some weeks	(*164*)
MGIA (Fig. 13)	Competitive, direct[b]	Yes	Human IgG	13 μg	?	?	(*53*)
Optical detection of the Ag Ab reaction at the glass–liquid interface (Fig. 14)	Direct[b]	?	Methotrexate	270 nM l^{-1}, SD: 6%	30	?	(*41*)
	Two-site[b]	?	Human IgG	30 nM l^{-1}, SD: 0.4 U	?	?	(*42*)
Energy-transfer immunoassay by stopped flow injection analysis (Fig. 15)	Homogeneous[a]	No	HSA	5×10^{-5}–10^{-7} M l^{-1}, VC: about 2.4%	—	—	(*170*)

Note: TELISA, thermistor enzyme-linked immunosorbent assay; HSA, human serum albumin; HCG, human chorionic gonadotropin; PIMIA, potentiometric ionophore-modulation immunoassay; MGIA, microgravimetric immunoassay; VC, variation coefficient; SD, standard deviation;

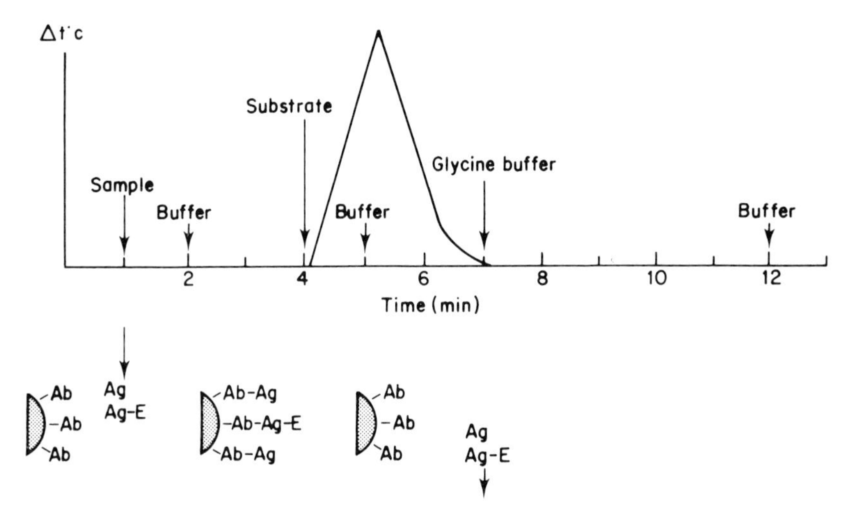

Fig. 11. Schematic presentation of a reaction cycle in the TELISA procedure. The arrows indicate changes in the perfusing medium (flow rate 0.8 ml min^{-1}). The cycle starts with potassium phosphate buffer pH 7.0 (0.2 M). At this time the thermistor column contains only immobilized antibodies. At the arrow "sample" a mixture of antigen and catalase-labelled antigen is introduced. The system is then washed with potassium phosphate buffer for two minutes. Now the sites on the antibodies of the column are occupied by antigen as well as by catalase-labelled antigen. The amount of catalase bound is measured by registering the heat produced during a one-minute pulse of 1 mM H_2O_2. After the heat pulse is registered, the system is washed with glycine/HCl (0.2 M, pH 2.2) to split the complex. After five minutes of washing, phosphate buffer is introduced, and the system is ready for another assay (taken from ref. (*49*)).

determination of human serum albumin. The principle is shown briefly in Fig. 11. The system is based on the competition of albumin (HSA) and catalase-conjugated HSA for the binding to immobilized anti-HSA antibodies. After the application of both, H_2O_2 is pulsed and the exothermic reaction of H_2O_2 to O_2 and H_2O (catalysed by catalase) is measured. Depending on the quantity of added conjugate and immobilized antibody, the range can be varied. A detection limit of 10^{-13} *M* HSA was achieved. Borrebaeck *et al.* (*159*) stated that this system could be used repeatedly up to 100 times.

Similar systems, based on measuring the change of O_2 partial pressure, were published by Aizawa *et al.* (*160*), Boitieux *et al.* (*161*), Haga *et al.* (*162*), and Karube *et al.* (*163*). The systems of Aizawa *et al.* (*160*) and of Haga *et al.* (*162*) were competitive immunoassays, the other two were two-site immunoassays (as shown in equations (25)–(28)). Normally, there must be an incubation interval between each step:

Aizawa *et al.* (*160*)

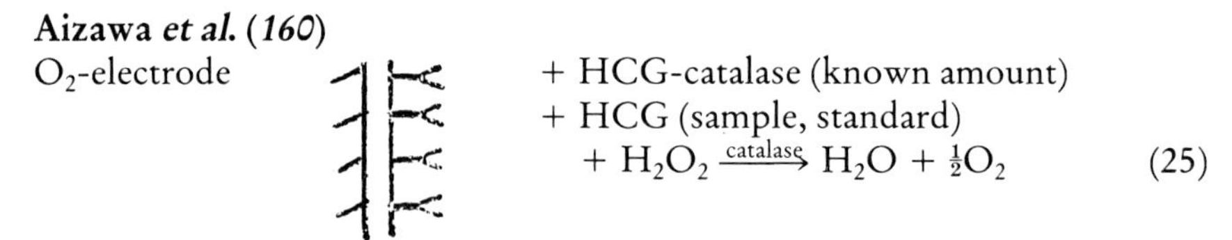

O_2-electrode

$$+ \text{HCG-catalase (known amount)} + \text{HCG (sample, standard)} + H_2O_2 \xrightarrow{\text{catalase}} H_2O + \tfrac{1}{2}O_2 \qquad (25)$$

antibody-membrane

Haga *et al.* (*162*)

O_2-Pt-electrode

$$+ \text{Ab} + \text{Ag-catalase (known amount)} + \text{Ag (sample, standard)} + H_2O_2 \xrightarrow{\text{catalase}} H_2O + \tfrac{1}{2}O_2 \qquad (26)$$

antigen-coated-membrane

Ab . . . Fab′-β-D-galactosidase;

Ag . . . theophylline

Boitieux *et al.* (*161*)

O_2-electrode

$$+ HB_sAg + \text{Anti-}HB_sAg\text{-glucose-oxidase} + \text{glucose} + O_2 \xrightarrow{\text{glucose oxidase}} \text{gluconic acid} + H_2O \qquad (27)$$

Anti-HB_sAg-IgG-membrane

Karube *et al.* (*163*)

O_2-electrode

$$+ \text{HSA} + \text{Anti-HSA-catalase} + H_2O_2 \xrightarrow{\text{catalase}} H_2O + \tfrac{1}{2}O_2. \qquad (28)$$

Anti-HSA-IgG-membrane

These four systems show the disadvantages of washing, elution, and long incubation steps, because they were not constructed for use in flow systems. Details are shown in Table XI.

Keating and Rechnitz (*164*) published the so-called PIMIA system, which is based on the change of the membrane potential caused by the antigen–antibody reaction. Details are shown in Fig. 12 and Table XI. The main disadvantage of this system, which was applied to the determination of antidigoxin antibodies for example, is the necessity of keeping constant the background K^+ concentration because of the sensitivity of the sensor to changes in the K^+ concentration (see also Section 3). The main advantage is the lack of unspecific reactions and the direct measurement principle. Only elution buffer is required.

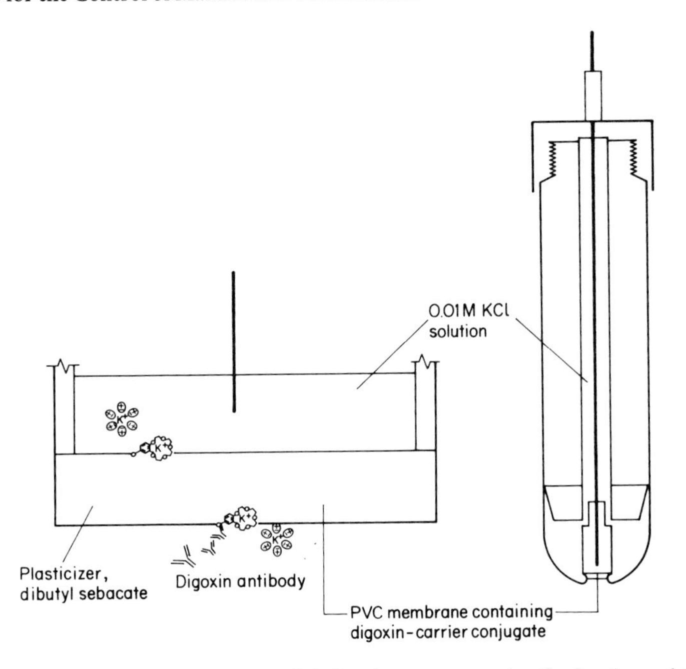

Fig. 12. Potentiometric ionophore-modulation immunoassay (antibody electrode) (taken from ref. (*164*)).

Another system, MGIA, in which the transducer is a surface acoustic device, was published by Roederer and Bastiaans (*53*). This sensor is based on the measurement of small changes of mass on the surface of a quartz piezoelectric crystal (Fig. 13). The comparison of a reference device with the antibody-coupled detecting device, on which the antigen–antibody reaction has taken place, is used for the determination of the analyte. The main disadvantages are the poor sensitivity, the non-specific adsorption, and the impossibility of application to the direct detection of haptens.

The last interesting system is based on measurement of the change of the transmitted light in a waveguide, caused by the change of the evanescent wave at the interface between two optical media by a macromolecule or complex bound to the glass surface (*41, 42*) (Fig. 14(*a*), 14(*b*)). The advantages are high sensitivity and the fact that separation steps, except for the elution, are not necessary. It is applicable in a direct or two-site mode (Table XI).

Apart from the above-mentioned detection principles, many other quite similar principles have been published, using other enzyme-conjugates, such as chloroperoxidase (*166*), adenosine deaminase, asparaginase, and urease

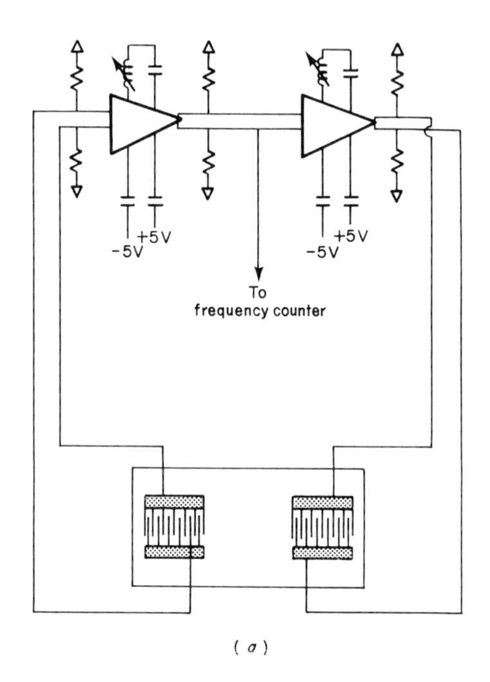

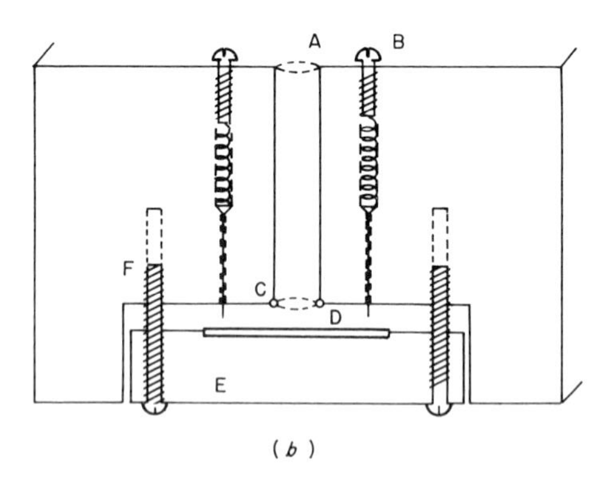

Fig. 13. Microgravimetric immunoassay (taken from ref. (*53*)). (*a*) Electrical oscillator circuit with SAW crystals. (*b*) Cut-away view of one SAW crystal in delrin detector cell: A, sample introduction well; B, screw-spring-electrode assembly; C, O-ring seal; D, crystal; E, delrin positioning insert; F, placement screws.

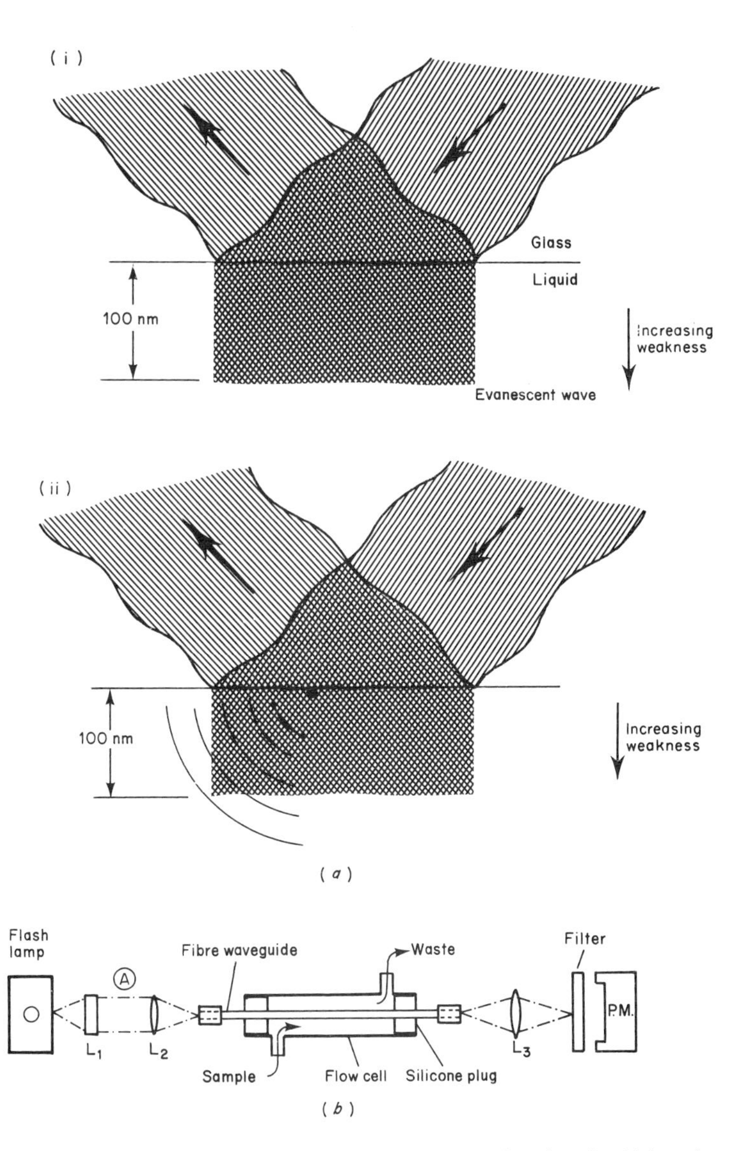

Fig. 14. Optical detection of antigen–antibody reaction at the glass–liquid interface. (*a*) Evanescent light. This occurs when light is totally reflected at a refractive index interface, such as that between a waveguide and the surrounding medium. (i) A light wave is a rapidly oscillating electromagnetic field (like a radio wave). When totally reflected, the wave induces a corresponding field which decreases exponentially at the other side of the interface. This short-range field is called the evanescent wave. (ii) When large molecules adhere to the interface, the evanescent field is disturbed and total reflection breaks down allowing light to escape from the waveguide. Detectors based on this system are consequently sensitive to the number and size of surface-adherent particles (taken from ref. (*165*)). (*b*) Diagram of a fibre-optic assembly with flow cell and light coupling optics (L_{1-3} are lenses; A, end of the fibre) (taken from ref. (*41*)).

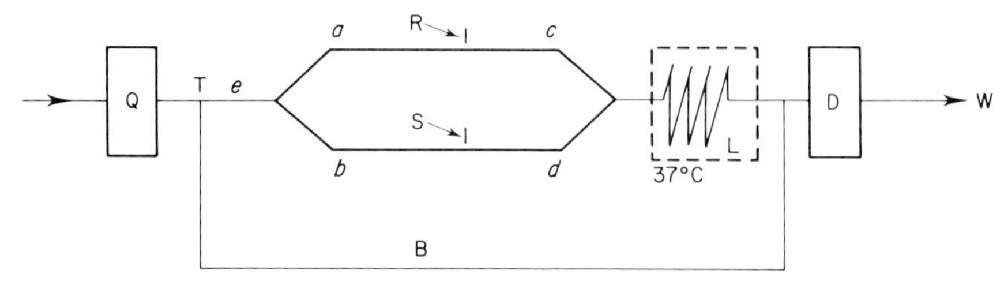

Fig. 15. Flow-injection analysis arrangement: B, bypass; D, fluorimetric detector; I, injection valve; L, reaction coil; Q, pump; R, reagent; S, sample, T, 3-way valve; W, waste; $a = b = 250$ mm, $c = d = 150$ mm, $e = 100$ mm (taken from ref. (*170*)).

conjugates (*167*), or other detection systems (e.g., luminescent immunoassay (*168, 169*)). The main disadvantages of all the systems mentioned are as follows. Frequently, non-specific binding of proteins to the devices takes place, which can be a major problem in the application of these devices in animal cell technology. Ionic interaction can occur with the sample solution. In most cases, except for the optical-fibre-based fluorescence assay, long incubation steps, especially for the two-site immunoassays, are necessary. Consequently, the application of energy-transfer immunoassay by stopped FIA with merging zones (Fig. 15, Table XI), working in a homogeneous mode, can be of great interest (*170*). The system is based on the use of fluorescein (donor fluorescent label) coupled antigen and rhodamine (acceptor fluorescent label) coupled antibody. The emission spectrum of one (donor) overlaps the excitation spectrum of the other (acceptor). When the antigen–antibody reaction takes place, energy transfer from the donor to the acceptor can occur, resulting in quenching of the donor fluorescence and enhancement of the acceptor fluorescence. In the presence of unlabelled sample (Ag) these effects are reversed. As shown in Fig. 15, the same amounts of the sample (S-value) and the reagent (equivalent mixture of rhodamine–antibody and fluorescein–antigen) (R-value) are applied to the system. After an incubation time of only 6 min, the fluorescence is detected by a fluorimeter. The advantages are that no unspecific reaction or unspecific binding of proteins can occur, that there are no possibilities for ionic interactions, and that no separation, elution, and washing steps are necessary. The main disadvantage is the necessity for reagent application.

In summary, there are many different possibilities for detecting a product of animal cell fermentation, but there is none, except the optical-fibre-based fluorescence assay and the energy transfer immunoassay, which can be employed in an elution-free mode. The application in a flow system is the only possibility for the use of these assays in an on-line mode in a continuous or semi-continuous way. Much developmental work is still needed for the optimization of affinity systems, such as the affinity constant or elution methods and for the optimization of membranes in electrode-based systems.

4.5. Determination of Enzymes

The detection of enzymes, produced as "product" or as "waste" by cells, is also of interest. When the enzyme is the desired product, optimal production has to be realized and controlled. But when the enzyme is only "waste", as is the case with proteases produced by hybridomas, detection is important for the production itself and the subsequent downstream processing (*110, 173*). Unfortunately, not much has been done so far in the development of enzyme-detecting biosensors. Nevertheless, some applications have been published. Guilbault (*85*), who reviewed the field of biosensors, made some points with respect to enzyme-detecting devices. The main problem is that it is difficult to achieve a truly self-contained, enzyme-sensing "substrate-electrode", because the substrate is small and is therefore immobilized with difficulty. The second point is that it is really better to monitor an enzyme using a kinetic, non-equilibrium method. Guilbault (*85*) gave practical examples of enzyme-sensing electrodes.

The application of FIA for the detection of enzyme concentration/activity in the liquid of a bioprocess has already been made several times. Kroner and Kula (*71*) and Recktenwald *et al.* (*174*) determined alkaline phosphatase and pullulanase, and formate dehydrogenase and L-leucine dehydrogenase, respectively. They used an FIA system with continuous diluted enzyme flow (out of the fermentation vessel or downstream process), which was to be detected, equipped with a discrete injection of the appropriate substrate (and finally a reversed FIA, according to van der Linden (*81*)). The colour change was detected subsequently. Danielsson and Mosbach (*175*) determined some enzymes, including urease and alkaline phosphatase, by the aid of a thermistor. The calorimeter was equipped with a fixed amount of enzyme, to be detected. The enzyme sample was mixed with a substrate solution of high concentration to guarantee substrate excess (and, finally, a competitive assay), determined by the thermistor. But both systems provide only an endpoint measurement or the measurement after a certain fixed time. A kinetic measurement, claimed by Guilbault (*85*) not to be necessary, was not applied. However, because of the relative ease of the endpoint determination, this method will be the method of choice for application in the control of a downstream process.

5. CONCLUSIONS

Many possibilities exist for determination of the cell count, the nutrients and metabolites, and the product concentration. Unfortunately, most of these methods were developed for the determination of the biomass of bacteria and yeasts or for the determination of biochemicals in medical laboratories. In such situations, in contrast to our criteria, the following criteria are of major importance: fast response, short lifetime (disposability), cheapness, sterility not a necessity (in general, except for *in vivo* applications). The fermentation industries demand the following criteria: long lifetime, lack of drift, recalibration possibilities, sterile application, cheapness. To achieve these conditions, much additional development work is needed.

REFERENCES

1. Merten, O.-W., Reiter, S., and Katinger, H. (1985). Stabilizing effect of reduced cultivation temperature on human × mouse hybridomas. *Dev. Biol. Stand.* **60**, 509–512.
2. Lloyd, D., Bohátka, S., and Szilágyi, J. (1985). Quadrupole mass spectrometry in the monitoring and control of fermentations. *Biosensors* **1**, 179–212.
3. Suzuki, S., and Karube, I. (1981). Bioelectrochemical sensors based on immobilized enzymes, whole cells, and proteins. *Appl. Biochem. Bioeng.* **3**, 145–174.
4. Karube, I., and Suzuki, S. (1984). Amperometric and potentiometric determinations with immobilized enzymes and microorganisms. *ISE Rev.* **6**, 15–58.
5. Fleischacker, R. J., Weaver, J. C., and Sinskey, A. J. (1981). Instrumentation for process control in cell culture. *Adv. Appl. Microbiol.* **27**, 137–167.
6. Schönherr, O. T., van Gelder, P. T. J. A., van Hees, P. J., van Os, A. M. J. M., and Roelofs, H. W. M. (1987). A hollow fibre dialysis system for the *in vitro* production of monoclonal antibodies replacing *in vivo* production in mice. *Dev. Biol. Stand.* **66**, 211–222.
7. Emery, N. (1986). Growth of hybridomas and secretion of monoclonal antibodies *in vitro*. Presented at: Society of Chemical Industry Symposium on "Large Scale Production of Monoclonal Antibodies" (London, 9 Dec. 1986).
8. Chang, T. H., Steplewski, Z., and Koprowski, H. (1980). Production of monoclonal antibodies in serum free medium. *J. Immunol. Method.* **39**, 369–375.
9. Murakami, H., Masui, H., Sato, G. H., Sueoka, N., Chow, T. P., and Kano-Sueoka, T. (1982). Growth of hybridoma cells in serum-free medium: Ethanolamine is an essential component. *Proc. Natl. Acad. Sci. U.S.A.* **79**, 1158–1162.
10. Darfler, F. J., and Insel, P. A. (1982). Serum-free culture of resting, PHA-stimulated, and transformed lymphoid cells, including hybridomas. *Exp. Cell Res.* **138**, 287–295.
11. Kovar, J., and Franek, F. (1985). Hybridoma cultivation in defined serum-free media: Growth supporting substances. I. Transferrin. *Folia Biologica* **31**, 167–175.
12. Kovar, J., and Franek, F. (1984). Serum-free medium for hybridoma and parental myeloma cell cultivation: A novel composition of growth supporting substances. *Immunol. Lett.* **7**, 339–345.
13. Cole, S. P. C., Vreeken, E. H., and Roder, J. C. (1985). Antibody production by human × human hybridomas in serum-free medium. *J. Immunol. Method.* **78**, 271–278.

14. Kawamoto, T., Sato, J. D., Le, A., McClure, D. B., and Sato, G. H. (1983). Development of a serum-free medium for growth of NS-1 mouse myeloma cells and its application to the isolation of NS-1 hybridomas. *Anal. Biochem.* **130**, 445–453.
15. Fazekas de St. Groth, S. (1983). Automated production of monoclonal antibodies in a cytostat. *J. Immunol. Method.* **57**, 121–136.
16. Reuveny, S., Velez, D., Miller, L., and Macmillan, J. D. (1986). Comparison of cell propagation methods for their effect on monoclonal antibody yield in fermenters. *J. Immunol. Method.* **86**, 61–69.
17. Rutzky, L. P., Tomita, J. T., Calenoff, M. A., and Kahan, B. D. (1977). Matrix perfusion cultivation of human choriocarcinoma and colon adenocarcinoma cells. *In Vitro* **13**, 191.
18. Butler, M. (1986). Nutrition of hybridoma cells. Presented at: Society of Chemical Industry Symposium on "Large Scale Production of Monoclonal Antibodies" (London, 9 Dec. 1986).
19. Butler, M. (1985). Growth limitations in high density microcarrier cultures. *Dev. Biol. Stand.* **60**, 269–280.
20. Hu, W. S., Dodge, T. C., Frame, K. K., and Himes, V. B. (1987). Effect of glucose and oxygen on the cultivation of mammalian cells. *Dev. Biol. Stand.* **66**, 279–290.
21. Roberts, R. S., Hsu, H. W., Lin, K. D., and Yang, T. J. (1976). Amino acid metabolism of myeloma cells in culture. *J. Cell Sci.* **21**, 609–615.
22. Polastri, G. D., Friesen, H. J., and Mauler, R. (1984). Amino acid utilization by Vero cells in microcarrier culture. *Dev. Biol. Stand.* **55**, 53–56.
23. Seaver, S. S., Rudolph, J. L., and Gabriels, J. E. Jr. (1984). A rapid HPLC technique for monitoring amino acid utilization in cell culture. *Bio Techniques* **2**, 254–260.
24. Merten, O.-W., Palfi, G. E., and Steiner, J. (1984). Unpublished results.
25. Lowe, C. R. (1984). Biosensors. *Trends Biotechnol.* **2**, 59–65.
26. Lowe, C. R. (1985). An introduction to the concepts and technology of biosensors. *Biosensors* **1**, 3–16.
27. Gronow, M., Kingdon, C. F. M., and Anderton, D. J. (1985). Biosensors. *In* "Molecular Biology and Biotechnology" (J. M. Walker and E. B. Gingold, eds.), pp. 295–324. The Royal Society of Chemistry, Burlington House, London.
28. Nilsson, H., Akerlund, A.-C., and Mosbach, K. (1973). Determination of glucose, urea and penicillin using enzyme-pH-electrodes. *Biochim. Biophys. Acta* **320**, 529–534.
29. Caras, S. D., Petelenz, D., and Janata, J. (1985). pH-based enzyme field effect transistors. Part 2. Glucose enfet. *Anal. Chem.* **57**, 1920–1923.
30. Caras, S. D., and Janata, J. pH based enzyme field effect transistors. Part 3. Penicillin enfet. *Anal. Chem.* **57**, 1924–1925.
31. Danielsson, B., Lundström, I., Mosbach, K., and Stiblert, L. (1979). On a new enzyme transducer combination: The enzyme transistor. *Anal. Lett.* **12** (B11), 1189–1199.
32. Sibbald, A. (1985). A chemical-sensitive integrated-circuit: The operational transducer. *Sensors and Actuators* **7**, 23–38.
33. Sibbald, A. (1983). Chemical-sensitive field-effect transistors. *IEE Proc.* **130**, 233–244.
34. Appelqvist, R., Marko-Varga, G., Gorton, L., Torstensson, A., and Johansson, G. (1985). Enzymatic determination of glucose in a flow system by catalytic oxidation on the nicotinamide coenzyme at a modified electrode. *Anal. Chim. Acta* **169**, 237–247.
35. Schelter-Graf, A., Schmidt, H.-L., and Huck, H. (1984). Determination of the substrates of dehydrogenases in biological material in flow injection systems with electrocatalytic NADH oxidation. *Anal. Chim. Acta* **163**, 299–303.
36. Cass, A. E. G., Davis, G., Francis, G. D., Hill, H. A. O., Aston, W. J., Higgins, I. J., Plotkin, E. V., Scott, L. D. L., and Turner, A. P. F. (1984). Ferrocene-mediated enzyme electrode for amperometric determination of glucose. *Anal. Chem.* **56**, 667–671.

37. Wingard, L. B. Jr. (1984). Cofactor modified electrodes. *Trends Anal. Chem.* **3**, 235–238.
38. Lowe, C. R., Goldfinch, M. J., and Lias, R. J. (1983). Some novel biomedical biosensors. *In* "Biotech 83", pp. 633–641. On line Publications, Northwood.
39. Schultz, J. S., Mansouri, S. and Goldstein, I. J. (1982). Affinity sensor: A new technique for developing implantable sensors for glucose and other metabolites. *Diabetes Care* **5**, 245–253.
40. Liedberg, B., Nylander, C., and Lundström, I. (1983). Surface plasmon resonance for gas detection and biosensing. *Sensors and Actuators* **4**, 299–304.
41. Sutherland, R. M., Dähne, C., Place, J. F., and Ringrose, A. S. (1984). Optical detection of antibody–antigen reactions at a glass–liquid interface. *Clin. Chem.* **30**, 1533–1538.
42. Sutherland, R. M., Dähne, C., Place, J. F., and Ringrose, A. S. (1984). Immunoassays at a quartz–liquid interface: Theory, instrumentation and preliminary application to the fluorescent immunoassay of human immunoglobulin G. *J. Immunol. Method.* **74**, 253–265.
43. Smith, J. B., Burke, S. E., Lefer, A. M., and Freilak, A. (1984). Continuous measurement of ATP secretion in vitro. *Pharm. Res.* **1**, 40–43.
44. Carrea, G., Bovara, R., Mazzola, G., Girotti, S., Roda, A., and Ghini, S. (1986). Bioluminescent continuous-flow assay of adenosine 5′-triphosphate using firefly luciferase immobilized on nylon tubes. *Anal. Chem.* **58**, 331–333.
45. Girotti, S., Roda, A., Ghini, S., Grigolo, B., Carrea, G., and Bovara, R. (1984). Continuous flow analysis of NADH using bacterial bioluminescent enzymes immobilized on nylon. *Anal. Lett.* **17**(B1), 1–12.
46. Roda, A., Girotti, S., Ghinh, S., Grigolo, B., Carrea, G., and Bovara, R. (1984). Continuous flow determination of primary bile acids, by bioluminescence, with use of nylon-immobilized bacterial enzymes. *Clin. Chem.* **30**, 206–210.
47. Mandenius, C. F., Danielsson, B., and Mattiasson, B. (1980). Enzyme thermistor control of the sucrose concentration at a fermentation with immobilized yeast. *Acta Chem. Scand.* **B34**, 463–465.
48. Scheller, F., Siegbahn, F., Danielsson, B., and Mosbach, K. (1985). High-sensitive enzyme thermistor determination of L-lactate by substrate recycling. *Anal. Chem.* **57**, 1740–1743.
49. Mattiasson, B., Borrebaeck, C., Sanfridson, B., and Mosbach, K. (1977). Thermometric enzyme linked immunosorbent assay: TELISA. *Biochim. Biophys. Acta* **483**, 221–227.
50. Guilbault, G. G. (1980). Use of the piezoelectric crystal detector in analytical chemistry. *ISE Rev.* **2**, 3–17.
51. Alder, J. F., and McCullen, J. J. (1983). Piezoelectric crystal for mass and chemical measurements. *Analyst* **108**, 1169–1189.
52. Guilbault, G. G. (1983). The piezoelectric crystal as an air pollution monitor. *Anal. Chem. Symp. Ser.* **17**, 637–643.
53. Roederer, J. E., and Bastiaans, G. J. (1983). Microgravimetric immunoassay with piezoelectric crystal. *Anal. Chem.* **55**, 2333–2336.
54. McLaughlin, J. K., Meyer, C. L., and Papoutsakis, E. T. (1985). Gas chromatography and gateway sensors for on-line state estimation of complex fermentations (Butanol-acetone fermentations). *Biotechnol. Bioeng.* **27**, 1246–1257.
55. Rousseau, I. (1984). Development of automatic on-line HPLC in fermentation processes. *Pers. Commun.*
56. Dinwoodie, R. C., and Mehnert, D. W. (1985). A continuous method for monitoring and controlling fermentations: Using an automated HPLC-system. *Biotechnol. Bioeng.* **27**, 1060–1062.
57. Enfors, S. O., and Cleland, N. (1983). Calibration of oxygen- and pH-based enzyme electrodes for fermentation control. *Anal. Chem. Symp. Ser.* **17**, 672–676.

58. Powell, M. S., and Slater, N. K. H. (1983). The deposition of bacterial cells from laminar flows onto solid surfaces. *Biotechnol. Bioeng.* **25**, 891–900.
59. Clarke, D. J., Calder, M. R., Carr, R. J. G., Blake-Coleman, B. C., Moody, S. C., and Collinge, T. A. (1985). The development and application of biosensing devices for bioreactor monitoring and control. *Biosensors* **1**, 213–320.
60. Danielsson, B., Gadd, K., Mattiasson, B., Mosbach, K. (1977). Enzyme thermistor determination of glucose in serum using immobilized glucose oxidase. *Clin. Chim. Acta* **81**, 163–175.
61. Davies, P., and Mosbach, K. (1974). The application of immobilized NAD^+ in an enzyme electrode and in model enzyme reactors. *Biochim. Biophys. Acta* **370**, 329–338.
62. Ruzicka, J., and Hansen, E. H. (1979). Stopped flow and merging zones—A new approach to enzymatic assay by slow injection analysis. *Anal. Chim. Acta* **106**, 207–224.
63. Hansen, E. H., Ruzicka, J., and Rietz, B. (1977). Flow injection analysis. Part VIII. Determination of glucose in blood serum with glucose dehydrogenase. *Anal. Chim. Acta* **89**, 241–254.
64. Rydevik, U., Nord, L., and Ingman, F. (1982). Automatic lactate determination by flow injection analysis. *Int. J. Sports Med.* **3**, 47–49.
65. Zabriskie, D. W., and Humphrey, A. E. (1978). Continuous dialysis for the on-line analysis of diffusible components in fermentation broth. *Biotechnol. Bioeng.* **20**, 1295–1301.
66. Cleland, N., and Enfors, S.-O. (1984). Externally buffered enzyme electrode for determination of glucose. *Anal. Chem.* **56**, 1880–1884.
67. Cleland, N., and Enfors, S.-O. (1984). Monitoring glucose consumption in an *Escherichia coli* cultivation with an enzyme electrode. *Anal. Chim. Acta* **163**, 281–285.
68. Mandenius, C. F., Danielsson, B., and Mattiasson, B. (1984). Evaluation of a dialysis probe for continuous sampling in fermenters and complex media. *Anal. Chim. Acta* **163**, 135–141.
69. Schmidt, W. J., Meyer, H.-D., Schügerl, K., Kuhlmann, W., and Bellgardt, K.-H. (1984). On-line analysis of fermentation media. *Anal. Chim. Acta* **163**, 101–109.
70. Chotani, G., and Constantinides, A. (1982). On-line glucose analyzes for fermentation applications. *Biotechnol. Bioeng.* **24**, 2743–2745.
71. Kroner, K. H., and Kula, M.-R. (1984). On-line measurement of extracellular enzymes during fermentation by using membrane techniques. *Anal. Chim. Acta* **163**, 3–15.
72. Wang, H. Y. (1984). Sensor development for fermentation monitoring and control. *Biotechnol. Bioeng. Symp.* **14**, 601–610.
73. Parker, C. P., Gardell, M. G., and DiBiasio, D. (1986). A complete system for fermentation monitoring. *Int. Biotechnol. Lab.* June, 33–40.
74. Ghoul, M., Ronat, E., and Engasser, J.-M. (1986). An automatic and sterilizable sampler for laboratory fermenters: Application to the on-line control of glucose concentration. *Biotechnol. Bioeng.* **28**, 119–121.
75. Tolbert, W. R., and Feder, J. (1983). Large-scale cell culture technology. *In* "Annual Reports on Fermentation Processes," Vol. 6 (G. T. Tsao, ed.), pp. 35–74. Academic Press, New York.
76. Tolbert, W. R., Feder, J., and Kimes, R. C. (1981). Large-scale rotating filter perfusion system for high density growth of mammalian suspension cultures. *In Vitro* **17**, 885–890.
77. Rebsamen, E., Goldinger, W., Scheirer, W., Merten, O.-W., and Pálfi, G. E. (1987). Use of a dynamic filtration method for separation of animal cells. *Dev. Biol. Stand.* **66**, 1273–1277.
78. Vick Roy, T. B., Mandel, D. K., Dea, D. K., Blanch, H. W., and Wilke, C. R. (1983). The application of cell recycle to continuous fermentative lactic acid production. *Biotechnol. Lett.* **5**, 665–670.
79. Tolbert, W. R., Lewis, C. Jr., White, P. J., and Feder, J. (1985). Perfusion culture systems

for production of mammalian cell biomolecules. *In* "Large-Scale Mammalian Cell Culture" (J. Feder and W. R. Tolbert, eds.), pp. 97–123. Academic Press, London.
80. Snyder, L. R. (1980). Continuous-flow analysis: Present and future. *Anal. Chim. Acta* **114**, 3–18.
81. van der Linden, W. E. (1986). Flow injection analysis in on-line process control. *Anal. Chim. Acta* **179**, 91–101.
82. Schifreen, R. S., Hanna, D. A., Bowers, L. D., and Carr, P. W. (1977). Analytical aspects of immobilized enzyme columns. *Anal. Chem.* **49**, 1929–1939.
83. Watson, B., and Keyes, M. H. (1976). A dedicated instrument for the analysis of blood urea nitrogen using an immobilized enzyme reactor. *Anal. Lett.* **9**, 713–725.
84. Gray, D. N., Keyes, M. H., and Watson, B. (1977). Immobilized enzymes in analytical chemistry. *Anal. Chem.* **49**, 1067A–1078A.
85. Guilbault, G. G. (1984). "Analytical Uses of Immobilized Enzymes." Dekker, New York.
86. Bourdillon, C., Thomas, V., and Thomas, D. (1982). Electrochemical study of D-glucose oxidase autoinactivation. *Enzyme Microb. Technol.* **4**, 175–180.
87. Decristoforo, G., and Knauseder, F. (1984). Rapid determination of cephalosporins with an immobilized enzyme reactor and sequential subtractive spectrophotometric detection in an automated flow-injection system. *Anal. Chim. Acta* **163**, 73–84.
88. Mattiasson, B., Danielsson, B., and Mosbach, K. (1976). A split-flow enzyme thermistor. *Anal. Lett.* **9**, 867–889.
89. Merten, O.-W., Palfi, G. E., and Steiner, J. (1986). On line determination of biochemical/physiological parameters in the fermentation of animal cells in a continuous or discontinuous mode. *In* "Advances in Biotechnological Processes," Vol. 6 (A. Mizrahi and A. T. van Wezel, eds.), pp. 111–178. Alan R. Liss, New York.
90. Clarke, D. J., Kell, D. B., Morris, J. G., and Burns, A. (1982). The role of ion-selective electrodes in microbial process control. *ISE Rev.* **4**, 75–131.
91. Turner, A. P. F., and Pickup, J. C. (1985). Diabetes mellitus: Biosensors for research and management. *Biosensors* **1**, 185–238.
92. Romette, J.-L., Froment, B., and Thomas, D. (1979). Glucose-oxidase electrode. Measurements of glucose in samples exhibiting high variability in oxygen content. *Clin. Chim. Acta* **95**, 249–253.
93. Gough, D. A., Lucisano, J. Y., and Tse, P. H. S. (1985). Two-dimensional enzyme electrode sensor for glucose. *Anal. Chem.* **57**, 2351–2357.
94. Greenfield, P. F., Kittrell, J. R., and Laurence, R. L. (1975). Inactivation of immobilized glucose oxidase by hydrogen peroxide. *Anal. Biochem.* **65**, 109–124.
95. Guilbault, G. G., and Nagy, G. (1973). Improved urea electrode. *Anal. Chem.* **45**, 417–419.
96. Tsuchida, T., and Yoda, K. (1981). Immobilization of D-glucose oxidase onto a hydrogen peroxide permselective membrane and application for an enzyme electrode. *Enzyme Microb. Technol.* **3**, 326–330.
97. Tsuchida, T., Takasugi, H., Yoda, K., Takizawa, K., and Kabayashi, S. (1985). Application of L-+-lactate electrode for clinical analysis and monitoring of tissue culture medium. *Biotechnol. Bioeng.* **27**, 837–841.
98. Ho, M. H., and Asouzu, M. U. (1984). Use of immobilized enzymes in flow injection analysis. *Ann. N.Y. Acad. Sci.* **434**, 526–528.
99. Mullen, W. H., Keedy, F. H., and Churchouse, S. J. (1986). Glucose enzyme electrode with extended linearity. Application to undiluted blood measurements. *Anal. Chim. Acta* **183**, 59–66.
100. Bourdillon, C., Hervagault, C., and Thomas, D. (1985). Increase in operational stability of immobilized glucose oxidase by the use of an artificial cosubstrate. *Biotechnol. Bioeng.* **27**, 1619–1622.

101. Yao, T., Sato, M., Kobayashi, Y., and Wasa, T. (1984). Flow injection analysis for glucose by the combined use of an immobilized glucose oxidase reactor and a peroxidase electrode. *Anal. Chim. Acta* **165**, 291–296.
102. Durliat, H., and Comtat, M. (1984). Amperometric enzyme electrode for determination of glucose based on thin-layer spectroelectrochemistry of glucose oxidase. *Anal. Chem.* **56**, 148–152.
103. Miyawaki, O., and Wingard, L. B. Jr. (1984). FAD and glucose oxidase immobilized on carbon. *Ann. N.Y. Acad. Sci.* **434**, 520–522.
104. Mullen, W. H., Churchouse, S. J., and Vadgama, P. M. (1985). Enzyme electrode for glucose based on the quinoprotein glucose dehydrogenase. *Analyst* **110**, 925–928.
105. Jeannesson, P., Manfait, M., and Jardillier, J.-C. (1983). A technique for laser raman spectroscopic studies of isolated cell populations. *Anal. Biochem.* **129**, 305–309.
106. Chapman, A. G., Fall, L., and Atkinson, D. E. (1971). Adenylate energy charge in *Escherichia coli* during growth and starvation. *J. Bacteriol.* **108**, 1072–1086.
107. Siro, M.-R., Romar, H., and Lövgren, T. (1982). Continuous flow method for extraction and bioluminescence assay for ATP in baker's yeast. *Eur. J. Appl. Microbiol. Biotechnol.* **15**, 258–264.
108. Hill, B. T., and Whatley, S. (1975). A simple, rapid microassay for DNA. *FEBS Lett.* **56**, 20–23.
109. Himmler, G., Palfi, G., Rüker, F., Katinger, H., and Scheirer, W. (1985). A laboratory fermenter for agarose immobilized hybridomas to produce monoclonal antibodies. *Dev. Biol. Stand.* **60**, 291–296.
110. Merten, O.-W., unpublished results.
111. Meyer, C., and Beyeler, W. (1984). Control strategies for continuous bioprocesses based on biological activities. *Biotechnol. Bioeng.* **26**, 916–925.
112. Luong, J. H. T., and Carrier, D. J. (1986). On-line measurement of culture fluorescence during cultivation of *Methylomonas mucosa*. *Appl. Microbiol. Biotechnol.* **24**, 65–70.
113. Scheper, T., Gebauer, A., Sauerbrei, A., Niehoff, A., and Schügerl, K. (1984). Measurement of biological parameters during fermentation processes. *Anal. Chim. Acta* **163**, 111–118.
114. Beyeler, W., Einsele, A., and Fiechter, A. (1981). On-line measurements of culture fluorescence: Method and application. *Eur. J. Appl. Microbiol. Biotechnol.* **13**, 10–14.
115. Leist, C., Meyer, H.-P., and Fiechter, A. (1986). Process control during the suspension culture of a human melanoma cell line in a mechanically stirred loop bioreactor. *J. Biotechnol.* **4**, 235–246.
116. Blake-Coleman, B. C., Calder, M. R., Carr, R. J. G., Moody, S. C., and Clarke, D. J. (1984). Direct monitoring of reactor biomass in fermentation control. *Trends Anal. Chem.* **3**, 229–235.
117. Koch, A. L. (1961). Some calculations on the turbidity of mitochondria and bacteria. *Biochim. Biophys. Acta* **51**, 429–471.
118. Mallette, M. F. (1969). Evaluation of growth by physical and chemical means. *In* "Methods in Microbiology," Vol. 6B (J. R. Norris and D. W. Ribbons, eds.), pp. 319–566. Academic Press, London.
119. Harris, C. M., and Kell, D. B. (1985). The estimation of microbial biomass. *Biosensors* **1**, 17–84.
120. Merten, O.-W., Palfi, G. E., Stäheli, J., and Steiner, J. (1987). Invasive infrared sensor for the determination of the cell number in a continuous fermentation of hybridomas. *Dev. Biol. Stand.* **66**, 357–360.
121. Matsunaga, T., Karube, I., and Suzuki, S. (1979). Electrode system for the determination of microbial populations. *Appl. Environ. Microbiol.* **37**, 117–121.

122. Matsunaga, T., Karube, I., and Suzuki, S. (1980). Electrochemical determination of cell populations. *Eur. J. Appl. Microbiol. Biotechnol.* **10**, 125–132.
123. Ishimori, Y., Karube, I., and Suzuki, S. (1981). Determination of microbial populations with piezoelectric membranes. *Appl. Environ. Microbiol.* **42**, 632–637.
124. Clarke, D. J., Blake-Coleman, B. C., Calder, M. R., Carr, R. J. D., and Moody, S. C. (1984). Sensors for bioreactor monitoring and control—a perspective. *J. Biotechnol.* **1**, 135–158.
125. Matsushita, T., Brendzel, A. M., Shotola, M. A., and Groh, K. R. (1982). Electrical determination of viability in saline-treated mouse myeloma cells. *Biophys. J.* **39**, 41–47.
126. Miller, S. J. O., Henrotte, M., and Miller, A. O. A. (1986). Growth of animal cells on microbeads. I. *In situ* estimation of numbers. *Biotechnol. Bioeng.* (in press).
127. Al-Hitti, I. K., Moody, J. G., and Thomas, J. D. R. (1984). Glucose oxidase membrane systems based on poly(vinyl) chloride) matrices for glucose determination with an iodide ion-selective electrode. *Analyst* **109**, 1205–1208.
128. Al-Hitti, I. K., Moody, G. J., and Thomas, J. D. R. (1984). 5. Immobilisation of enzymes in membranes for use with electrochemical sensors. *J. Biomed. Eng.* **6**, 178–180.
129. Mell, L. D., and Maloy, J. T. (1975). A model for the amperometric enzyme electrode obtained through digital simulation and applied to the immobilized glucose oxidase system. *Anal. Chem.* **47**, 299–307.
130. Gorton, L., and Bhatti, K. M. (1979). Potentiometric determination of glucose by enzymatic oxidation in a flow system. *Anal. Chim. Acta* **105**, 43–52.
131. Mansouri, S., and Schultz, J. S. (1984). A miniature optical glucose sensor based on affinity binding. *Bio/Technology* **2**, 885–890.
132. D'Costa, E. J., Higgins, I. J., and Turner, A. P. F. (1986). Quinoprotein glucose dehydrogenase and its application in an amperometric glucose sensor. *Biosensor* **2**, 71–87.
133. El Degheidy, M. M., Wilkins, E. S., and Soudi, O. (1986). Optimization of an implantable coated wire glucose sensor. *J. Biomed. Eng.* **8**, 121–129.
134. Keyes, M. H., Semersky, F. E., and Gray, D. N. (1979). Glucose analysis utilizing immobilized enzymes. *Enzyme Microb. Technol.* **1**, 91–94.
135. Clarke, L. C., Jr. (1970). Membrane polarographic electrode system and method with electrochemical compensation. US Patent 3,539,455.
136. Guilbault, G. G. (1982). Ion-selective electrodes applied to enzyme systems. *ISE Rev.* **4**, 187–231.
137. Aston, W. J., and Turner, A. P. F. (1984). Biosensors and biofuel cells. *Biotechnol. Genet. Eng. Rev.* **1**, 89–120.
138. Updike, S. J., and Hicks, G. P. (1967). The enzyme electrode. *Nature (London)* **214**, 986–988.
139. Updike, S. J., and Hicks, G. P. (1967). Reagentless substrate analysis with immobilized enzymes. *Science* **158**, 270–272.
140. Enfors, S.-O. (1981). Oxygen-stabilized enzyme electrode for D-glucose analysis in fermentation broths. *Enzyme Microb. Technol.* **3**, 29–32.
141. Enfors, S.-O., and Molin, N. (1978). Enzyme electrodes for fermentation control. *Process Biochem.* 9–11, 24.
142. Shichiri, M., Kawamori, R., Goriya, Y., Yamasaki, Y., Nomura, M., Hakui, N., and Abe, H. (1983). Glycaemic control in pancreatectomized dogs with a wearable artificial endocrine pancreas. *Diabetologia* **24**, 179–184.
143. Shichiri, M., Kawamori, R., Yamasaki, Y., Hakui, N., and Abe, H. (1982). Wearable artificial endocrine pancreas with needle-type glucose sensor. *Lancet* **2**, 1129–1131.
144. Abel, P., Müller, A., and Fischer, U. (1984). Experience with an implantable glucose sensor as a prerequisite of an artificial beta cell. *Biomed. Biochim. Acta* **43**, 577–584.

145. Kessler, M., Höper, J., Volkholz, H.-J., Sailer, D., and Demling, L. (1984). A new glucose electrode for tissue measurements. *Hepato-gastroenterol.* **31**, 285–288.
146. Massom, M., and Townshend, A. (1984). Determination of glucose in blood by flow injection analysis and an immobilized glucose oxidase column. *Anal. Chim. Acta* **166**, 111–118.
147. Wieck, H. J., Heider, G. H. Jr., and Yacynych, A. M. (1984). Chemically modified reticulated vitreous carbon electrode with immobilized enzyme as a detector in flow-injection determination of glucose. *Anal. Chim. Acta* **158**, 137–141.
148. Lundbäck, H., and Olsson, B. (1985). Amperometric determination of galactose, lactose, and dihydroxyacetone using galactose oxidase in a flow injection system with immobilized enzyme reactors and on-line dialysis. *Anal. Lett.* **18**, 871–889.
149. Mascini, M., Fortunati, S., Moscone, D., Palleschi, G., Massi-Benedetti, M., and Fabietti, P. (1985). An L-lactate sensor with immobilized enzyme for use in *in vitro* studies with an endocrine artificial pancreas. *Clin. Chem.* **31**, 451–453.
150. Mascini, M., Moscone, D., and Palleschi, G. (1984). A lactate electrode with lactate oxidase immobilized on nylon net for blood serum samples in flow systems. *Anal. Chim. Acta* **157**, 45–51.
151. Racine, P., Engelhardt, R., Higelin, J. C., and Mindt, M. (1975). An instrument for the rapid determination of L-lactate in biological fluids. *Med. Instrum.* **9**, 11–14.
152. Mizutani, F., Tsuda, K., Karube, I., Suzuki, S., and Matsumoto, K. (1980). Determination of glutamate pyruvate transaminase and pyruvate with an amperometric pyruvate oxidase sensor. *Anal. Chim. Acta* **118**, 65–71.
153. Fogt, E. J., Cahalan, P. T., Jeyne, A., and Schwinghammer, M. A. (1985). Simplified procedure for forming polymer-based ion-selective electrodes. *Anal. Chem.* **57**, 1155–1157.
154. Guilbault, G. G., and Hrabankova, E. (1971). New enzyme electrode probes for D-amino acids and asparagine. *Anal. Chim. Acta* **56**, 285–290.
155. Romette, J. L., Yang, J. S., Kusakabe, H., and Thomas, D. (1983). Enzyme electrode for specific determination of L-lysine. *Biotechnol. Bioeng.* **25**, 2557–2566.
156. White, W. C., and Guilbault, G. G. (1978). Lysine specific enzyme electrode for determination of lysine in grains and foodstuffs. *Anal. Chem.* **50**, 1481–1486.
157. Fung, K. W., Kuan, S. S., Sung, H. Y., and Guilbault, G. G. (1979). Methionine selective enzyme electrode. *Anal. Chem.* **51**, 2319–2324.
158. Havas, J., and Guilbault, G. G. (1982). Tyrosine-selective enzyme probe and its application. *Anal. Chem.* **54**, 1991–1997.
159. Borrebaeck, C., Börjeson, J., and Mattiasson, B. (1978). Thermometric enzyme linked immunosorbent assay in continuous flow system: Optimization and evaluation using human serum albumin as a model system. *Clin. Chim. Acta* **86**, 267–278.
160. Aizawa, M., Morioka, A., Suzuki, S., and Nagamura, Y. (1979). Enzyme immunosensor. III. Amperometric determination of human chorionic gonadotropin by membrane-bound antibody. *Anal. Biochem.* **94**, 22–28.
161. Boitieux, J.-L., Thomas, D., and Desmet, G. (1984). Oxygen electrode-based enzyme immunoassay for the amperometric determination of hepatitis B surface antigen. *Anal. Chim. Acta* **163**, 309–313.
162. Haga, M., Ikuta, M., Kato, Y., and Suzuki, Y. (1984). Enzyme immunosensor using hapten-bound membrane and Fab′-β-D-galactosidase complexes. *Chem. Lett.* 1313–1316.
163. Karube, I., Matsunaga, T., Suzuki, S., Asano, T., and Itoh, S. (1984). Immobilized antibody-based flow type enzyme immunosensor for determination of human serum albumin. *J. Biotechnol.* **1**, 279–286.
164. Keating, M. Y., and Rechnitz, G. A. (1984). Potentiometric digoxin antibody measurements with antigen-ionophore based membrane electrodes. *Anal. Chem.* **56**, 801–806.

165. North, J. R. (1985). Antibody-based biosensors. *Trends Biotechnol.* **3**, 180–186.
166. Fonong, T., and Rechnitz, G. A. (1984). Homogeneous potentiometric enzyme immunoassay for human immunoglobulin G. *Anal. Chem.* **56**, 2586–2590.
167. Gebauer, C. R., and Rechnitz, G. A. (1982). Deaminating enzyme labels for potentiometric enzyme immunoassay. *Anal. Biochem.* **124**, 338–348.
168. Ikariyama, Y., Suzuki, S., and Aizawa, M. (1984). Solid-phase luminescent catalyst immunoassay for human serum albumin with hemin as labeling catalyst. *Anal. Chim. Acta* **156**, 245–252.
169. Téronanne, B., Carrié, M.-L., Nicolas, J.-C., and Crastes de Paulet, A. (1986). Bioluminescent immunosorbent for rapid immunoassays. *Anal. Biochem.* **154**, 118–125.
170. Lim, C. S., Miller, J. N., and Bridges, J. W. (1980). Automation of an energy-transfer immunoassay by using stopped-flow injection analysis with merging zones. *Anal. Chim. Acta* **114**, 183–189.
171. Ruzicka, J., and Hansen, E. H. (1980). Flow injection analysis. Principles, applications and trends. *Anal. Chim. Acta* **114**, 19–44.
172. Schügerl, K. (1985). Sensor-Messtechniken in der biotechnologischen Forschung und Industrie. *Naturwissenschaften* **72**, 400–407.
173. Schlaeger, E. J., Eggimann, B., and Gast, A. (1987). Proteolytic activity in the culture supernatants of mouse hybridoma cells. *Dev. Biol. Stand.* **66**, 403–408.
174. Recktenwald, A., Kroner, K.-H., and Kula, M.-R. (1985). On-line monitoring of enzymes in downstream processing flow injection analysis (FIA). *Enzyme Microb. Technol.* **7**, 607–612.
175. Danielsson, B., and Mosbach, K. (1979). Determination of enzyme activities with the enzyme thermistor unit. *FEBS Lett.* **101**, 47–50.

6

The Design of Bench-scale Reactors

NORMAN A. DE BRUYNE
Techne Inc,
3700 Brunswick Pike,
Princeton, NJ 08540, U.S.A.

ANIMAL CELL BIOTECHNOLOGY VOL. 3
ISBN 0-12-657553 3

1. INTRODUCTION

It was my friend, Professor Leonard Weiss who got me interested in equipment for cell culture. He told me of his difficulties with "heatless" cell culture systems which had to be used in a refrigerator because they pre-empted the temperature control in an incubator. The outcome was a low-pressure pneumatic stirrer (Fig. 1) (*9*). This led to the Techne ACA pneumatic stirrer

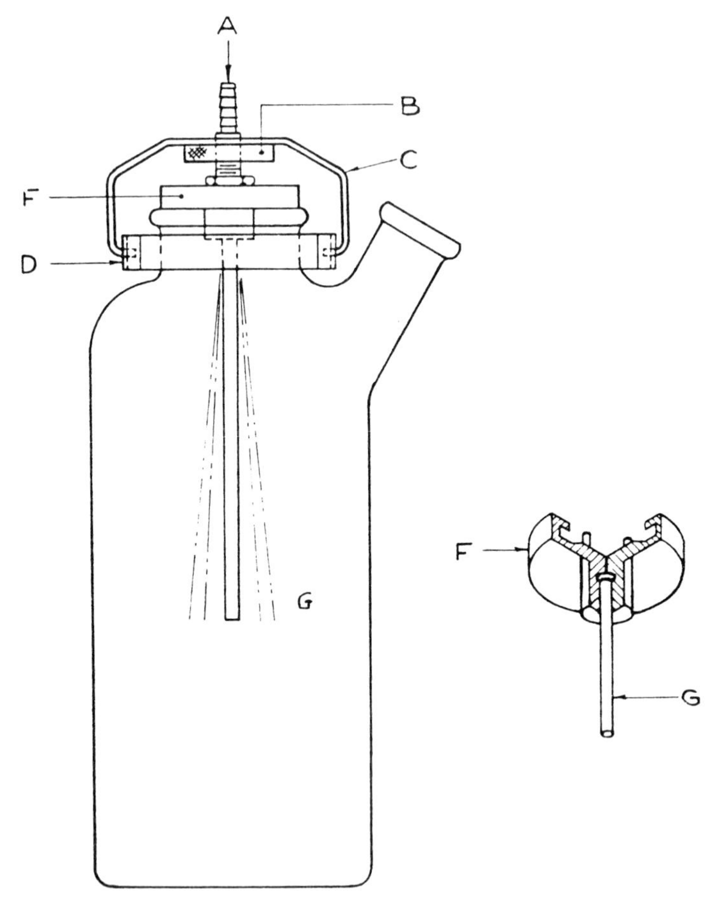

Fig. 1. The pneumatic stirring device. A Kontes 500 ml spinner culture bottle is fitted with a loosely fitting stainless steel collar D with two ears carrying a stainless steel wire loop C, which when tightened up by the knurled disc B forces the periphery of the rubber diaphragm F against the top of the bottle. Tubulation A which gives access to the top surface of the diaphragm is connected by a flexible tube to a source of alternating current air. The bottom surface of the diaphragm has a central rubber knob into which the glass stirrer rod G is fitted, as shown in cutaway sketch (right).

which employed three diaphragms 120° apart to give an orbiting action (the analogue of three-phase electrical supply) and which had a smooth motion at the low speeds required by microcarriers (Fig. 2).

Pharmacia of Uppsala, Sweden, tested this stirrer in comparison with a conventional cell-culture stirrer and flask (Fig. 3) and the results were gratifying, as exemplified by Fig. 4. Finally, the electromagnetic stirrer outlined in Fig. 5 was produced, which has a stirring action identical with that of the ACA stirrer but which can give the slow start and interval stirring needed in the preliminary plating operation of cells on microcarriers (*8, 10, 11*). This electromagnetic stirrer (Techne MCS stirrer) has been widely

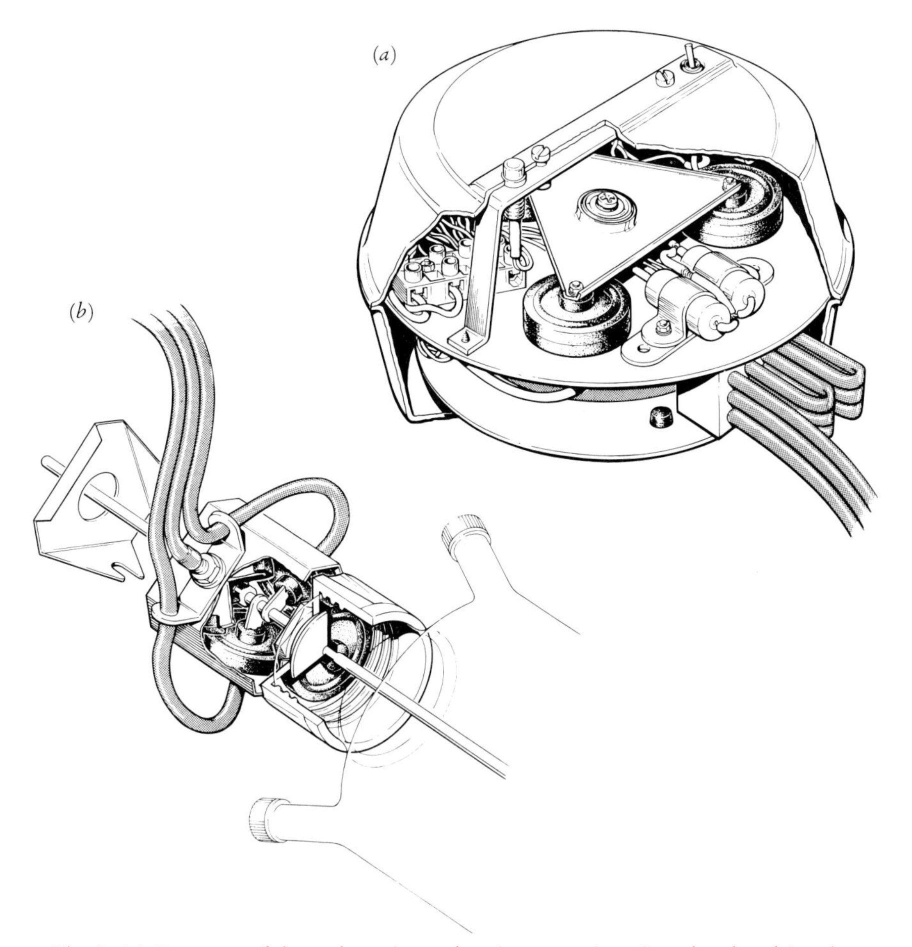

Fig. 2. (*a*) Generator of three-phase air supply using a nutating triangular plate driven by an electric motor. (*b*) Three-phase air stirrer fitted to screw cap of vessel.

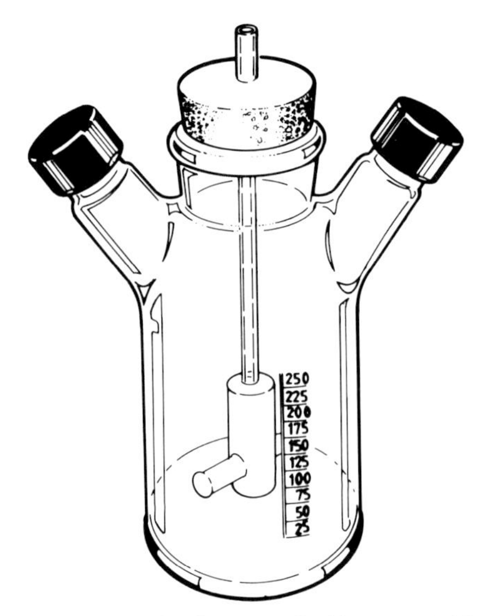

Fig. 3. Conventional culture vessel with suspended impeller.

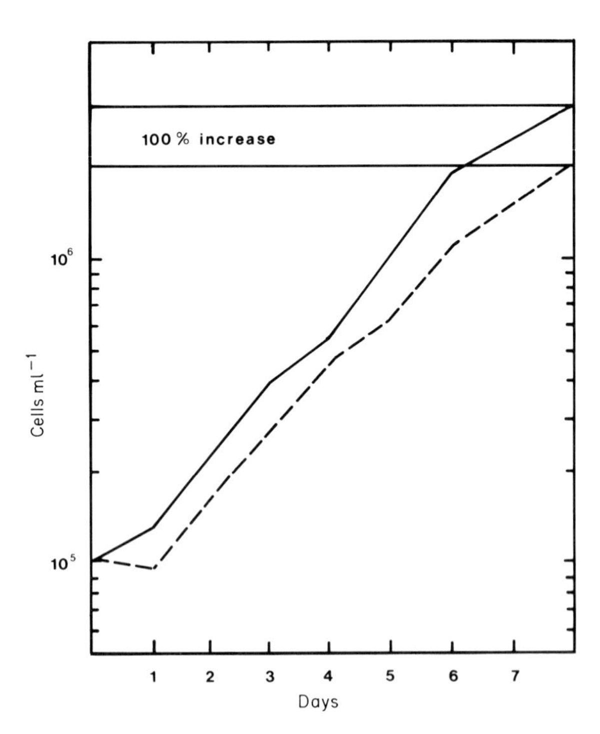

Fig. 4. Comparison of performance of an alternating current air (ACA) stirrer, shown by continuous curve, with the dotted curve of the conventional stirrer shown in Fig. 3.

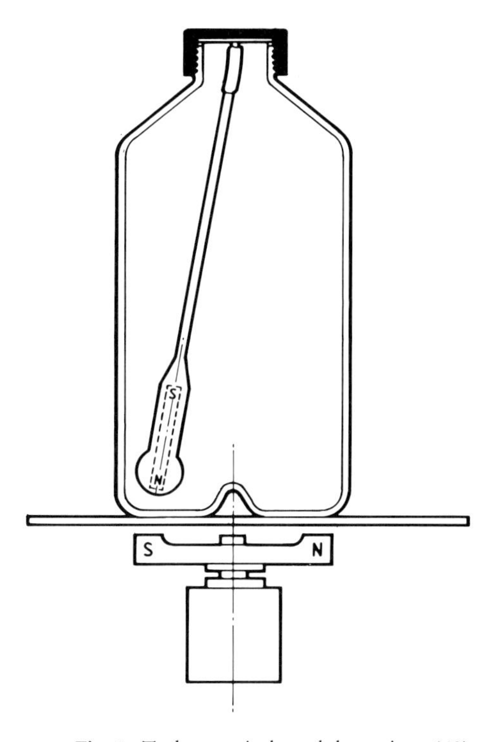

Fig. 5. Techne conical pendulum stirrer (*43*).

accepted in Europe and the U.S.A. However, there is now a market for bioreactors holding up to three litres of medium with provision for oxygenation and control by microprocessor or computer; this chapter is largely an account of what has been achieved to solve the problems involved.

2. SHEARING AND TURBULENCE

2.1. Shear

A lap joint can be made between two strips of metal by riveting, welding, or glueing, and by pulling the free ends of the strips we create a shear stress in the joint. The shear stress in a solid is the product of the shear strain and the modulus of shear rigidity.

A liquid can also sustain a shear stress for a time less than its relaxation time (*28*), but the shear stress will disappear at longer times unless sustained by sliding. Relaxation times for mobile liquids like water are extremely small but can be measured (*7*). The shear stress in a liquid is the product of the rate of shear strain multiplied by the viscosity. Quite a small shear stress would produce enormous rates of strain were it not for the instability that we call turbulence. Kühn (*23*) has pointed out that a force of 1 g applied tangentially to a plate of 1 cm^2 in water and 1 cm distant from a parallel fixed plate, would cause the plate to move with a velocity of 10^5 cm s^{-1} or over 2200 miles an hour, were it not for the onset of turbulence. Conversely, at reasonable speeds shear stresses are extremely small.

Laminar shear motion can only exist below an appropriate Reynolds number, which for rotating fluids in Nd^2/ν where N is the speed of rotation, d is the diameter of the impeller and ν is the kinematic viscosity. Makers of viscometers are forced to use very small clearances, as exemplified by the Couette (1890) viscometer (*4*), or tubes of very small diameter, as exemplified by capillary viscometers, in order to get sufficiently high rates of shear; but makers of bioreactors can, and should, avoid such small clearances. Nevaril *et al.* (*31*) and Augenstain *et al.* (*1*) were able to kill cells by shear, but they did not use bioreactors; they were forced to use what were (in effect) viscometers to do so!

2.2. Turbulence

Biologists frequently express fear about shear, but cells may be more at risk from turbulence or air bubbles than from shear. What is turbulence? "One may visualise turbulence as a haphazard and ever changing system of eddies superposed on the mean motion of that due to the mean flow alone" (*18*).

3. REVOLUTIONS PER MINUTE, TIP SPEED, TORQUE, INTEGRATED SHEAR FACTOR (ISF)

It is common practice to quote revolutions per minute (RPM) of the impeller when assessing the possibility of damage to cells in suspension or on microcarriers. In classical tissue culture, stirring at 60 RPM was the golden rule. A better parameter is the tangential velocity of the tips of the impeller. This "tip speed", however, is flawed because it takes no account of the influence of the stationary wall of the vessel on the shear stress produced in the liquid.

The torque T (newton-metres, N m) required to rotate a vertical shaft in an unbounded viscous liquid is (*24*)

$$T = 4\pi\eta\omega l a^2, \tag{1}$$

where η is the dynamic viscosity in Pascal-seconds (Pa s) for water at 37°C; a

is the radius of the shaft in metres; ω is the angular velocity in radians ($\omega = 2\pi N/60$, where N is RPM); l = length in metres of the immersed shaft. η, a^2, ω, l have combined dimensions ML^2T^{-2}, which are also the dimensions of torque.

Let $\eta = 0.0007$ Pa s, $a = 0.02$ m ($a^2 = 0.0004$ m^2), $\omega = 10.47$ rad s^{-1} (100 RPM), $l = 0.02$ m. Then

$$T = -4\pi \times 0.0007 \times 0.02 \times 0.0004 = 0.737 \times 10^{-6} \text{ N m}.$$

If we now surround the vertical shaft with a fixed concentric cylinder of radius b, the torque will be (*24*)

$$T = 4\pi\eta\omega l a^2 b^2/(b^2 - a^2), \tag{2}$$

so that the torque will be increased by $b^2/(b^2 - a^2)$, or $b^2/(b + a)(b - a)$. If $a = 0.02$ m, $b = 0.03$ m, the torque will be increased 1.8 times by the presence of the stationary flask wall.

Integrated Shear Factor (ISF)

Let D_t be the internal diameter of the flask and D_i be the diameter of the impeller (both in metres). The integrated shear factor (ISF) is $\omega D_i/(D_t - D_i)$. The "tip speed" of the impeller is $2\pi(N/60) \times (D_i/2)$, where N = revolutions per minute (RPM) and $D_i/2$ is the radius of impeller. $(D_t - D_i)/2$ is the gap between impeller tip and flask. Strictly speaking, the torque on the revolving impeller (regarding it as a cylinder) will be inversely proportional to the gap only if the curvature (the reciprocal of the radius) is small.

Nevertheless, in practice the integrated shear factor gives good correlation, as found by its originators Sinskey *et al.* (*35*) and confirmed by Hu and Wang (*17*), and is simple to use. It is possible to use a "torque factor" $\omega a^2 b^2/(b^2 - a^2)$ derived from equation (2) in place of the ISF; this also gives good correlation using Hu and Wang's experimental results (*17*).

4. THE FLOATING STIRRER

4.1. James Thomson and the Thomson Secondary Flow

A feature of the Techne BRO6 Bioreactor for animal cell culture is the floating impeller, which has its origin in a discovery announced 130 years ago by James Thomson, elder brother of Lord Kelvin, at a meeting in Dublin of the British Association. He used it to explain the grand circulation of the atmosphere and illustrated it by stating that, after the removal of a spoon from

Fig. 6. Photographs of aluminium flakes in water rotated by an impeller 500 mm above the base of a cylindrical vessel.

stirring a cup of tea, the leaves ended up in a heap in the middle of the bottom of the cup, apparently in defiance of centrifugal or "apocentric" force (*40*).*

> A secondary flow can significantly alter the primary flow motion through a slight redistribution of angular momentum and vorticity. The secondary flow is often produced by viscous boundary layers and the control of the main motion is accomplished by vortex line stretching and the conservation of angular momentum. This is the interaction involved in vortex descent, whether it be a hurricane dissipating itself over land, or a stirred cup of tea to rest. (*16*)

If instead of withdrawing the teaspoon we continue to stir with the tip of the spoon, in the air–liquid interface one can see the whole Thomson motion in action. The tea leaves ascend in a tight spiral around the vertical axis of the flask and are flung out against the wall and descend to the bottom in a wider spiral. If, on the other hand, one stirs in the usual way, with the spoon scraping the bottom of the cup, the Thomson motion can be destroyed. This simple fact at once suggests that the proper place for an impeller in a bioreactor for cell culture is in the interface and not at the bottom of the flask, which is where it has always been put in conventional cell culture.

A surprising feature of the Thomson secondary flow is the height to which particles can be levitated by it.

Figure 6 is a series of photographs taken, at the times shown on the digital clock, of aluminium flakes in water rotated by an impeller (not visible) 500 mm above the base of the tubular flask. After an initial lag or "pull up time" the vertical speed of ascent is about 7 mm s^{-1}. The photograph on the extreme right was given an exposure of $\frac{1}{4}$ s to show the direction of motion of the flakes. A similar tubular flask, but fitted with a draught tube operated by an air lift, produced considerable foam in which the particles became immobilized.

4.2. Research Work at the National Engineering Laboratory

Research work is in progress by A. J. P. Spragg and his colleagues at the National Engineering Laboratory, East Kilbride, Glasgow, U.K., on mathematical and laboratory analysis of stirred-tank mixing. As a first step towards more complex situations, they investigated the flow patterns around a horizontal disc rotated by a vertical shaft in a concentric cylindrical vessel.

* He delighted in the invention of new words and he gave us "radian", "interface", "poundal", "numeric", "apocentric" and "torque". The secondary motion discovered by James Thomson is familiar to hydrodynamicists, but I have never seen any acknowledgement of its discoverer. After its announcement in 1857, the subject was "before his mind during the rest of his life . . . when in the last years of his life the affliction of partial blindness came upon him . . . he set himself leisure to complete his work" (Obituary Notice in *Proc. Roy. Soc.* Vol. LIII (1893)). I have made it a practice to call the phenomenon "the Thomson secondary motion".

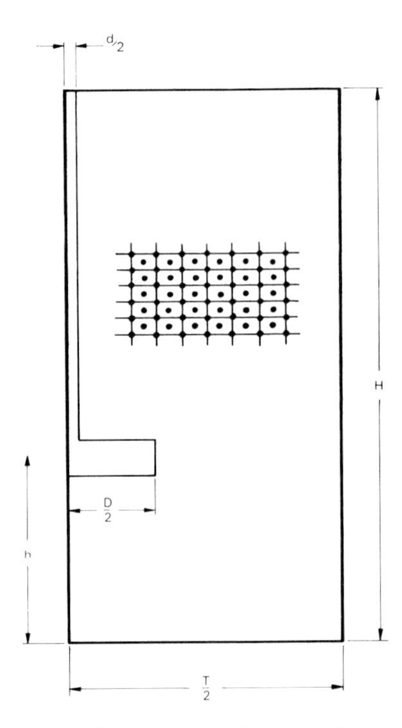

Fig. 7. Two-dimensional grid. (Reproduced by kind permission of The National Engineering Laboratory, East Kilbride, Glasgow.)

The speed of rotation was chosen to give a low Reynolds number (about 60) so that the flow was laminar. The mathematical predictions were in good agreement with laboratory results obtained by laying the two-dimensional grid shown in Fig. 7 over a photo of the vertical cross-section of the tank in such a fashion that all physical boundaries correspond to grid lines, and all three computational variables are defined and stored at the intersections of these grid lines (*38*). The partial differential equations governing the flow are reduced to sets of non-linear algebraic simultaneous equations by integration over "control volumes" using appropriate interpolating approximations ("up-wind" differencing for convective terms, central differencing for diffusive terms). These sets of equations are then solved by point Gauss–Seidel over relaxation. Boundary conditions are incorporated explicitly for swirl velocity and steam function, and by Couette flow analysis of the vorticity equation to derive additional equations for boundary vorticity values.

For the mixing calculations, a calculated flow field is taken as input; again the partial differential equation governing dispersion of a tracer species is

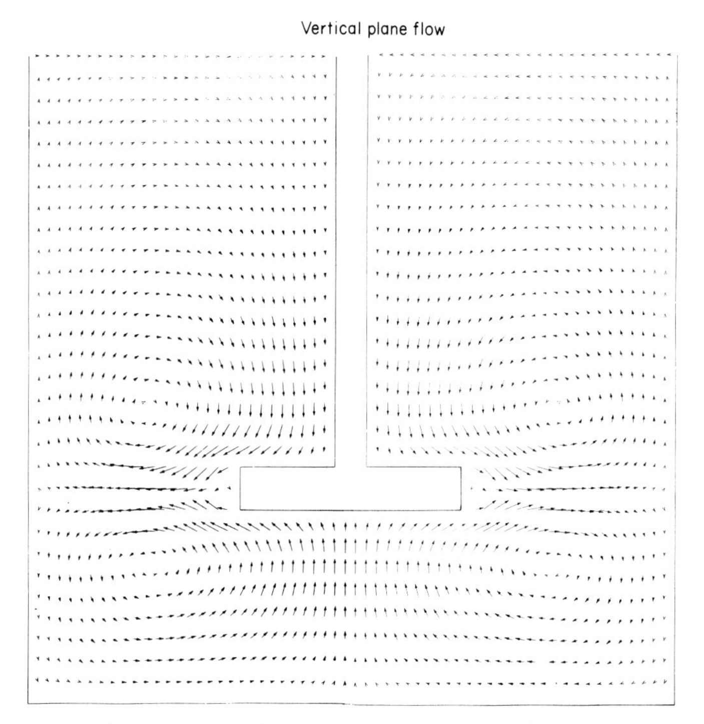

Fig. 8. Picture of vertical plane flow induced by a rotating disc. (Reproduced by kind permission of The National Engineering Laboratory, East Kilbride, Glasgow.)

reduced to a set of algebraic equations by entirely the same procedure used for the flow field calculations. The only basic difference is that the problem is time-dependent, so an "explicit time-stepping" procedure is used to predict a "new" set of concentration values from an "old" set, and the concentration values are defined and stored on a different staggered grid to ensure strict overall conservation of the tracer species. The model of a rotating disc has a striking resemblance to the Techne bioreactor BRO6, and inspection of the vertical plane flow picture (Fig. 8) shows that the direction of flow below the disc corresponds to a Thomson secondary motion; presumably if the disc were in the interface, the correspondence would be complete.

4.3. Stability

The simplest way of ensuring that the impeller is in the interface is to make it buoyant (*45*). But how to rotate it and to ensure that it stays in position on the vertical axis of the flask?. The rotation can be achieved by fixing a permanent magnet inside the buoyant impeller and by having a powerful rotating master magnet underneath the flask. This arrangement of magnets can give rise to at least two different kinds of instability. One instability becomes manifest when the liquid interface is at a low level; the master magnet will then pull the floating impeller onto the bottom. This can be overcome by suspending the flask on a spring which will cause the flask and its contents to rise upwards as the liquid level falls.

Another instability is wandering of the buoyant impeller laterally in the interface. A one-dimensional analysis of the lateral forces acting on the magnet in the buoyant impeller is simple, but it produces a complicated formula which can only be interpreted numerically (by computer). This was done by Dr. David Dreyfuss (Techne Inc., Princeton), who obtained a surprisingly simple result. One can expect stable operation of an unconfined floating magnetic stirrer if the magnets are sufficiently strong to overcome viscous drag, and if the separation between the magnets is more than half the length of the larger magnet (see Fig. 9).

However, we found that swirling or gross disturbance could displace the impeller outside of the range of lateral stability, so that it would still be

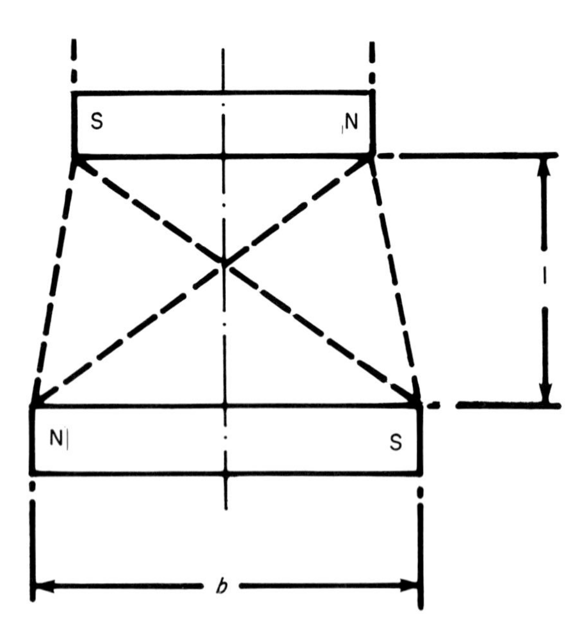

Fig. 9. Interaction between poles of floating and master magnets.

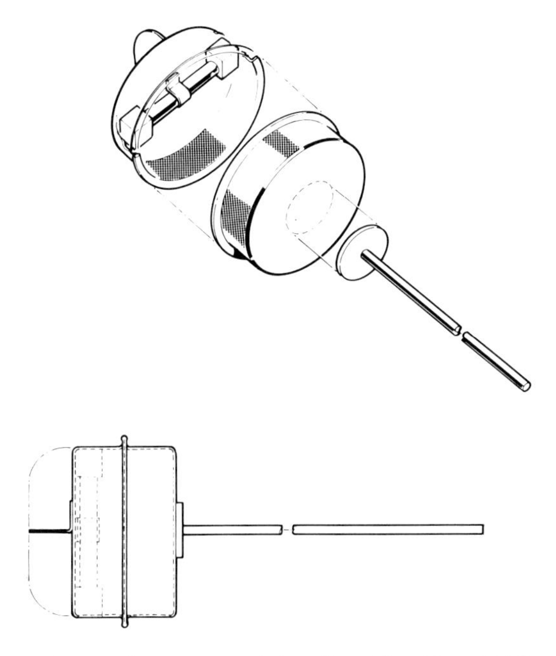

Fig. 10. Floating magnetic impeller with guide rod (*46*).

necessary to provide some mechanical stabilization. A bearing in the liquid medium is undesirable, because of the risk of damage to microcarriers or to cells. The best solution to the problem is shown in Fig. 10. A vertical rod attached to the top of the floating impeller permits only vertical and rotary motion in a sleeve above the liquid (devised by Dr. G. J. MacMichael, Techne, Princeton).

5. CONTINUOUS CULTURE

5.1. The Advantages of Continuous Culture

The obvious advantage of continuous culture is its economy. In a batch process for, say, the production of monoclonal antibodies (MABs) from hybridomas, the amount produced is limited by the lifetime of the hybridomas. In continuous culture the amount of MABs obtainable is theoretically infinite and in practice is large compared with a single batch.

> Batch culture is a discontinuous system used for culturing micro-organisms, . . . characterized by a single loading or inoculation into fresh medium at the start of the process and a single harvesting stage at the end, usually when the substrate has been exhausted. Organisms grown in batch systems generally show a sigmoid growth curve, undergoing exponential growth for a few generations following an initial lag phase. Growth then ceases, and the cells may lyze, die or stop growing. (*Macmillans Dictionary of Biotechnology* (1986).)

To the research worker, an advantage perhaps even more important than that of high production is that in continuous production the system is in a steady state:

> A batch culture is a closed system in which a given number of cells are inoculated into a limited amount of nutrient medium. As cell multiplication proceeds, nutrients are consumed and metabolites accumulate, thereby changing the environment of the culture. These changes in turn affect cell metabolism and lead ultimately to cessation of cell multiplication. A closed batch culture consists of a series of transient states difficult to define and even more difficult to control. Such difficulties can, however, be overcome by the use of an open system such as continuous-flow culture, in which there is input of substrate and output of cells and cell products. Such open systems offer the possibility of obtaining a steady state in which constant conditions can be maintained indefinitely (*41*)

5.2. Continuous Culture with the de St.Groth Cytostat

The classic paper of de St.Groth (*12*) has demonstrated the practicality and the value of the chemostat for culturing mammalian cell lines. His "Cytostat" is in essence six chemostats incorporated into one instrument, for studying growth kinetics and for the production of secretions, such as immunoglobulins. To date, his laboratory at the Basel Institute for Immunology has successfully used the Cytostat with 83 different cell lines.

We have now had considerable experience at Techne (Princeton) with the de St.Groth Cytostat. Our version incorporates four chemostats; therefore, four cell lines can be cultured simultaneously or the same cell lines can be cultured under four different sets of growth parameters (see Fig. 11). The only significant change from the original Cytostat is the use of floating impellers, as shown in Fig. 12, which are rotated by a magnetic field produced by four stationary electromagnets and which gently circulate the medium in a Thomson flow pattern and maximize gas transmission from the head-space. The chemostats are kept at constant temperature by their immersion in a water bath. Most of our work has been with hybridomas and has been successful.

Mouse hybridoma 14-4-4s (ATCC HB32) have been routinely maintained in mid-log phase at populations of 1.25×10^6 cells ml^{-1} with doubling times

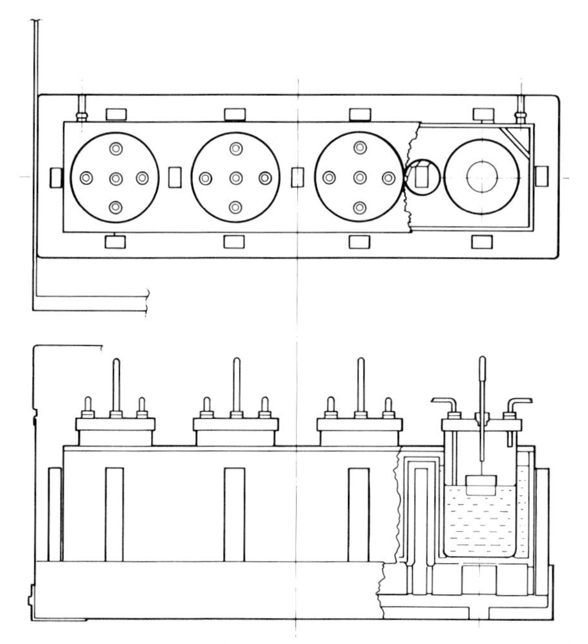

Fig. 11. Four Chemostats in water bath showing vertical pole pieces, which produce rotating magnetic fields.

(DT) as low as 13 hr. The doubling time, however, is not as important in continuous production as it is in batch production, because (quoting from de St.Groth),

> The strict correlation between cell number and yield, points up the irrelevance of doubling time and the importance of the density limit for cytostatic production of monoclonal antibodies. If cells can be maintained in exponential growth of high concentrations, the yield will also be high irrespective of whether the cells are growing fast or not. In fact a slower rate of growth at a given cell density is entirely advantageous as it means that the same quantity of antibody is recovered in a smaller volume, i.e. more cheaply and with less bulk for the subsequent isolation of the antibodies (*13*).

Currently our laboratory cytostat is being used to optimize the concentration of constituents of the media, and thus to maximize immunoglobin production. For example, when using the Dulbecco's Modified Eagle's (DME) as a base medium, glucose and foetal bovine serum (FBS) were shown to be limiting in HB32 cultures doubling times and monoclonal antibody production rates (MPR). In addition, the instrument demonstrated that FBS was superior to both new-born bovine serum and a commercial serum substitute, with MPRs of 827, 612 and 463 Ab $cell^{-1}$ s^{-1}, respectively.

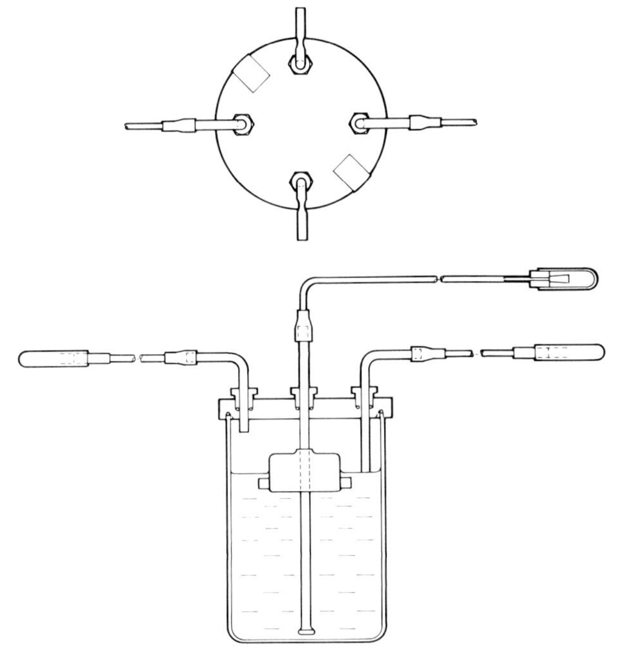

Fig. 12. Construction of Chemostat flask.

5.3. Continuous Culture with the BRO6

Used as a chemostat, the Techne BRO6 Bioreactor with its 3000 ml working volume will produce significant quantities of secreted products and eukaryotic biomass. For example, with HB32 it has produced 2.51 litres of effluent per day at a titre of 16.7 mg IgG/l^{-1}. As in the Cytostat, when a steady state of continuous production of immunoglobulin has been attained, it will be found that automatic control systems become superfluous. During the culturing of the hybridoma HB32, all monitored parameters remained constant without the need for pH and oxygen controllers. The cells were cultured in DME with 22.5 mM glucose (feed concentration), and 10% FBS. The medium was buffered with sodium bicarbonate in equilibrium with a 5% CO_2 atmosphere. The culture was maintained at 1.25×10^6 cells ml^{-1} for 240 hr. DT was 19.9 hr, and MPR was 990 Ab $cell^{-1}$ s^{-1} with a titre of 17.0 μg ml^{-1}. The pH, oxygen consumption, glucose and lactate concentrations were 6.9, 4.8×10^{-12} g $cell^{-1}$ hr^{-1}, 12.2 mM and 10.2 mM, respectively. The entire system was stabilized simply by maintaining steady-state growth. This was achieved by a daily cell count which by reference to de St.Groth's equation numbered (6) in his analysis determined the medium flow rate, which was held constant at 87 ± 1.0 ml hr^{-1}.

5.4. Application of de St.Groth's Equation

If the cells are counted every 24 hr (1440 min) and if the volume of liquid is 600 ml (the maximum working volume in each cytostat flask) then equation [6] of de St.Groth's analysis (*12*) can be written:

$$\text{Required increment} = m^{+} - m = \frac{600}{1400} \times \frac{C_t^{+}}{C_0^{-}}$$

where V is the volume in millilitres, C_t^{+} is the concentration per ml of live cells at $t = 1400$ min, C_0^{+} is concentration of cells at $t = 0$ (i.e. yesterday). From this we get the following simple relationship (*13*) tabulated below.

Observed change in concentration ratio	Required change in input in ml/min
0.6	−0.21
0.7	−0.15
0.8	−0.9
0.9	0.04
1.0	0.00
1.1	0.04
1.2	0.08
1.3	0.11
1.4	0.14
1.5	0.17
1.6	0.20

Thus if the concentration rose in 24 hr from 1.85×10^6 ml^{-1} to 2.4×10^6 ml^{-1} the ratio is 1.30 and hence the input has to be increased by 0.11 ml min^{-1}. If the readings were in the opposite order, the ratio 1.85/2.4 = 0.77 would indicate that the input of medium had to be decreased by about 0.11 ml min^{-1} by interpolation between the ratios of 0.7 and 0.8, i.e. between the corresponding increments of −0.15 and −0.09.

Alternatively, one can use the data in Table I to determine what changes must be made in the rate of input to medium.

TABLE I The Incremental Change Required in the Influx (ml min^{-1}) Required to Produce a Steady State when the Concentration Ratio $C(24)/C(0)$ is Known from Measurements at Intervals of 24 hr

Cell concentration	Volume							
$C(24)/C(0)$	250	500	750	1000	1500	2000	2500	3000
0.10	−0.40	−0.80	−1.20	−1.60	−2.40	−3.20	−4.00	−4.80
0.20	−0.28	−0.56	−0.84	−1.12	−1.68	−2.24	−2.79	−3.35
0.30	−0.21	−0.42	−0.63	−0.84	−1.25	−1.67	−2.09	−2.51
0.40	−0.16	−0.32	−0.48	−0.64	−0.95	−1.27	−1.59	−1.91
0.50	−0.12	−0.24	−0.36	−0.48	−0.72	−0.96	−1.20	−1.44
0.60	−0.09	−0.18	−0.27	−0.35	−0.53	−0.71	−0.89	−1.07
0.70	−0.06	−0.12	−0.19	−0.25	−0.37	−0.50	−0.62	−0.74
0.80	−0.04	−0.08	−0.12	−0.15	−0.23	−0.31	−0.39	−0.47
0.90	−0.02	−0.04	−0.05	−0.07	−0.11	−0.15	−0.18	−0.22
1.00	0.00	0.00	0.00	0.00	0.00	0.00	0.00	0.00
1.10	0.02	0.03	0.05	0.07	0.10	0.13	0.17	0.20
1.20	0.03	0.06	0.09	0.13	0.19	0.25	0.32	0.38
1.30	0.05	0.09	0.14	0.18	0.27	0.36	0.46	0.55
1.40	0.06	0.12	0.18	0.23	0.35	0.47	0.58	0.70
1.50	0.07	0.14	0.21	0.28	0.42	0.56	0.70	0.84
1.60	0.08	0.16	0.24	0.33	0.49	0.65	0.82	0.98
1.70	0.09	0.18	0.28	0.37	0.55	0.74	0.92	1.10
1.80	0.10	0.20	0.31	0.41	0.61	0.82	1.02	1.22
1.90	0.11	0.22	0.33	0.45	0.67	0.89	1.11	1.33
2.00	0.12	0.24	0.36	0.48	0.72	0.96	1.20	1.44
2.20	0.14	0.27	0.41	0.55	0.82	1.10	1.37	1.64
2.40	0.15	0.30	0.46	0.61	0.91	1.22	1.52	1.82
2.60	0.17	0.33	0.50	0.66	1.00	1.33	1.66	1.99
2.80	0.18	0.36	0.54	0.72	1.07	1.43	1.79	2.15
3.00	0.19	0.38	0.57	0.76	1.14	1.53	1.91	2.29
3.50	0.22	0.43	0.65	0.87	1.30	1.74	2.17	2.60
4.00	0.24	0.48	0.72	0.96	1.44	1.93	2.41	2.89

6. METHODS OF OXYGENATION

Oxygen can become the rate-limiting substrate in conventional culturing and techniques; when this happens, the mammalian cells become dependent on glycolysis to meet their energy demands, with up to 80–90% conversion of glucose to lactic acid. Under such conditions, the pH rapidly drops below optimum levels, and growth slows. If the deficiency is severe, the cells will enter a death phase which can be irreversible.

TABLE II Oxygen Consumption Rates of Various Cell Lines[a]

Cell type	Consumption (moles/10^6 cells/hr)	Consumption (10^{12} g/cell/hr)	Reference
Leucocyte	0.04	1.28	(*14*)
FS.4	0.05	1.6	(*15*)
Lymphoblastoid	0.05	1.6	(*20*)
Skin fibroblast	0.06	1.28–10.56	(*6*)
Hepatic	0.07–0.28	2.24–8.96	(*29*)
HeLa	0.10–0.39	0.32–1.28	(*33*)
CEF fibroblast	0.10	3.2	(*35*)
Liver	0.11	3.52	(*19*)
Hybridoma 14-4-48	0.13	4.16	(*39*)
W1-38	0.17–0.50	5.44–16	(*2*)
LS fibroblast	0.18	5.76	(*22*)
BHK	0.20	6.4	(*34*)
Long To	0.24	7.68	(*32*)
Hybridoma NBI	0.33	10.56	(*3*)
Detroit 6 (marrow)	0.43	13.76	(*32*)
HeLa	0.47	15.04	(*6*)

[a]This table is based on Table 1 of Spier and Griffiths (*36*).

An excellent review of the various methods of obtaining an effective rate of oxygen solution in the medium has been published by Spier and Griffiths (*36*) (see Table II).

There are many ways of dissolving oxygen, but we will confine our discussion to three ways in which advances have been made in recent years.

6.1. Transfer by Diffusion from Head-space

Oxygen can be transferred by diffusion from the head-space through the interface into the liquid medium, which must be well stirred. Flasks for cell culture are usually filled to not more than half their volume in order to leave the upper half as a head-space or reservoir for oxygen and other gases. For moderate volumes, up to about 2 litres, this type of oxygenation is effective.

To maximize the rate of diffusion, the flask should have a high ratio of exposed surface area A to volume V and this is best achieved with a spherical vessel half-filled with the medium. The virtues of spherical flasks are well known to organic chemists who call them boiling flasks. They are excellent also from a hydrodynamic point of view.

The variation of A/V with h the height of water, in a six litre spherical flask (three litre working volume) is shown in Fig. 13 together with corresponding ratios in a cylindrical flask.

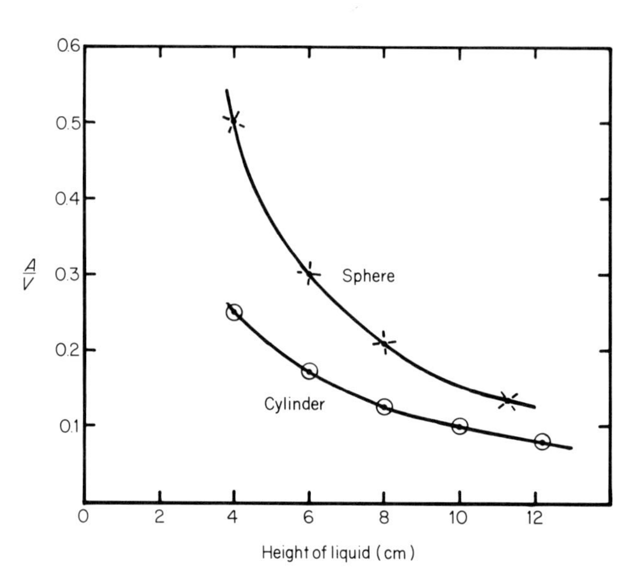

Fig. 13. Variation of the ratio of area to volume (A/V) with height of liquid in spherical and cylindrical flasks.

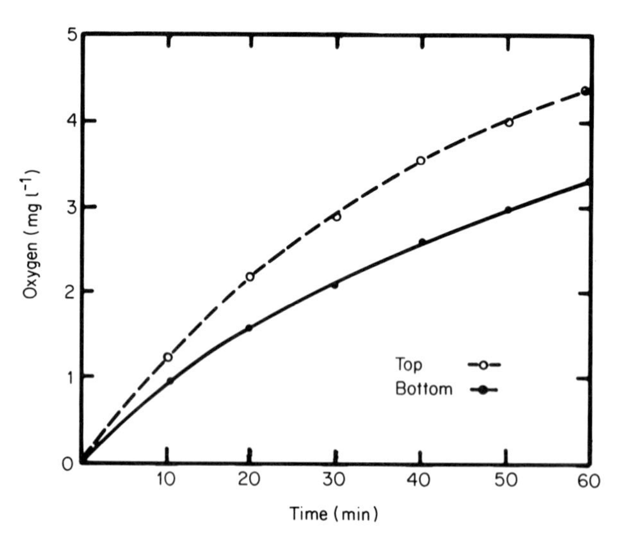

Fig. 14. The dotted curve shows oxygen intake obtained with the impeller at top of the liquid compared with its performance when it is at the bottom.

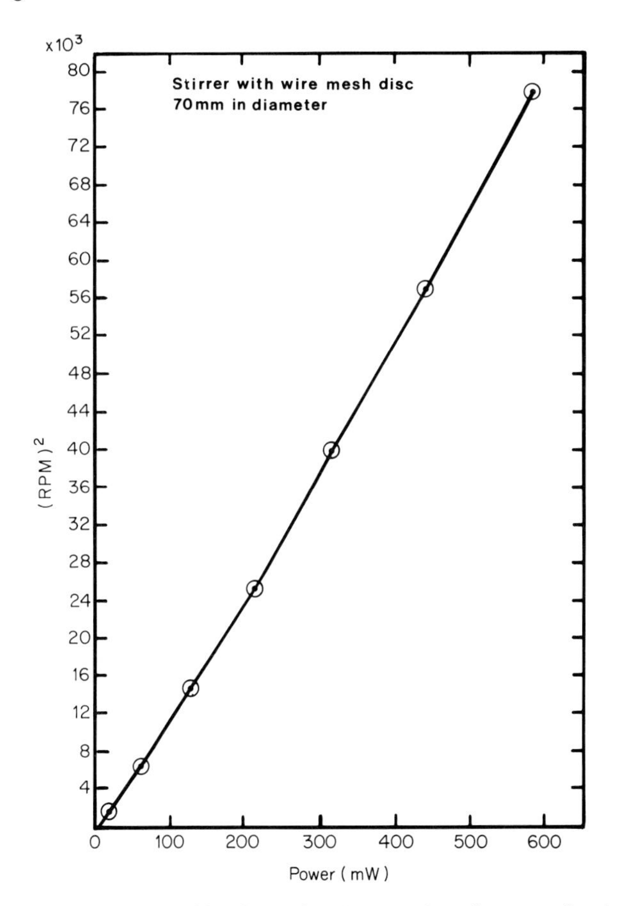

Fig. 15. The power consumed by the peplum is approximately proportional to the square of the speed of rotation.

The position of the stirrer is important: a location near the interface is better than one at the bottom of the flask (Fig. 14). The value of the power is very nearly proportional to the square of the speed of the impeller (Fig. 15).

Diffusion from the head-space can be increased by agitation of the liquid surface, as has been shown by Hu and Wang (*17*). A floating stirrer will produce surface agitation, which can be increased by adding a peplum of stainless steel wire mesh of 70 mm diameter, as shown in Fig. 16. The peplum revolves in the interface, and was designed by Mr. G. Feldscher. The enhancement of oxygen transmission is proportional to the area of the peplum (Fig. 17) and to its RPM. Diffusion from the head-space can also be improved

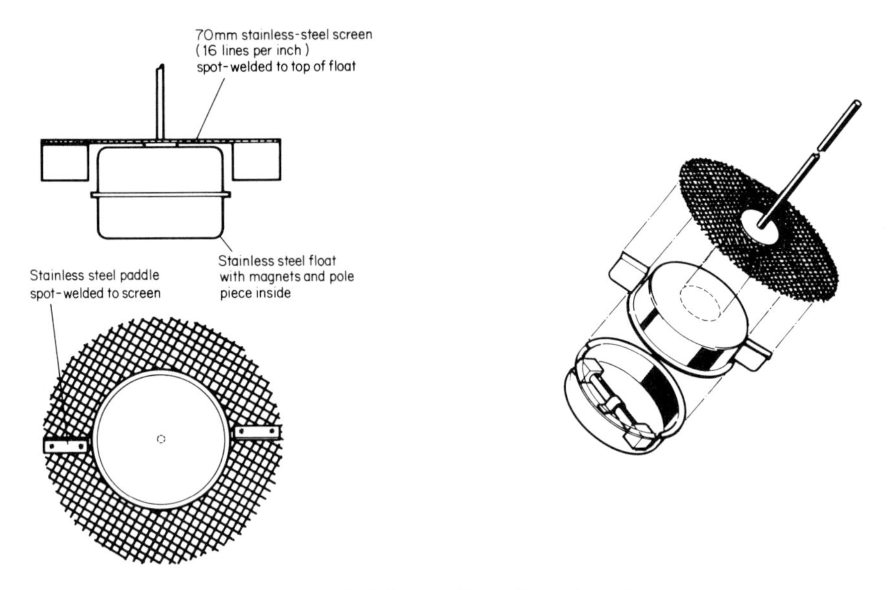

Fig. 16. Exploded view of impeller with peplum.

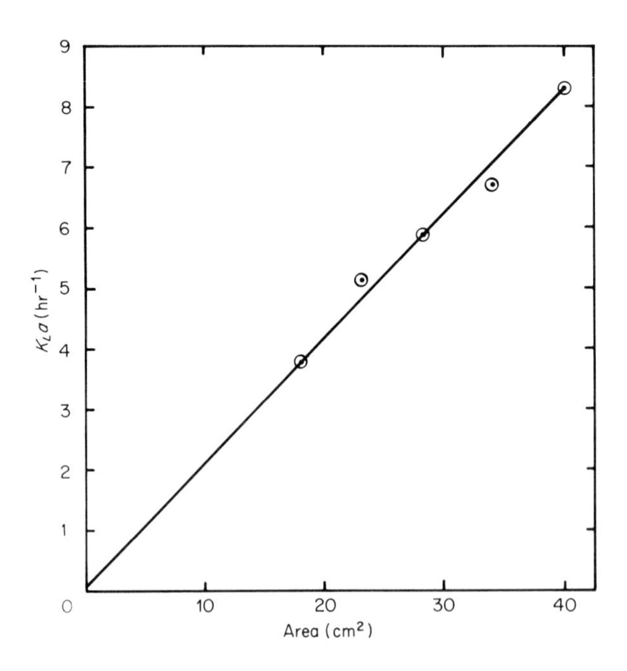

Fig. 17. The value of $K_L a$ is proportional to the area of the peplum.

by increasing the mole fraction of oxygen. As Henry's law of partial pressure states, the mole ratio of a gas dissolved in a liquid is directly proportional to the mole ratio in the gaseous phase above the liquid. But the high levels of oxygen can be cytotoxic.

6.2. Oxygenation by Air Bubbles

Oxygenation by air bubbles in an air-lift and/or by sparging with a sinter, is simple and gives high K_La values. But, at least in a laboratory, it is best avoided. Foaming is inevitable if there are proteins in the medium: this will trap microcarriers and cells. Aerosols that are produced can be toxic. Anti-foaming agents are toxic to many cell lines. Figure 18 shows a comparison between the BR-O6, a disposable air-lift, and a T75 flask. The other problem with sparging is that the bubbles generated by the sinter may damage cells. Kilburn and Webb (*21*) state that the mechanism of damage is purely physical: "The initial damage to the cell involves an increase in its permeability to the

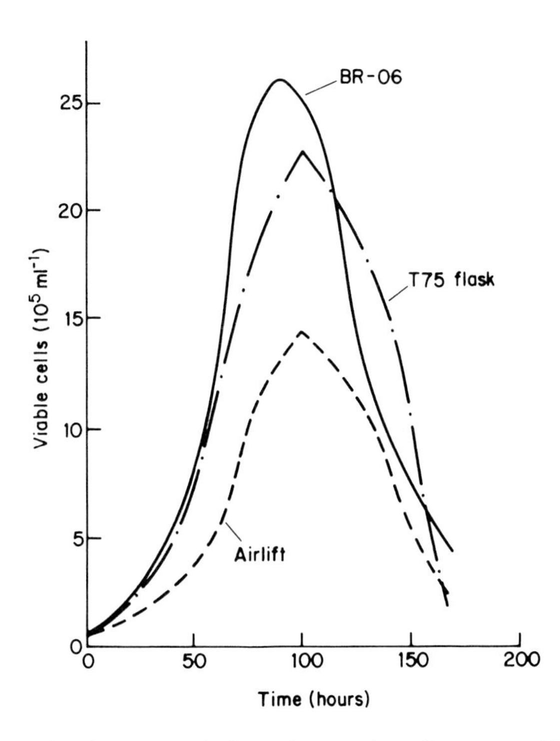

Fig. 18. A comparison between antibody production achieved in BR-O6, airlift and T75 flask.

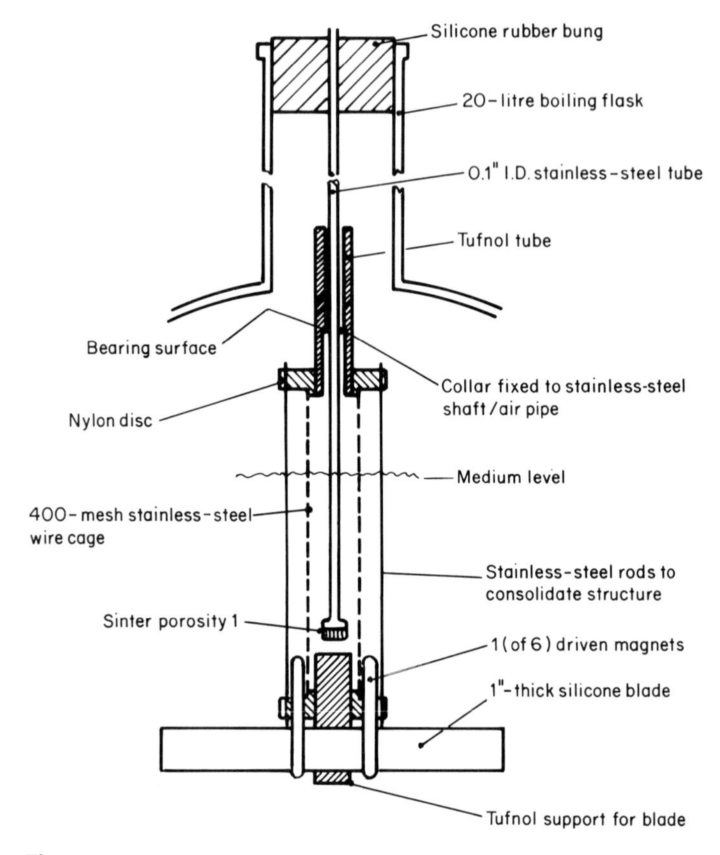

Fig. 19. Sparging cage. (Reproduced by kind permission of Spier and Whiteside.)

dye trypan blue without immediate lysis." It seems probable that surface-active forces at the bubble interface drastically modify or strip a surface layer from the cell.

In the short-term tests, the detrimental effects of bubbling could be prevented by increasing the serum content of the medium from 2% to 10% or by adding 0.02% Pluronic F68 (Wyandotte Chemicals Corp, Wyandotte, MI), a non-ionic surface-active polymer of poly(oxyethylene) and poly-(oxypropylene). Both serum and Pluronic probably stabilize the cell–liquid interface by forming a highly condensed interfacial structure of adsorbed molecules. The interfacial tension may also have been reduced, but this effect cannot, in itself, account for the protective action, since a number of non-toxic, low molecular weight surface-active compounds, which would also reduce interfacial tension, did not protect the cells.

It may be pertinent to note that the pressure (p) inside an air bubble is given by the equation $p = y/R$ Pa, where y is the surface tension (which is 7.1 N m for water) and R is the radius. Let $R = 100\ \mu m = 10^{-4}$ m (diameter is 200 μm = 0.2 mm). Then $p = 14.2 \times 10^4$ Pa or 20.6 lb in.$^{-2}$ (psi). Bubbles of diameter less than 100 μm will of course have greater internal pressure. If I were a cell with a diameter between 1 and 10 μm, I would be chary of contacting such a potential bomb.

In a sparging cage (see Fig. 19), the sinter is enclosed in a 400-mesh stainless-steel wire cage which is closed at the top and bottom only partially submerged in the medium. The air bubbles in the foam are disrupted as the foam passes outwards through the mesh into the head-space and there can be no entrapment of microcarriers. (Figure 19 is reproduced from Spier and Whiteside (*37*).)

6.3. Permeable Membranes

A method that holds some promise for effective oxygen transfer is the use of gas-permeable membranes (permeators) (*30*). By incorporating gas-permeable tubing into the reactor vessel, one can greatly increase the gas–liquid interface without producing air bubbles. The rate of diffusion through the membrane is directly proportional to the gas pressure applied to the tubing inlet.

Though permeators can increase $K_L a$, and have been successfully used for culturing mammalian cells (*27*), they do have some drawbacks. The tubing must be wrapped around a support that allows for maximum surface exposure, and this can be cumbersome. For this reason, the use of permeators can set limits to reactor design.

If the tubing becomes pinched at any point along its length, the permeator becomes inoperable. Also, the tubing must be attached to the gas inlet and outlet ports without leaving tailing slack. Finally, after a run is completed, the tubing must either be thoroughly cleaned or discarded.

B. Braun Melsungen AG have a German patent 29 40 446 (British patent 2 059 436) for the use of a silicone membrane for suspension or microcarrier cell culture. In their MC culture vessel (Fig. 20), silicone rubber tubing is wound round vertical rods to form a cylinder whose diameter is somewhat less than the inside diameter of the vessel, which has a working volume of 1.5–2 l. Air at a pressure of 0.8 bar (about 12 psi), applied to the inside of the silicone rubber tubing, supplies oxygen at a rate of at least 30–40 mg l^{-1} hr^{-1}.

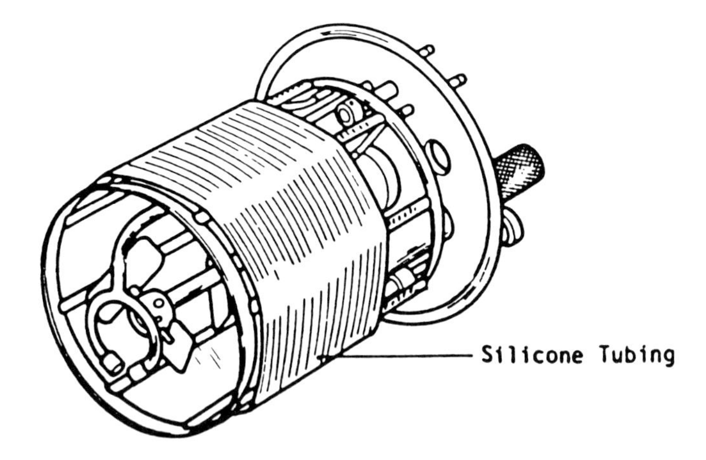

Fig. 20. Reproduced by kind permission of Braun, Melsungen A.G.

Lehmann *et al.* (*25*) have used a porous polypropylene tubing, developed by ENKA AG of Wuppertal, West Germany, called Accurel. Its properties are shown in Table III. The production and sale of bioreactors with Accurel aerators is in the hands of Diessel GmbH & Co. of Hildesheim, West Germany, and Fig. 21 shows the Accurel tubing for a 150-litre bioreactor. (This photo was kindly supplied by Diessel GmbH & Co.) Our tests made with pure oxygen, and correcting for differing surface areas of Accurel and silicone rubber tubing, showed that silicone rubber had 53% of the permeability of Accurel. Tests were made in the BRO5 reactor with impeller speed of 160 RPM in 2500 ml of saline.

TABLE III Properties of Accurel Membranes

Accurel hollow fibre (Enka AG) $d_o = 2.6$ mm. $d_i = 1.8$ mm	
Pore volume	75%
Medium pore size	0.3 m
Burst pressure	6.5 bar
Implosion pressure	1.5 bar
Specific surface, outside	81.7 $cm^2 m^{-1}$
Specific surface, inside	56.5 $cm^2 m^{-1}$
Specific volume, outside	5.3 $cm^3 m^{-1}$
Specific volume, inside	2.5 $cm^3 m^{-1}$

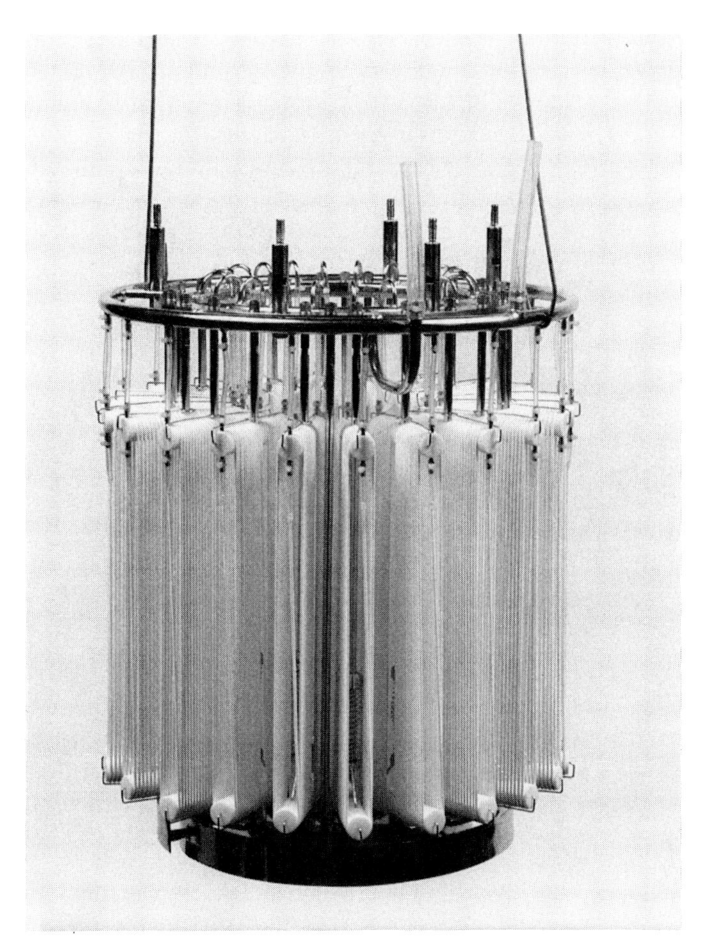

Fig. 21. (Reproduced by kind permission of Diessel GmbH & Co.) Accurel® tubing used in a 150 litre bioreactor.

6.4. Measurement of Dissolved Oxygen

Since the introduction of rugged "galvanic" oxygen meters by Mackereth (*26*), and a general realization of the importance of an adequate amount of dissolved oxygen, measurement of dissolved oxygen has become a routine operation.

Experiment shows that the rate of solution of O_2 in a liquid is proportional to the difference between the saturation concentration C^* and the actual concentration C (in milligrams of oxygen per litre). The constant of proportionality is written as $K_L a$ and chemical engineers call it a volumetric mass transfer coefficient; a is a function of the ratio of exposed liquid area to its volume (V). As with all mass transfer coefficients, the simple symbols are symbolic of a mass of ignorance about what is really going on (5), but are nevertheless useful. I find it more intelligible to write the transmission coefficient as K/L where L = Volume/Area = Height of liquid in a vessel of cylindrical or other constant cross-section.

The foregoing can be stated concisely in the following simple mass balance equation:

$$\text{Oxygen transfer rate (OTR)} = \frac{dc}{dr} = -K_L a(C^* - C_t). \tag{3}$$

From which we see that when $C_t = 0$, OTR $= -K_L aC^*$. This is a quick, rough way of finding $K_L a$. The negative sign indicates that the rate decreases as C_t increases. The integrated version below is the best way of finding the numerical value of $K_L a$.

$$K_L a = -\frac{1}{t} \ln \left(\frac{C^*}{C^* - C_t} \right). \tag{4}$$

As an example, we will use equation (4) to find $K_L a$ in 1000 ml of saline in a BRO6 flask with a floating stirrer fitted with a 70 mm diameter disk surface agitator of stainless-steel wire mesh, rotated at 150 RPM. The saline is at 37°C. The results are tabulated in Table IV and shown graphically in Fig. 22, and summarized in Table V.

TABLE IV $K_L a$ in a BRO6 Flask Containing 1 litre Saline Stirred at 150 RPM

Time (min)	C^*	C_t	$C^* - C_t$	$C^*/(C^* - C_t)$	$2.3 \log_{10}[C^*/(C^* - C_t)]$
5	6.2	1.55	4.65	1.333	2.3 × 0.1249 = 0.2873
10	6.2	3.35	2.85	2.175	2.3 × 0.3375 = 0.7763
15	6.2	4.45	1.75	3.543	2.3 × 0.5494 = 1.2636
20	6.2	5.10	1.10	5.636	2.3 × 0.7510 = 1.727
25	6.2	5.55	0.65	9.538	2.3 × 0.9795 = 2.2528
30	6.2	5.75	0.45	13.778	2.3 × 1.1392 = 2.620

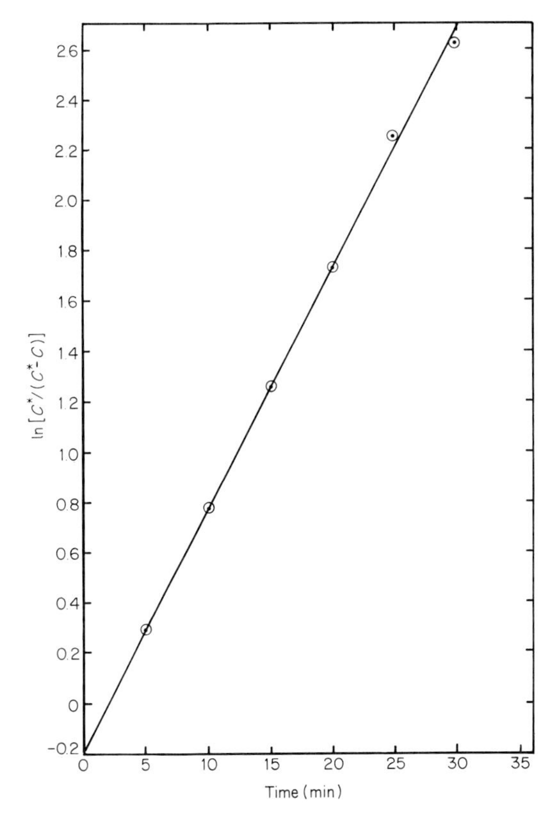

Fig. 22. Determination of $K_L a$.

TABLE V Summary of Data on $K_L a$ from Fig. 22

t (min)	$\ln [C/(C^* - C_t)]$
30	2.65
5	0.28
25	2.37
Hence in 60 min	$\frac{2.37 \times 60}{25} = 5.69$
	$K_L a = 5.7\,\text{hr}^{-1}$

6.5. Measurement of Rate of Oxygen Consumption

The textbook method for the dynamic determination of the rate of oxygen is to sparge the system to saturation (C^*) and then to turn the sparger off. The consumption of oxygen by cells reduces the concentration to a value (C_t) over a set time period (t). The rate of consumption is determined from equation (3).

In our laboratory, we have no spargers. We switch the impeller off at the beginning of the test and on again at the conclusion (Fig. 23).

7. DESIGN PRINCIPLES FOR BR-O6

1. Follow the aircraft rules "Keep it simple and add more lightness".
2. Use a standard, catalogued, borosilicate vessel, not "a prize product of the glass blower's art" (a phrase due to Professor de St.Groth).
3. The bioreactor must be able to cover the range of volumes between 500 ml and 3000 ml without changing the vessel.
4. All tubing, probes, and sensors should enter through the ten ports in a flat stainless steel lid. There should be no motor above the lid. (Eventually, we found it possible to dispense with the use of a motor entirely.)
5. Use an external heating pad on which the vessel rests. This dispenses with the need to use a circulator and jacketed vessel and facilitates easy removal of the vessel for sterilization.

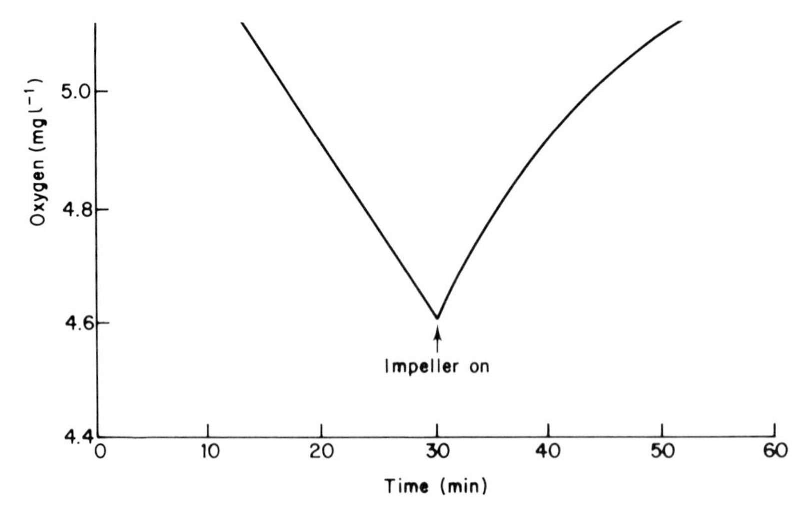

Fig. 23. Method of determining the rate of oxygen utilization in HB hybridoma cells by switching off impeller. Impeller speed 300 RPM; cell concentration 5.5×10^5 ml^{-1}) gas O_2 20%; N_2 75%; CO_2 5%: volume of medium 2500 ml.

6. Provide a tray deep enough to retain 3 l in case of an accident.
7. The electric supply to the heater should be at a low voltage for safety (e.g. 25 V).
8. Controls for gas flow, input and output, speed, and temperature are mandatory. They are all that is needed for continuous production. Other controls such as those for pH and pO_2 control and control by microprocessor or computer can be supplied to meet a customer's requirements. Automatic controls are unnecessary in continuous production; all that is required is a daily cell count and adjustment of the input as shown in Table I.
9. Electronic circuits of modular construction for ease of testing and servicing. Crystal oscillator as a reference for the counter and variable oscillator for setting speed.

In designing the BR-O6, we have not adapted a fermenter to make it suitable for cell culture but have allowed the mammalian cells to set the parameters, rather than hoping that they will tolerate those of a preconceived design.

Because both circulation and oxygenation increase with the speed of the impeller, one of the first questions we asked was "will the HB32 cells be

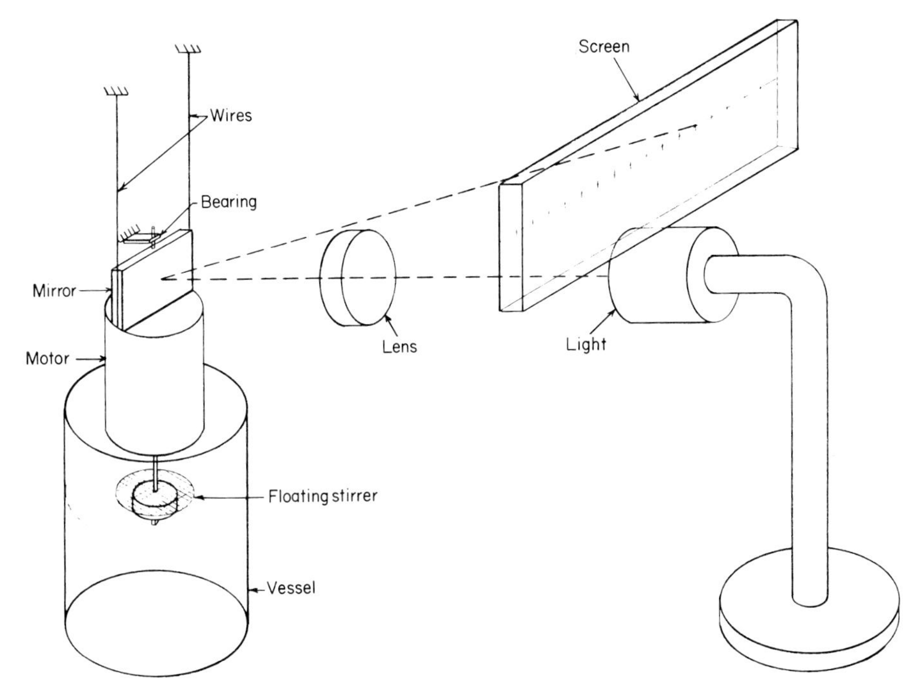

Fig. 24. Measurement of torque and power required for floating stirrer.

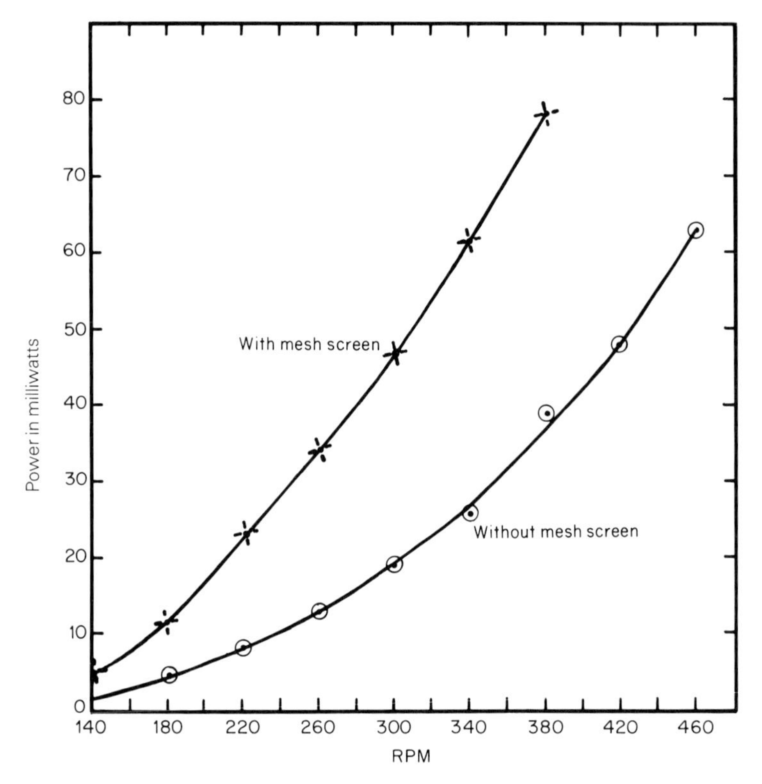

Fig. 25. Power required to rotate impeller with and without peplum mesh screen.

harmed if we run the impeller at 300 RPM?" The answer was "not only do the cells tolerate this increased stress but actually exhibited increase in growth rates of about 10%. Also the viable population number increased over 25%."

Another important question was "How much power do we need for the impeller?" We got an answer with the setup shown in Fig. 24. A small electric motor was hung up on its two conducting wires to form a bifilar suspension; this gave us the torque which, multiplied by the angular velocity, gives the power. As expected from previous work, the power was so small that a magnetic drive from a remote rotating magnetic field seemed feasible (Fig. 25). If this magnetic field were produced by stationary electromagnets switched electronically, then there would be no need for a conventional electric motor with its bearings, noise, and limited life.

What is the correct wattage for the heating pad? This is easily found by pouring 3 litres of water at about 40°C into a 6-litre flask and then plotting the temperature against time as the water cools. The water must, of course, be well-stirred. From this graph or recording we can find the rate of loss of heat

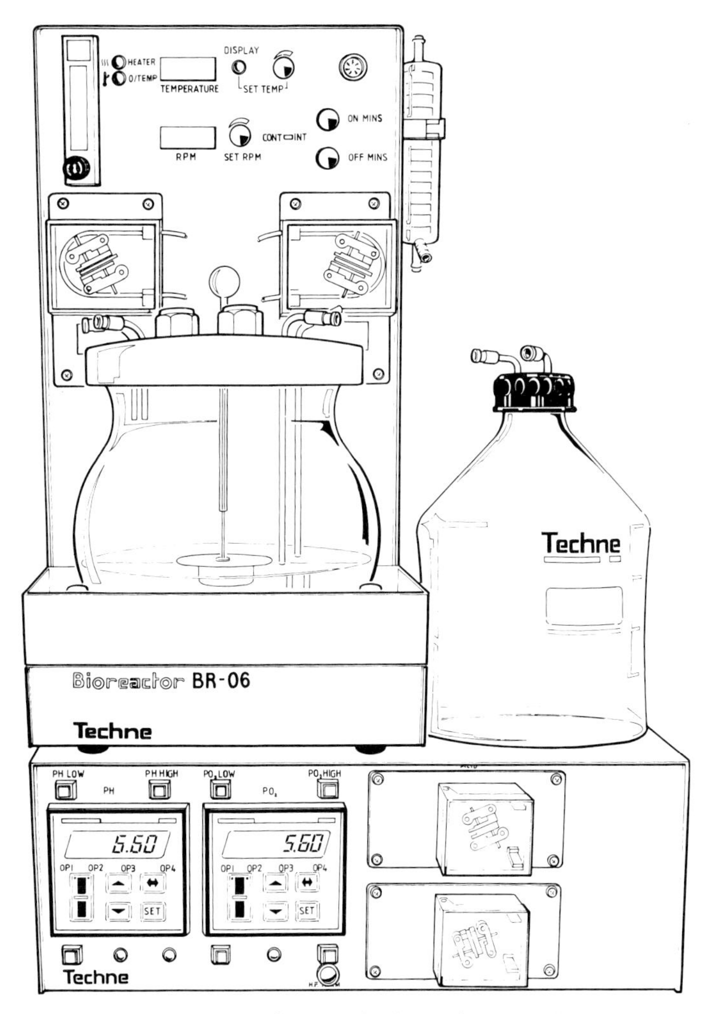

Fig. 26. BR-O6 Bioreactor, Techne (Cambridge) Ltd., mounted on control consule.

at 37°C, which is the operating temperature of the BR-O6. Multiply the loss of heat (in watts) by 2 and that is the wattage of the heater. It is in fact about 50 W. If we place the heating pad centrally under the spherical flask, the resultant convection will assist the Thomson flow, but will it give a uniform temperature throughout the liquid? We must try it and see. Designing, like politics, is the art of the possible.

The result was the BR-O6 (Fig. 26), having been tested for continuous production, for microcarrier culture (*42*), for batch operation and for operations between 500 ml and 3000 ml.

Acknowledgements

It is a pleasure to thank another friend, Professor Stephen Fazekas de St.Groth for all his help, humour and enlightenment in launching the de St.Groth Cytostat. It opened our eyes to the simplicity and effectiveness of continuous production.

It is a pleasure to record the help of Ann Noden and other members of the drawing office staff at Duxford, Cambridge, England, and of all those who managed to get the manuscript typed.

Finally, I thank Dr. Gregory J. MacMichael, Biotech Development Manager at Techne, for his considerable assistance in writing this chapter and for the whole of the biological investigations.

REFERENCES

1. Augenstein, D. C., Sinsey, A. J., and Wang, D. I. C. (1971). *Biotechnol. Bioeng.* **13**, 409–418.
2. Balin, A. K., Goodman, D., Rasmussen, H. and Cristofolo, V. J. (1976). The effect of oxygen tension on the growth and metabolism of WI-38 cells. *J. Cell Physiol.* **89**, 235–250.
3. Boraston, R., Thompson, P. W., Garland, S., and Birch, D. (1982). Growth and oxygen requirements of antibody producing mouse hybridoma cells in suspension culture. *Dev. Biol. Stand.* **55**, 103–111.
4. Couette, M. N. (1898). *Ann. de Chimie et Physik.* **XX1**, 433.
5. Cussler, C. F. (1984). "Diffusion—Mass Transfer in Fluid Systems". Cambridge University Press, Cambridge.
6. Danes, B. S., Broadfoot, M. M. and Paul, J. (1963). A comparative study of respiratory metabolism in cultured mammalian cell strains. *Expl. Cell Res.* **30**, 369–378.
7. de Bruyne, N. A. (1941). Note on viscous and plastic flow. *Proc. Phys. Soc.* **53**, 251–257.
8. de Bruyne, N. A. (1961). Intelligent use of thermostatically controlled Bath. *R&D* Sept. (U.K.). [Out of print. Copies can be supplied by Techne (CAMBRIDGE) Ltd., Duxford, Cambridge, CB2 4PZ. *Note:* On p. 2 the numerical values under "Power Required" are incorrect.]
9. de Bruyne, N. A., Weiss, L., and Huber, D. (1974). A pneumatic stirrer for suspension cultures of mammalian cells. *In Vitro* **10**, 184–187.
10, de Bruyne, N. A. (1979). *Dev. Biol. Stand.* **42**, 135.
11. de Bruyne, N. A., and Morgan, B. J. (1981). Stirrers for suspension cell cultures. *American Laboratory*, June.
12. de St.Groth, S. F. (1983). Automated production of monoclonal antibodies in a cytostat. *J. Immunol. Methods* **57**, 121–136.
13. de St.Groth, S. F. (1986). Methods of Enzymology.
14. Documenta Geigy (1972). Ciba Geigy Ltd, P. 618 Basle.
15. Fleischaker, R. J., and Sinskey, A. J. (1981). Oxygen demand and supply in cell culture. *Eur. J. Appl. Microbiol. Biotechnol.* **12**, 193–197.
16. Greenspan, H. P. (1980). "The Theory of Rotating Liquids", p. 3. Cambridge University Press, Cambridge.
17. Hu, W. S., and Wang, D. I. C. (1986). Use of surface aerators to improve oxygen transfer. *In* "Mammalian Cell Culture" (W. G. Thilly, ed.), pp. 182–184. Butterworth, London.

18. Hunter Rouse (1978). "Elementary Mechanics of Fluids", pp. 176–180. Dover, New York.
19. Jensen, M. D. (1979). *In* "Practical Tissue Culture Applications", Chapter 8, pp. 115–136. Academic Press, London.
20. Katinger, H. W. D., Scheiner, W., and Krömer, E. (1978). Der Blasen Säulenfermenter für die Massensuspensionskultur tierischer Zellen. *Chem. Ing. Tech.* **50**, 472.
21. Kilburn, D. C., and Webb, F. C. (1968). The cultivation of animal cells at controlled dissolved oxygen partial pressure. *Biotechnol. Bioeng.* **X**, 801–813.
22. Kilburn, D. C., and Lilly, M. D. (1969). The effect of dissolved oxygen partial pressure on the growth and carbohydrate metabolism of mouse LS cells. *J. Cell. Sci.* **4**, 25–37.
23. Kühn, W. (1939). *Z. Phys. Chem. B* **42**, 1.
24. Lamb, H. (1932). "Hydrodynamics", 6th edn, p. 558. Dover, New York.
25. Lehmann, J., Piehl, G. W., and Schulz, R. (1985). Bubble free cell culture aeration with porous moving membranes. *Biotech. Forum* **2**, 112–117.
26. Mackereth, F. J. H. (1964). An improved galvanic cell for determination of oxygen concentration in fluids. *J. Sci. Instrum.* **41**, 38–41. [Also U.K. Patent 999 909 (1965).]
27. Maroudas, N. (1973). New methods for large-scale culture of anchorage-dependent cells. *In* "New Techniques in Biophysics and Cell Biology" (R. H. Payne and J. B. Smith, eds.), Chapter 3, pp. 67–86. Wiley, New York.
28. Maxwell, J. C. (1867). *Phil. Trans. R. Soc. London* **157**, 52.
29. Mclimans, W. F., Blumenson, L. E., and Tunnah, K. V. (1968). Kinetics of gas diffusion in mammalian cell culture. *Biotechnol. Bioeng.* **10**, 741–763.
30. Munder, P. G., Modobell, M., and Wallach, D. F. H. (1971). Cell propagation of films of polymeric fluoro-carbon as a means to regulate pericellular pH & pO_2 in cultured monolayers. *FEBS Lett.* **15**, 191.
31. Nevaril, C. G., Lynch, E. C., Alfrey, C. P., and Hellows, J. D. (1968). *J. Lab. Clin. Med.* **71**, 784–790.
32. Phillips, H. J., and Andrews, R. V. (1960). *Proc. Soc. Exp. Biol. Med.* **103**, 160–163.
33. Phillips, H. J., and McCarthy, H. L. (1956). Oxygen uptake and lactate formation of Hela Cells. *Proc. Soc. Exp. Biol. Med.* **93**, 573–576.
34. Radlett, P. J., Telling, R. C., Whiteside, J. P., and Maskell, M. A. (1972). The supply of oxygen to submerged cultures of BHK21 cells. *Biotechnol. Bioeng.* **14**, 437–445.
35. Sinskey, A. J., Fleischaker, R. J., Tyo, M. A., Giard, D. J., and Wang, D. I. C. (1981). Production of cell derived products, virus and interferon. *Ann. N.Y. Acad, Sci.* **369**, 47–60.
36. Spier, R. E., and Griffiths, B. (1984). An examination of the data and concepts germane to the oxygenation of cultured animal cells. (ESACT 1982.) *Dev. Biol. Stand.* **55**, 81–92.
37. Spier, R. E., and Whiteside, J. P. (1984). Description of a device which facilitates the oxygenation of microcarrier cultures. (ESCAT 1982.) *Dev. Biol. Stand.* **55**, 151–152.
38. Spragg, A. J. P., Maskell, S. J., and Patrick, M. A. (1985). "5th European Conference on Mixing BHRA". Cranfield, Bedford, U.K.
39. Techne (1986). Unpublished.
40. Thomson, J. (1857). On the ground currents of atmosphere circulation. *Trans. Sections. Br. Assoc. Sci.* 38–39. [Reprinted in "Collected Papers in Physics and Engineering" (Larmor and Thomson, eds.). Cambridge University Press, 1912.]
41. Tovey, M. B. (1980). The cultivation of animal cells in the chemostat: application to the study of tumour cell multiplication. *Adv. Cancer Res.* **33**, 1–37.
42. Whiteside, J. P., Farmer, S., and Spier, R. E. (1985). The use of caged aeration for the growth of animal cells on microcarriers. (Joint ESACT/IABS meeting 1984.) *Dev. Biol. Stand.* **60**, 283–290.
43. U.S. Patent 4 382 685. [The MCS with Pearson flask.]

44. U.S. Patent 4 465 377. [N. A. de Bruyne Floating Stirrer driven by rotating magnetic field at side of flask.]
45. U.S. Patent 4 498 785. [N. A. de Bruyne Floating Stirrer. Filing date 30th Sept. 1982.]
46. Patent applied for.

PART III

CULTURE SYSTEMS

7

Overview of Cell Culture Systems and their Scale-up

J. B. GRIFFITHS
Vaccine Research and Production Laboratory, PHLS CAMR,
Porton Down,
Salisbury, Wiltshire, U.K.

1. INTRODUCTION

In this section of the book an attempt is made to review the many culture systems which have been developed in response to the need for mass cultivation of cells and their products. Each contribution is presented by an expert on that particular culture who will underline the advantages, and limitations, of using a particular technique or reactor. The range of systems

ANIMAL CELL BIOTECHNOLOGY VOL. 3
ISBN 0-12-657553 3

now available seems at first sight to be enormous, but in fact many are divergences of a common technical development, often made 20–30 years ago. Also, many distinctively different systems do run on the same set of conceptual factors. One such parameter is that if cells are to be maintained at a high density and in a controlled environment, then a medium perfusion system has to be used. The problems to be overcome in scale-up are well known. They include maintaining a proper supply of nutrients (including oxygen) without causing any physicochemical damage to the cells and, at high densities, removing toxic metabolites. The problem is one of transfer between the environment and the cell. The various means of solving this problem have given rise to the diversity of culture systems that are currently in use. For anchorage-dependent cells there is the additional problem of increasing the surface area for cell attachment and this again has been solved in many different and ingenious ways.

In this overview chapter, the general problems involved in scale-up are discussed and a historical development of culture systems is presented. This has not only been done out of interest to show the developmental themes which have led to current culture reactors, but also because some of the original or "old" technology used ingenious solutions to the problem which may stimulate today's innovators of culture reactors. The range of available equipment and techniques is confusing for newcomers who have to decide which system will best fit their needs, if indeed they can be persuaded to go beyond multiple roller bottles or low-density suspension cultures. Resources do not usually allow a multiple choice to be made and it is difficult to evaluate the relative performance of competing culture types and to decide whether inherent complexities would make the system risky for a commercial process. Therefore, in this section an attempt is made to give as objective a summary as possible of the strengths and weaknesses, the scale-up capabilities, and the potential applications of the culture systems reviewed.

It is recognized that no single cell-culture system can be universally used, owing to the variability in cell types, product expression kinetics, and process requirements (listed in Table I). Suspension culture is the scale-up method of

TABLE I Factors Affecting the Design of Cell Reactors and Production Processes

	Factors	Parameters
Growth mode	Suspension	Differences in fragility
	Anchorage	Differences in adhesion
Production	Product expression	Growing or stationary cells
	Product type	Intracellular/extracellular/lytic
	Product concentration	Feedback inhibition
	Culture type	Batch, fed-batch, continuous
Downstream processing	Product concentration	Ultrafiltration?
	Batch or continuous	

TABLE II Comparison of Substrate-attached and Suspension Culture Systems

Attached systems	Suspension systems
Easy to change medium	Easy and cheap to scale-up
Easy to perfuse	Unit scale-up to high volumes
Easy to change medium : cell ratio	Utilizes less space
Products often expressed better	Easy to monitor and control
More flexible (all cells)	Cells can be sampled
(Microcarrier culture has the advantages of both systems)	

choice because environmental monitoring and control allows homogeneous and near-linear increase in volume. However, many commercially important cell lines either will not grow in suspension culture, or have a significantly lower productivity, or express products with altered properties in suspension when compared to growth in anchorage-dependent cultures (*85*). Although suspension culture technology is predominant, and every means of using such cells is being exploited, it is still very important to increase the productivity and unit size of anchorage-dependent reactors. The comparative merits of suspension and anchorage-dependent cells are summarized in Table II. Similarly, every effort is being made to construct continuous, as opposed to batch, processes as a more cost-effective methodology (to avoid repetitive production of cell seed with the accompanying quality-assurance workload, and to avoid equipment downtime and turn-around costs). Again, this is not always possible, e.g. in production of lytic viruses, but the change in emphasis over the last few years on producing recombinant rather than native products encourages the use of continuous processes, even for vaccines.

These factors have to be taken into account when considering the various culture options reviewed on the following pages.

2. SCALING-UP

Scaling-up from small laboratory flasks (static or spinner) in a unit rather than multiple process has always been a prime objective for cell culturists. Initially it was to meet the demands of virologists both for research and the manufacture of vaccines. Currently there is a wide range of molecules that can be economically developed only if efficient large-scale culture methods are used. Animal cells, in comparison to bacteria, have low productivity as a consequence of a slow growth rate (doubling time of 12–24 hr compared to 0.3–3 hr for bacteria). Thus, on a cell weight basis (g biomass $litre^{-1}$ hr^{-1}) bacteria are 60- to 200-fold more efficient. In addition, animal cells need far more complex and expensive growth media (*31*) and more critical environmental control and process handling (to avoid the high risk of microbial

TABLE III Recombinant Animal Cell Products

Viruses	HBsAg, HSV 1 & 2, influenza, CMV, EBV,[a] rabies,[a] HIV,[a] FMDV,[a] Lassa fever[a], JEE,[a] RSV, VSV
Blood products	Factor VIII, Factor IX, Protein C, immunoglobulins
Hormones	hGH, hCG, insulin, erythropoetin, relaxin, LH
Others	t-PA, α & β-IFN, IL-2, PDGF, MCAb (e.g. OKT-3)

[a] Expressed via Vaccinia virus.

contamination, for example). Animal cells are only considered for an industrial process when there is no alternative, i.e. when prokaryote and other eukaryotic systems, even after recombinant DNA technology, cannot produce the required product in the correct configuration or with freedom from toxic components. Thus, comparisons between bacterial and animal cells are unproductive—the challenge is to increase the efficiency of existing cell systems. This has already been significantly achieved by means of hybridization techniques (DNA and cell) and by optimizing the environment in which cells grow. There already exist many recombinant products from animal cells (Table III). This chapter, however, is concerned with the increased productivity that can be achieved with the correct choice of culture reactor.

The need to achieve higher productivity is exemplified by the data in Table IV (*45*), which give an indication of the culture volumes needed to produce clinical doses of many important biologicals.

The basic problem in achieving these aims is the relative fragility of cells to mechanical stress brought about by stirring, sparging, etc. As only low-speed, non-turbulent, stirring can be used, mass transfer rates are low and means of

TABLE IV Culture Requirements for Producing Therapeutic Doses of Animal Cell Products[a]

Product	Cell requirement per dose	Culture volume (litres)
Polio	2×10^4	0.0001
Rabies	4×10^6	0.005
HSV	2×10^7	0.03
FMDV	2×10^7	0.01
IFN (anti-virol)	10^5 day^{-1}	0.1
(anti-tumour)	5×10^5 day^{-1}	0.5
t-PA	$>10^{10}$	1–10
MCAb	10^{12}	100
UK	10^{12}	500

[a] Data based on Katinger and Bleim (*3*).

aeration and maintaining optimum nutritional conditions in a non-damaging manner are difficult. There are four major areas in which key developments have occurred to improve cell culture performance: mixing, aeration, perfusion and substrate (for anchorage-dependent cells). These factors are all closely interlinked and should not be considered in isolation, but they do allow individual innovations to be conveniently described.

One major development which has evolved as a combined result of these innovations is the ability to maintain cells at high densities (over 10^8 cells ml^{-1}). Currently there is a choice between volumetric scale-up (increasing size whilst maintaining the same process intensity) and density scale-up (increasing the process intensity in relatively small volumes). The ideal which is being strived after is a combination of these techniques, and assessments are being made of all current working systems to find the technology with the potential to achieve this.

2.1. Relative Merits of Process Intensity

In terms of cell density within body tissues and organs, a dense cell population would be in the order of 1–2 $\times$ 10^9 cells cm^{-3}. In a culture reactor, assuming a spherical cell of 12 μm, a dense culture would be in excess of 5 $\times$ 10^8 cells ml^{-1} (*89*). This then is the upper limit that can be realistically aimed for; but can it be achieved? The supply of oxygen is perhaps the most critical parameter. It has been estimated that a 100 g organ contains 2 $\times$ 10^{11} cells and, as the average blood flow through tissues is 100 ml g^{-1} min^{-1}, then cells *in vitro* have a supply of 3.6 $\times$ 10^{12} g oxygen $cell^{-1}$ min^{-1} (*89*). This is within the range found for cells in culture (*86*). Blood delivers oxygen to cells through an extensive capillary system (calculated as having a total surface area of 1.9 $\times$ 10^6 cm^2/kg body weight). This can be mimicked in cell cultures with the use of fibres or thin-walled vinyl tubing. Calculations on how close together these "artificial capillaries" have to be is based on observation as five cell diameters (*3*), or computer models of oxygen diffusion rates which range from 10–30 cell diameters (*89*). Thus, a gap of 100–300 μm between oxygen sources should be adequate to maintain dense cells.

It is possible for cells to exist in small-scale culture in dense masses, but scale-up has to be by multiplicity. However, before committing too much of one's developmental resources to determining whether these systems will scale-up as unit processes, the advantages and disadvantages should first be assessed (Table V). The argument that, since a 1-litre culture supports 5 $\times$ 10^8 cells ml^{-1}, then this is equivalent to a 500–1000-litre reactor of conventional cell density (2–5 $\times$ 10^6 cells ml^{-1}), and that thus there is a reduced need for large volumes, is partly over-ridden by the requirement for a large-volume

TABLE V High-density Culture Systems

Advantages	Disadvantages
Smaller reactor volume	Volumetric scale-up difficult
More concentrated product	Mass transfer problems
Reduced serum requirement	Sterilization problems
Long run lengths	Higher degree of control sophistication needed

reservoir. A given volume of medium is capable of yielding a finite number of cells—for Eagle's MEM this is 2–4 × 10^5 cells ml^{-1} in a closed batch system and 1–2 × 10^6 cells ml^{-1} in an efficiently controlled recirculating perfusion system (*31*). Thus a 1-litre high-density culture theoretically needs a 500-litre reservoir. In fact, with a controlled sequential flow of medium from preparation through to product extraction, this value can be halved. In addition, a great saving in space can be obtained now that medium can be supplied in plastic bags (similar to blood transfusion bags) in units up to 500 litres (*116*). However, these calculations are based on the assumption that product expression per cell is the same in both dilute and dense systems. This is hardly ever true and the fact that one usually gets diminishing returns during scale-up still remains a prime problem. There are many environmental factors that contribute to this phenomenon (e.g. end-product repression and toxic metabolites), but there are also many factors intrinsic to the cell which can only be resolved by careful comparative studies of cell physiology at low and high densities. Proximal contact, together with chemical and neuronal signals, are the main regulatory agencies in animals (*29*) and the effect of close contact on surface receptors has to be established in dense cell culture.

Facilitated downstream processing owing to a high product concentration is often quoted as one of the advantages of high-density culture. However, for a particular product, one has to decide whether it is easier to concentrate a dilute supernatant or overcome the problem of cell debris contaminating a concentrated supernatant. This example serves as a warning that there are two sides to many of the arguments for and against a particular culture concept and one should not get carried away with intellectual considerations. Undeniably on the plus side for dense cultures is the fact that the requirement for expensive medium components (e.g. serum, growth factors) is significantly reduced (*10, 84*).

The history of using animal cells for the manufacture of pharmaceuticals has been typified by the concern not to use a transformed or malignant cell, and to prove that the cell line in use has not undergone such a transformation. One requirement was that human diploid cells must remain as a monolayer, i.e. multilayering was taken as a sign of such a change. It is fortunate that

Biological Control Agencies are able to relax this requirement, and even permit the use of continuous (heteroploid) cell lines (e.g. Vero) owing to improved biochemical techniques for assessing the biological hazards in the system. This change in attitude is very important for the future of dense cell systems as substrates for therapeutic, rather than just diagnostic, products (*108*).

2.2. Mixing

This topic has been covered in Chapter 6. The first stirred cultures relied on a magnetic bar which gave laminar mixing and became scale-up limiting at quite low volumes (2–5 litres). Modifications to increase the surface area of a bar magnet by attaching various designs of "paddle" (*27, 30, 96, 111*) allow a greater mixing capability without an increase in stirrer speed or, more importantly, to the tip velocity of the impeller (which should not exceed 1.5–2 m s^{-1}). The aim is to achieve non-turbulent, streamlined bulk flow patterns within the culture fluid so that mechanical stress damage is minimized. Turbine impellers are damaging to many cell types and the marine impeller has become the configuration of choice. Modifications to the marine impeller have subsequently been made to increase mixing efficiency at low speeds (*119*). Other mixing concepts that have been successfully used are the Vibromixer (*20*), airlift (*43, 44*), pump (Celligen fermenter) (*75*), and internal loop (*58*). All these systems are diagrammatically summarized in Fig. 1.

2.2.1. Airlift Reactors

Airlift, as an alternative to standard stirred-tank reactors, has been gaining popularity because of its gentler mixing action and suitability for shear-sensitive cells (*5, 43*). It works on the principle that the gas mixture introduced into the base of the draft tube within the culture establishes a circulation of medium owing to the density differential between the oxygenated air (in the draft tube) and the less oxygenated air in the outer zone. Oxygen levels and pH can be controlled by varying the composition of the gas mixture. The critical factors are the reactor height, aspect ratio, and gas flow velocity. The optimum configuration for mixing is to have the draft tube diameter essentially the same as the total outside (downflow) zones, and the most efficient gas velocity is 0.5–1.5 cm s^{-1} (*79*). In practice, air flow rates (AFR) of about 300 ml min^{-1} are employed. Scale-up is relatively simple as it is directly linear and 1000-litre reactors are in commercial use (*1, 5, 6*). It having been established that cells can withstand cyclical exposure to hydrostatic pressures of 25 psi, plans are well advanced in commissioning a 5000-litre vessel. In some respects it is easier to scale-up, as increases in height improve the mixing times

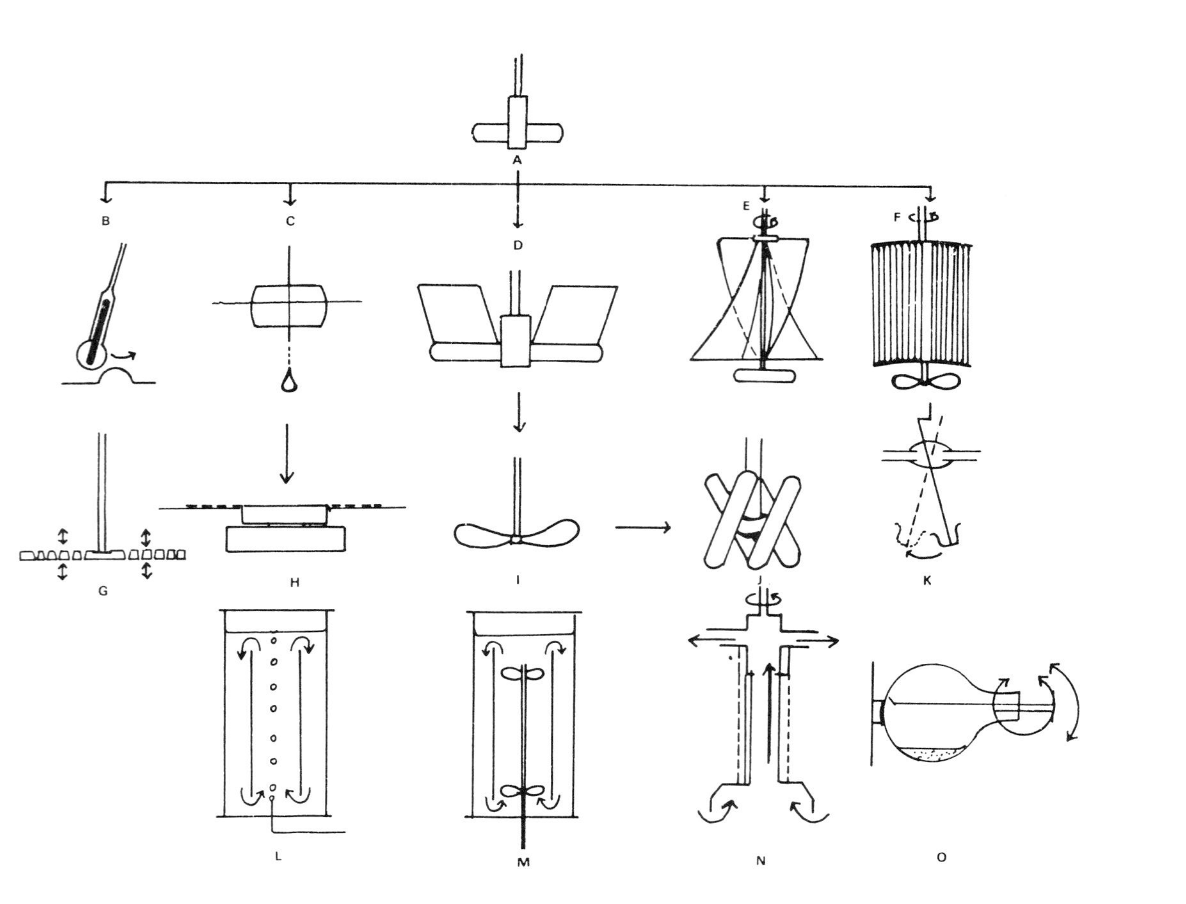

A
B
C
D
E
F
G
H
I
J
K
L
M
N
O

and mass transfer rate. Aspect ratios of between 6:1 and 12:1 are normally used.

The performance and productivity of the airlift system are similar to that of conventional stirred reactors, i.e. 2–3 $\times$ 10^6 cells ml^{-1} in the batch mode, with the possibility of using continuous or semi-continuous batch feed and perfusion options. It must be considered a low-process-intensity system with no obvious route to significantly increasing cell density, other than an external loop through a filtration unit and returning the cells to the airlift vessel. Although scale-up is straightforward, it does have the disadvantage of being dimensionally linear, requiring considerable vertical space.

The main advantages of airlift over stirred reactors, besides its ease of scale-up and suitability for shear-sensitive cells, is that no moving parts and mechanical seals are needed. This not only means less sophisticated engineering design, but it also eliminates a potential source of contamination. Also, the system makes it easy to meet the required oxygen transfer rates and it requires less power input than other types of culture.

2.3. Aeration

It has long been recognized that bubbling (sparging) in a cell culture vessel causes damage to many cell lines unless it is carried out at very low rates. The general guidelines for sparging are to use large-diameter bubbles (1–3 mm) at a low flow rate (5–10 $cm^3\ l^{-1}\ min^{-1}$). More recently it has been suggested that bubble damage occurs at the surface of the culture during bubble disengagement (*13, 36*) but that this can be minimized by increasing the aspect ratio of a culture vessel to 6–12:1 in order to reduce the residence time of cells at the surface. This is the basis of the airlift reactor. In addition, the inclusion of Pluronic F-68 (0.01%) very effectively protects cells against bubble damage (*13, 36*).

Small-scale cultures rely on oxygen transfer through the air–medium interface (surface aeration). However, the relatively low solubility of oxygen (7.6 mg ml^{-1}) and diffusion rates (static cultures) or mixing efficiency (stirred cultures) causes oxygen to become a limiting factor during scale-up (often at 5-litre scale). This is despite keeping a low aspect ratio of 1–2:1. Considerable efforts have been made to supply adequate oxygen in a non-harmful manner

---

Fig. 1. Agitation/mixing systems used for animal cell cultures. A, magnetic spinner bar; B, MCS system (*120*); C, cytostat surface stirrer (*15*); D, increased surface area spinner; E, sail impeller (*16*); F, membrane basket agitator (*59*); G, vibromixer (*20*); H, surface stirrer BRO6 (*121*); I, marine impeller; J, cell ascenseur (*119*); K, skull stirrer (*2*); L, airlift; M, internal loop (*58*); N, celligen (*75*); O, ZKA bioreactor (*114*).

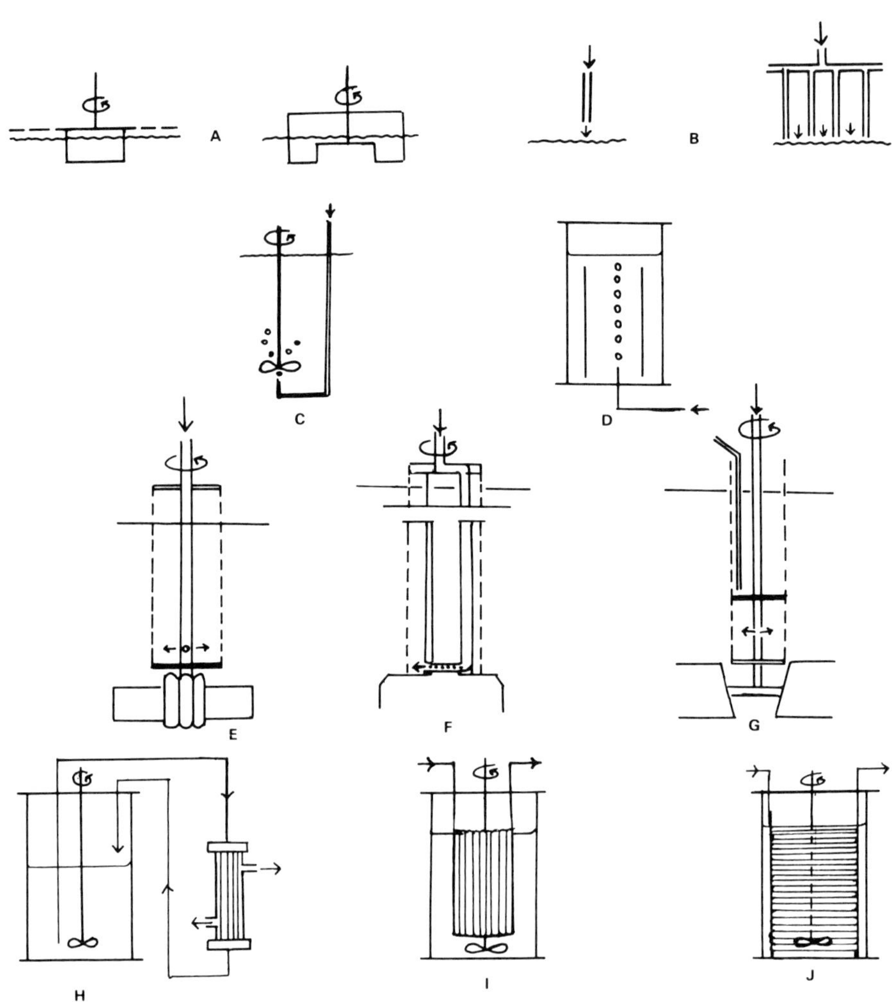

Fig. 2. Systems for aerating animal cell cultures. A, surface aerators (*40, 121*); B, surface aeration by air jets (*100*); C, sparging; D, airlift; E, caged aerator (*105*); F, celligen (*75*); G, spin exchanger (*35*); H, external loop oxygenator; I, membrane basket (*69*); J, membrane tubing oxygenation (*55*).

to cell cultures in order to allow scale-up in volume and intensity. The efficiency of surface aeration can be increased at least 4-fold with the use of a surface aerator (*40*), or by injecting air through multiple nozzles placed immediately above the medium surface (*100*). Examples of the many alternative methods used to increase aeration are shown diagrammatically in Fig. 2. One method is to circulate the medium from the culture through an

TABLE VI Aeration Methods of a 39-litre Stirred Bioreactor

Aeration method	O_2 ($mg\,l^{-1}\,hr^{-1}$)	No. cells/ml $\times 10^6$ supplied[a]
Air ($10\ ml\ l^{-1}\ min^{-1}$, 40 r.p.m.)		
Surface aeration	0.5	0.08
Direct sparging	4.6	0.76
Spin filter sparging	3.0	0.40
Perfusion (1 volume hr^{-1})	12.6	2.10
Spin filter sparging + perfusion	15.9	2.65
Oxygen ($10\ ml\ l^{-1}\ min^{-1}$, 80 r.p.m.)		
Spin filter sparging	51.0	8.50
Spin filter sparging + perfusion	92.0	15.00

[a] Assuming oxygen utilization rate of 2–6 μg/million cells/hr.

oxygenator and back into the culture, but for success a fast perfusion rate is needed. The oxygenator can be a capillary fibre device, silicone tubing (*37*), or a specially adapted reservoir vessel (*26*). The removal of cell-free medium has been achieved using both gravitational settling methods (*9*) and spin filters (*35*). The critical factor is keeping the cells from blocking the filter used for removing the medium. In microcarrier culture, where a mesh size of 60–100 μm can be used, this is relatively straightforward and perfusion rates of at least 2 volumes hr^{-1} can be achieved. It is more difficult for suspension cultures, as a filter mesh of under 10 μm is required and this only allows perfusion rates of 1–2 volumes day^{-1} without blockage occurring. The importance of perfusion in culture oxygenation is exemplified by the data in Table VI (*32*).

Two main methods have been employed to overcome the difficulties of sparging and the complexities of perfusion, apart from increasing the partial pressure of oxygen or the atmospheric pressure of the culture. These are aerating the medium in a cell-free compartment within the culture by means of a mesh screen (*33, 35, 88, 105*), or by using a bubble-free system based on diffusion through thin vinyl or polypropylene tubing (*18, 47, 66*). This latter method originally used tubing arranged around the vessel periphery, but owing to the great lengths involved it was rather inconvenient to use, although commercial models are available (*55*). A more practical means of organizing the tubing is by using the stirrer as a holder (membrane agitator) (*59*)—this allows both an efficient presentation of the tubing and a large-surface-area impeller. The disadvantages of this aeration method are that scale-up is linear (i.e. tubing length has to increase disproportionately during volume scale-up) and different gases vary in their diffusion rates so that gases cannot be mixed (e.g. oxygen and carbon dioxide). However, yields of 8×10^6 cells ml^{-1} on Cytodex microcarriers have been reported (*59*).

Oxygenation of cultures is a complex subject. If gas bubbles are used, then a high aspect ratio increases the efficiency owing to the higher pressure within the culture and the residence time of the bubbles—this is needed to offset the disadvantages of using large bubbles. If bubbling is not used, then a very low aspect ratio has to be used to increase surface aeration. Oxygen is toxic at concentrations above 21% partial pressure and care has to be taken if this concentration is increased in the gas mixture—in fact, it should only be used in efficiently mixed systems, and well into the culture cycle when the oxygen demand is high. Fortunately, cells seem to find a low oxygen concentration (about 10% partial pressure) optimal. The problem of foam has to be resolved in sparged cultures, especially when serum is present. Antifoam agents, such as silicone antifoam (Sigma, 6–30 p.p.m.) are effective, but should not be used over long culture periods as they do have detrimental effects on cells.

2.4. Perfusion

Perfusion is a long-established technique in cell culture (since 1912). Initially it was used to keep cells viable in very small chambers used for microscopical examination. However, the realization that cells are better suited to a continuous supply of nutrients and removal of toxic waste products than to the alternative feasting and fasting routines of a batch culture, has led to a widespread use of this technique. Small-scale applications probably culminated in the dual-rotary-circumfusion system (*78*), which allowed the progression and maintenance of differentiated foetal tissue to be microscopically examined in 24 replicate chambers. Medium replacement was carried out by the automatic gravitational filling and emptying of chambers as they revolved on a wheel, and not by pump action as used in subsequent systems.

The current use of perfusion is in response to the need to maintain high cell densities in a unit volume (or unit surface area). This technology originated in 1957 with the Cytogenerator (Fig. 3), a U-shaped tube with its side-arms constructed of sintered glass surrounded by the medium reservoir (*23, 24*). This development was of interest because (a) it allowed high-density growth of suspension cells (an unheard of 1–2×10^7 cells ml^{-1} in 1957) and (b) it had a very gentle agitation/mixing mechanism based on the flow of perfused medium and a tidal action created by alternating pulses of air to the tops of the U-tubes. The inevitable problem of filter blockage was delayed by reversing the direction of medium flow every 24 hr. The volume of the reactor, and the reservoir, was 300 ml. At a perfusion rate of 600 ml day^{-1} a density of 1×10^7

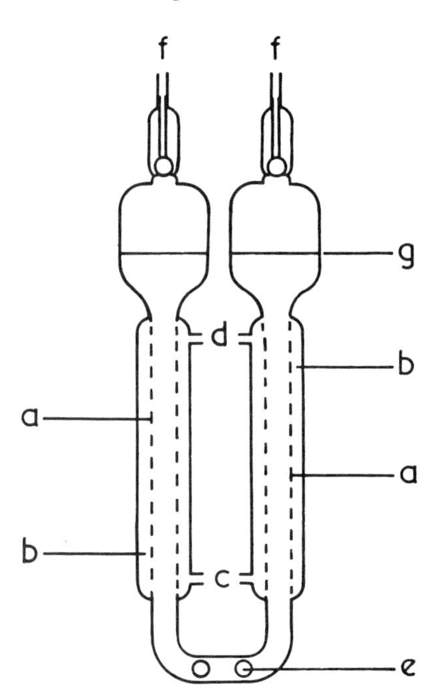

Fig. 3. Graff cytogenerator (*23*). A, sinter glass tube; B, outer medium reservoir; C, medium inlet from reservoir; D, medium outlet; E, sampling port; F, gas in/out; G, medium level when stationary.

BHK cells ml^{-1} was maintained for 1000 hr with 150 ml being withdrawn from the culture every 24 hr (*25*). The efficiency of the filter did decrease after an average of 650 hr. Unfortunately, the U-shaped configuration made the apparatus difficult to scale-up, but it did establish many useful principles and it is surprising that more use has not been made of this (for 1957) advanced technology.

The success of the Cytogenerator for suspension cells provided the impetus to develop perfusion systems for substrate-attached cells. The first of these was the glass helix perfusion chamber (*64*), of 40 ml capacity and used at an initial perfusion rate of 2 volumes day^{-1} but increasing to 400 volumes day^{-1} in the later stages of the culture cycle. Cell yields in the order of 10^6 ml^{-1} were obtained.

This technique has been developed further in many laboratories using glass pipes (*37*) and spheres (*7, 26, 77, 106, 107*) at scales up to at least 100 litres (*104*). In the 1960s, large-scale production of attached cells could only be achieved with multiple-batch roller bottle culture. By means of specially designed swivel caps, Kruse (*51*) was able to transfer to roller bottles his

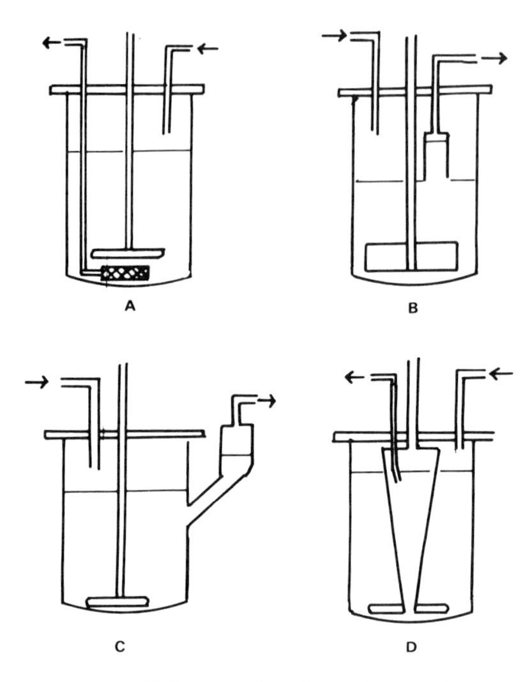

Fig. 4. Devices to separate cells from medium for perfusion culture. A, stainless steel mesh filter (*98*); B, gravitational settling (internal) (*9*); C, gravitational settling (external) (*16, 47*); D, centrifugal separation in stirrer shaft (*82*).

successful flask perfusion technique (*52, 53*), in which cell densities of up to 17 monolayer equivalents (45×10^6 cells cm^{-2}) with perfusion rates in the order of 10 volume day^{-1}. This development is commercially available (*118*), as are modifications which give larger surface areas (by packing the roller bottle with spaced glass tubing) (*11, 21, 81*). One modification rotates the bottle alternately clockwise and anti-clockwise through 360° to avoid the use of swivel caps (*112*). Many other developments to increase the surface area per unit volume have been made which rely on perfusion, and are described in Section 2.5.

Perfused suspension (and microcarrier) cultures continued to be developed (Fig. 4) using gravitational settling systems (*9, 16, 47, 96*), which permit only slow perfusion rates but do not become blocked, or by using the centrifugal force of an enlarged and hollow stirrer shaft (*82*), or by static filters (*98*). However, most current procedures are based on spin-filter devices or on filtration systems.

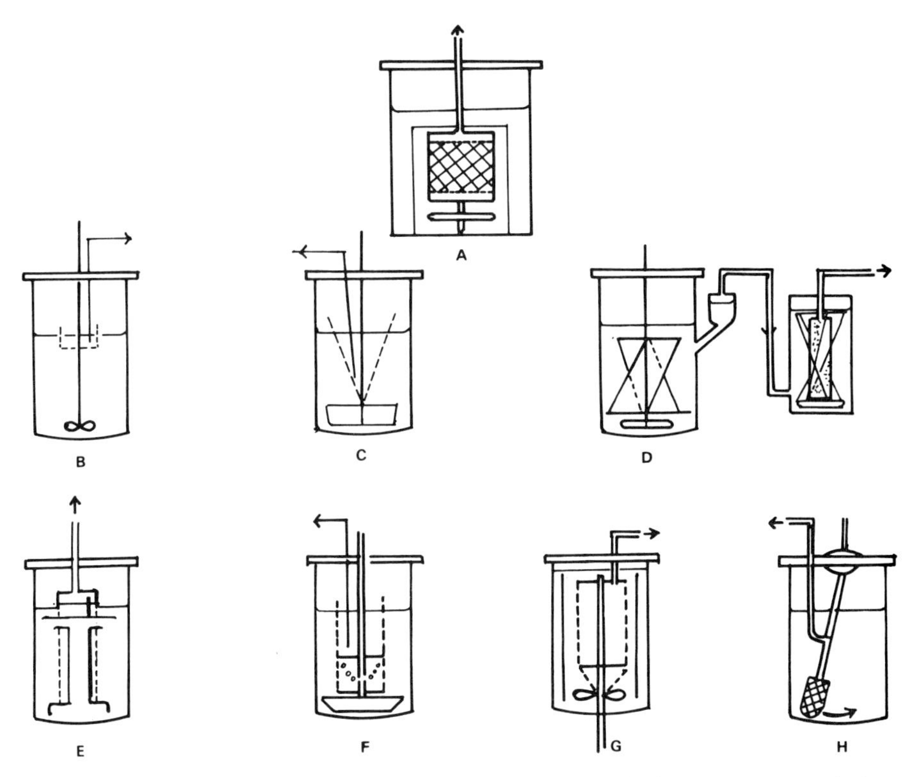

Fig. 5. Development of spin filters. A, Himmelfarb (*38*); B, wire basket (*100*); C, internal wire filter (*27*); D, external filter with spinning sails (*16*); E, celligen (*75*); F, spin exchanger (*35*); G, rotating wire cage (*103*); H, perfusion wand (*111*).

2.4.1. Spin-Filter Devices (Fig. 5)

All static filters within a culture become blocked sooner rather than later. This has always been a limiting factor in process intensity scale-up. There are two means of preventing (or delaying) this; they are to provide a large filter surface area (to reduce the flow rate per unit area), or to rotate the filter, thus creating a boundary effect and preventing cell attachment and filter clogging. This principle was first used by Himmelfarb in 1969 (*38*) and allowed cell densities of 6.5×10^7 ml^{-1} (L1210 mouse leukaemia cells) to be attained. The filter was a 3 μm porosity stainless-steel mesh rotated at 300 r.p.m. and used at volumes up to 1 litre. In later developments (*94*), microfibre filter tubes, ceramic, and fused glass bead filters were used. This technology was borrowed successfully for the growth of cells on microcarriers where the porosity is not so critical (*27, 28, 34, 100, 101*). The same units, made of stainless steel are also currently being used in many laboratories for the growth of hybridoma cells at intermediate cell densities (1–5 × 10^7 ml^{-1}) (*17, 76, 102*). Porcelain filters of 1 μm porosity have also been used (*95*). The limiting factor is that perfusion rates of only 1–2 volumes day^{-1} can be achieved with small-pore (5–10 μm) filters. However, a recent paper (*103*) describes the use of a filter with a 53 μm pore mesh which, by means of a suitably integrated design of fermenter geometry, impeller and cage construction, and stirring speed, allows a 95% retention of suspension cells. The cage is fitted into a draft tube and a perfusion rate of 0.12 volumes hr^{-1} is used at scales up to 15 litres. Another modification of this technique is to have rotating Teflon sails around the stationary filter in a satellite vessel (*17, 96*). The boundary effect of the sails on a 1 μm porcelain filter allowed sufficient perfusion rates to support Walker Rat carcinosarcoma cells at 3 × 10^7 ml^{-1}. The use of twin drives, one for stirring and the other for a high-speed spin filter, is another possible solution.

Spin filters provide a simple technical solution to scaling-up process intensity within the well-established stirred fermenters. They do, however, limit process intensity (especially for suspension cells) and can also bring to a process a certain degree of unpredictability as to when filter clogging begins to adversely affect, or cause a shutdown, of the culture.

2.4.2. Filtration Systems (Fig. 6)

Various cultures now exist that rely on the use of a range of filtration materials that allow the passage of low-molecular-weight compounds but retain the cells and even, in some examples, the cell product. These methods were pioneered by Gori (*22*), who developed the dialysis fermenter as a simpler, and more easily scaled-up system, to the cytogenerator. Dialysis membrane was chosen because other filter materials (e.g. sintered glass) encouraged cell growth on their surfaces and this led to severe blockage problems. The apparatus was used as a chemostat with the objective of maintaining maximum growth rates for long periods of time (over 40 days) rather than supporting high cell densities. However, a modern equivalent is available (the Bioengineering Membrane Laboratory fermenter) (*113*), which has a Cuprophan dialysis membrane (MW cutoff 10 000 daltons) forming an inner chamber. The reactor tank has two independent stirrer shafts, one inside the membrane and the other in the exterior reaction chamber. The unit was designed principally for microbial use, but can be adapted in a variety of ways for animal cells.

Hollow fibres are available in both ultrafiltration and filtration grades and, in conjunction with conventional stirred reactors, can be utilized for perfusion. Two approaches, illustrated in Fig. 6, have been used. In the first, the cells, medium and product are pumped through an external loop to the fibre

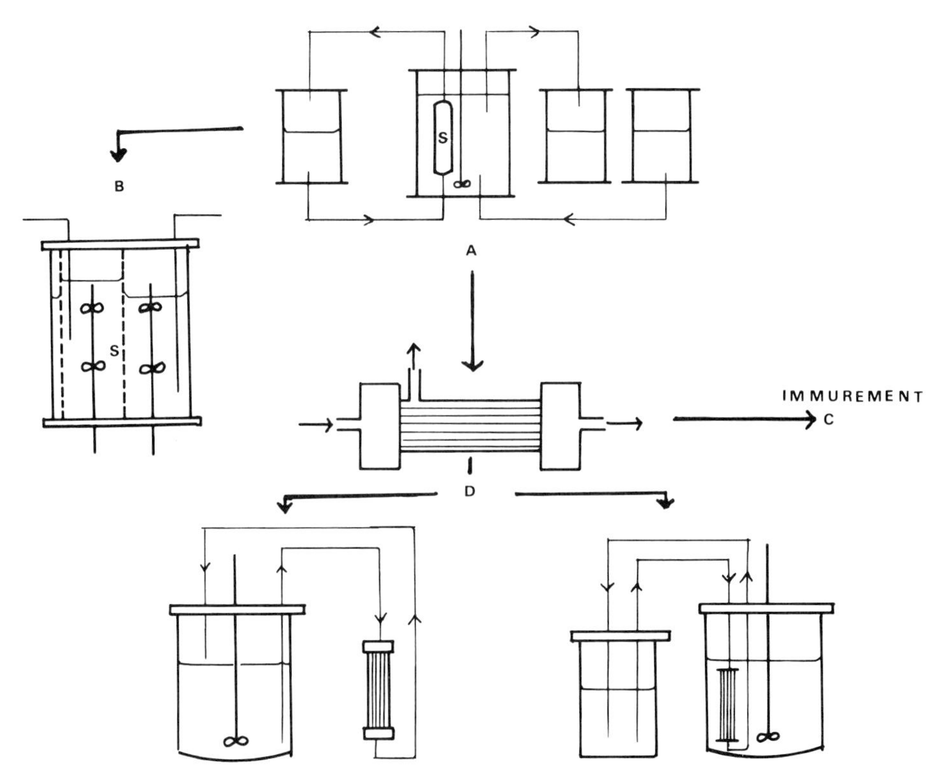

Fig. 6. Culture systems based on filtration membrane principles. A, dialysis fermenter of Gori (*22*); B, membrane laboratory fermenter (*113*); C, immurement techniques summarized in Fig. 7; D, hollow-fibre units used in an external loop (*109*) or internal filter (*13*); S, dialysis chamber.

cartridge and the cells are returned to the reactor (*109*). The second method is to perfuse medium through fibres situated in the reactor, i.e. an alternative to spin filters (*13*). The external loop method has the advantage that the product can be continuously harvested in a concentrated form and the cell density increases within the reactor as they are returned. The disadvantages are potential damage to cells as they are pumped around the loop and the loss of environmental control in the vinyl tubing during the cells' residence time in the loop. The internal system consists of a fibre bundle (MW 6000) only with the cartridge removed and has a greater potential for maintaining high cell densities owing to the ability to rapidly circulate fresh medium. The product has to be harvested in a more dilute concentration with the cells and medium and is thus a semi-batch operation. Both systems are capable of supporting 1–5 $\times 10^7$ cells ml^{-1} in the bioreactor and can be a simpler design solution than many other reactor configurations, including the spin filter.

2.4.3. Continuous-Flow (Chemostat) Culture

This methodology was comprehensively reviewed in a previous volume (*97*) and will not be described in detail here. In the context of a production process, it is only useful for growing (dividing) cells and thus has a limited application unless used as a two-step process. An example is monoclonal antibody production from hybridomas which is not growth-associated. It is a method which provides invaluable research and development data on growth and production kinetics, but has not found a ready application for the manufacture of cell products. One reason is that it is a low-density system providing a continuous trickle of cells and a low-titre product which is far from ideal for downstream processing. Batch culture is a popular concept as a manufacturing process because each batch is a definable entity for quality assurance purposes, initiating from a characterized seed lot. The developmental logic from batch is to a fed-batch, recirculation or perfusion system which allows higher cell densities and product concentration. Continuous-flow culture is only suitable for suspension cells in which the product is growth-dependent, it is low-yielding, and needs an extra downstream processing step to remove the cells. Also, continuous cell division could result in changes to the biological properties of cells and there are always difficulties with animal cells in precisely defining the growth-limiting factor.

2.5. Cell Immobilization

Density scale-up using spin filters, etc., is technically limited to a 10–50-fold increase over non-perfused cultures. To achieve higher cell densities, various immobilization techniques have been developed in order to achieve more efficient and critical control of the cell environment by means of medium perfusion. Basically, two concepts are being used. Firstly, immurement, or the retention of cells within a compartment which allows free passage of medium. Secondly, the entrapment approach, which provides a substrate with the physical configuration to capture and trap cells. Reliance is also placed upon suspension cells weakly attaching to or adsorbing onto the surface of the entrapment matrix. Anchorage-dependent cells benefit from the greatly increased surface area of such materials. Many immobilization materials protect cells against mechanical stresses caused by stirring and high perfusion rates. Although primarily aimed at providing independent high-density systems, they can, in addition, improve the efficiency of spin filter techniques as the support material is so much larger than individual cells.

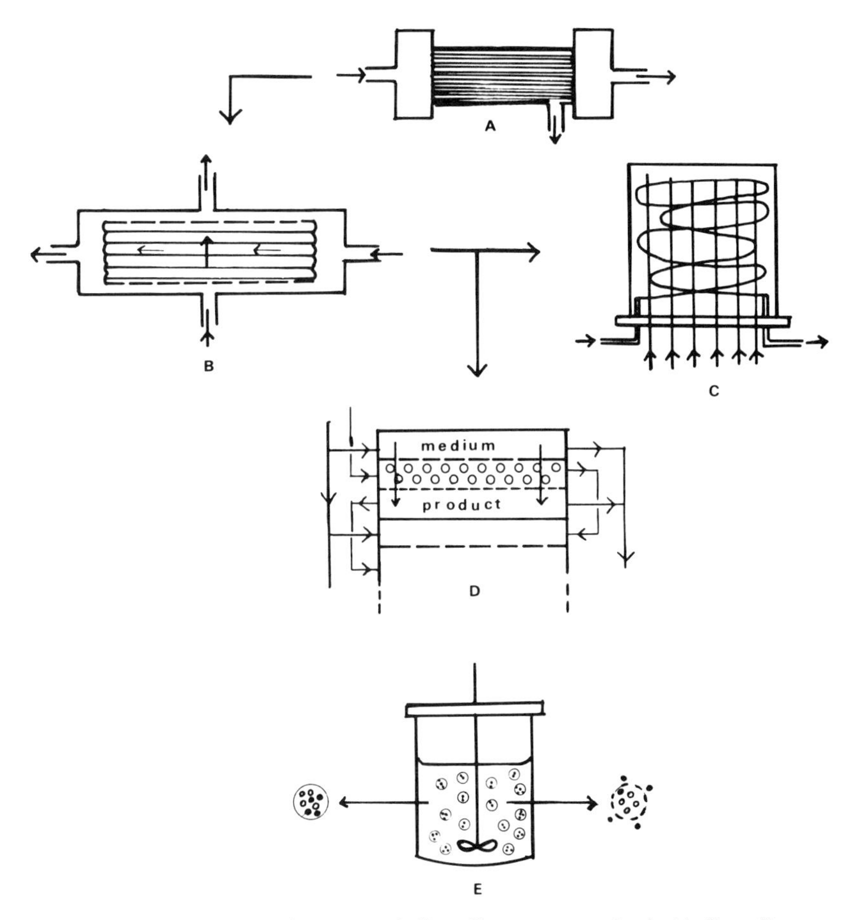

Fig. 7. Cell immurement techniques. A, hollow-fibre reactor; B, flat-bed hollow-fibre reactor (*54*); C, static maintenance reactor (*15*); D, membrane reactor (*48*); E, encapsulation (*74*).

2.5.1. Immurement Techniques (Fig. 7)

Modified Filtration Techniques. An efficient system for perfusion of high cell densities was developed by Knazek (*49*) in 1972 using ultrafiltration capillary fibres (cellulose acetate for medium diffusion and silicone polycarbonate for gas diffusion). These original cultures (Amicon Vitafiber) were available in 25, 250, and 2500 cm^2 sizes with extracapillary volumes of 2.5, 25, and 250 ml respectively. The fibres had a spongy wall (50–70 μm thick) and a central lumen of 200 μm diameter. The external surface was very porous, but the lumen lining was a thin ultrafiltration skin with either 1000, 5000, or

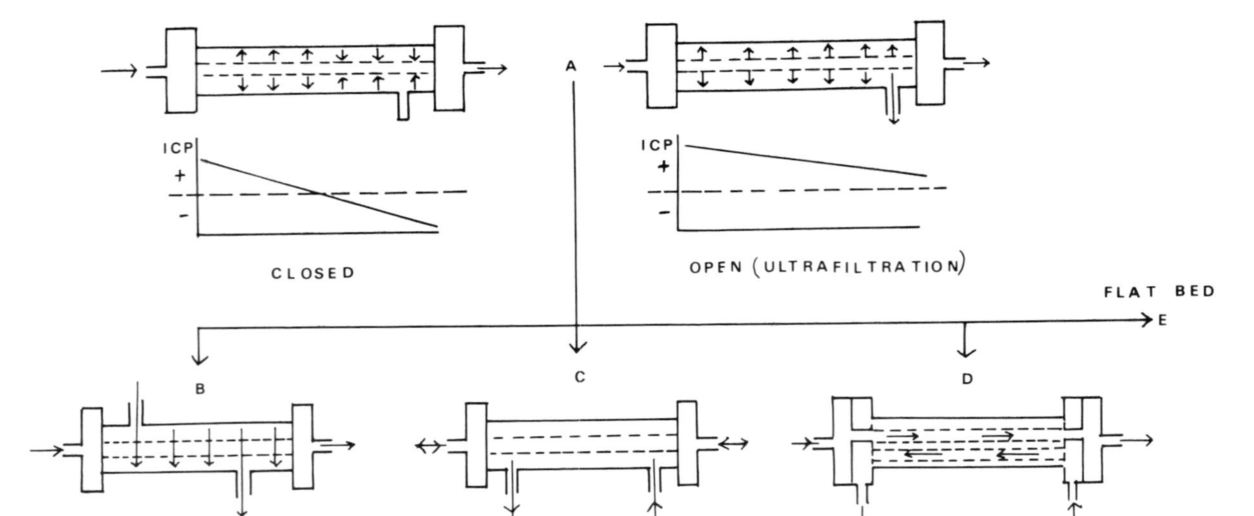

Fig. 8. Homogeneity and scale-up problems of hollow-fibre reactors, and possible solutions. A, flow path patterns in open (ultrafiltration) and closed systems and the pressure differential between inside (ICP) and outside the fibres which results in uneven growth (*91*); B, crossflow diffusion of nutrients, with gas flow through the fibres; C, cyclical pressure pulses between the inside and outside of the fibres (*115*); D, contraflow fibres of nutrients and oxygen; E, flat-bed reactor (illustrated in Fig. 7), an example of crossflow diffusion.

10 000 MW cut-offs. A unit consisted of thousands of fibres "potted" at either end in a cylindrical housing and was capable of supporting 1–10 × 10^6 cells cm^{-2} (10^8 cells ml^{-1}). This concept has been developed over the years and is now very widely used in many commercial applications (see Chapters 15 and 16), especially for producing monoclonal antibodies from hybridoma (suspension) cells. There is no questioning the efficiency of these capillary fibre cultures, in which over 10^8 cells ml^{-1} can be maintained for long periods (months) within the extracapillary compartment. However, there is a scale-up problem in that pressure and concentration gradients build up along the length of the cartridge (Fig. 8). This phenomenon has been explained in detail (*91*), and is due to transmembrane flux being directed back into the lumen for the latter half of the flowpath in the closed (re-cycle) mode (i.e. when medium passes through the lumen only). This results in much higher cell densities at the inlet end compared to the outlet or, with suspension cells, a packed cell mass accumulating at the outlet end. In the ultrafiltration mode (if the extracapillary volume is continuously being withdrawn) the pressure drops continuously along the length of the cartridge, again resulting in uneven growth. Both configurations are therefore severely limited for scaling-up. This particularly applies to reactors using ultrafiltration fibres, chosen because of their ability to compartmentalize the cell and product away from the bulk of the re-circulating medium. Some systems are based on filtration-grade membranes (0.1–0.22 μm) in order to achieve a greater flux rate (*10, 92*), i.e. only the cells not the product are retained in the extracapillary space. Scale-up is mainly limited to providing a series of replicate units, although attempts to fabricate larger units are being made. Methods of overcoming this gradient problem include the insertion of contraflow and aeration fibres (*92*), a cyclical flushing of medium either from the lumen into the extracapillary space or into the lumen from the extracapillary space (*115*), and the use of flat-bed reactors (*96, 16, 54*) rather than the cylindrical filtration cartridges (Fig. 8).

Tharakan and Chau (*92, 93*) have designed a unit that combines the high surface-area-to-volume ratio of hollow fibres with the efficient feed delivery of a radial flow reactor. This culture consists of a central perforated stainless-steel radial flow distributor, which gives a uniform delivery of nutrients along the total length of the fibres, and an annular bed of hollow fibres through which the gas mixture flows. Medium diffuses into these fibres and is carried away by the gas flow. This arrangement permits considerable scale-up whilst maintaining homogeneous conditions.

The concept of using pressure differentials to simulate *in vivo* arterial and venous flow is used in the Acusyst (*115*) system. The culture has a dual medium circuit, one passing through the lumen of the fibres, the other through the extracapillary space. By cyclically alternating the pressure between the two circuits, media is made to pass either into or out of the lumen. This allows a flushing of media through the cell compartment and overcomes to some extent the gradient problem. It also allows extra concentration of the product just before harvesting. The system is described fully in Chapter 16.

Another means of overcoming the concentration gradient limitation to scale-up is to use a flat-bed hollow fibre perfusion reactor (*16*). The fibres are used for aeration, and as an attachment surface for the cells; 3–6 layers of fibres are sandwiched between steel microphore filter plates (20 μm pores), between which the medium is perfused. Densities of 10^6 cells cm^2 have been maintained for periods of 60 days. The flat-bed configuration brought about a 4-fold increase in cell yield over the cartridge-type owing to the short nutrient flow path preventing any gradients or the need to use fast shearing flow rates. It provides a very good basis for scale-up and in fact it has allowed the development of the InVitron Static Maintenance Reactor (SMR) (Chapter 17) which uses silicone tubing for gas exchange and fibres for nutrients (*96*). The spacing of the fibres (5 mm) seems somewhat excessive for optimum conditions, but the reactor is for cell maintenance only—cells are grown to high densities in other cultures (either suspension or microcarrier), and mixed with the matrix material in the SMR. A similar concept, but using just the membrane sandwich without fibres, is the Membroferm Bioreactor (*48*) (see Chapter 11). In this system, membranes of different porosities can be used, so that one can have consecutive arrangements or combinations of medium, cell, and product compartments.

Hollow-fibre culture has mainly been used for suspension cells, particularly hybridomas, with fibres based on cellulose acetate. It is now becoming increasingly possible to grow anchorage-dependent cells in the same system using fibres based on polypropylene. A 1-litre cartridge, which supports over 10^{10} suspension cells, has a surface area of approximately 1 m^2. The normal expectation is a yield of $1\text{–}2 \times 10^5$ cells cm^{-2}, thus a cartridge would be expected to support $1\text{–}2 \times 10^9$ cells but it is claimed that the yield would be higher owing to the process control facilities of this type of equipment. Quantitative data on cell yields are difficult to determine with any degree of accuracy in this type of system, but figures of $1\text{–}2 \times 10^6$ cm^{-2} have been confirmed (*54*).

There are several means of overcoming the problem of poor attachment to fibres. One solution is to coat polysulfone 3S1000 fibres (Amicon) with poly-D-lysine which results in yields of over 10^7 fibroblast cells per cm^2 (*92*). Another approach is to pack the extracapillary space with microcarriers (*90*). Both polyacrylamide and dextran microcarriers have been successfully used.

Besides the scale-up problem, a serious disadvantage of filtration fibres is that many types cannot be steam-sterilized and ethylene oxide sterilization has to be used followed by careful aseptic assembly of the whole unit.

Polysulfone and Teflon fibres can be steam-sterilized. Also, many of the fibres have toxic components which have to be leached out before cells are added.

To summarize, hollow-fibre systems have large surface-area-to-volume ratios, continuous removal of waste products, high cell densities which allow a tissue architecture to be established, separation of the cells from the nutrient flow, thus eliminating the effects of shear, etc., and the possibility of concentrating the product in a small volume. Disadvantages are diffusional limitations, especially for homogeneity and scale-up, process control complexity, sterilization method, possibility of fibre blockage and membrane leakage, and sampling difficulties.

Microencapsulation. A totally different approach, but using similar principles, is the microencapsulation technique, originally developed by Mosbach and co-workers (*70*) and Lim and Sun (*60*), and currently being used for the commercial production of monoclonal antibodies (*12, 80*). There are basically three different methodologies currently in use. The first, Encapcel (*12, 80*) (Damon Biotechnology), traps the cells in sodium alginate spheres, which are then coated with polylysine to form a semi-permeable membrane. The internal gel is then solubilized with sodium citrate, which releases the cells into free suspension within the capsule. Cells grow within the sphere from a seed of 2×10^6 to a final intracapsular concentration of 500×10^6 ml^{-1} after several weeks. The second method, Geltrap (Karyon technology), is a simplified version of the preceding method using calcium alginate only, which allows products such as antibody to diffuse into the culture medium. A problem with this method is that spheres tend to be rather large (0.5–1 mm), which causes severe nutrient limitation in the centre. The third method uses agarose beads in which the cells are within a honeycombed matrix within the gel (*71, 73*). The agarose fragments, unfortunately, have a wide size distribution and a lower mechanical strength compared to alginate (*74*). Other materials that have been used are collagen and fibrin, mainly to promote encapsulation systems for anchorage-dependent cells.

The advantages of these techniques are that the cells and (with a suitable capsule membrane such as the Encapcel) product are compartmentalized into small (500 μm) capsules which can be stirred in conventional fermenter equipment. Thus volumetric scale-up possibilities are far greater with this technology than with the fibre systems. The fragile cells are protected from mechanical stresses (thus higher stirring speeds can be used) and can be kept in serum-free medium (serum is in the extracapsular medium). In the Encapcel method cell densities of over 10^8 ml^{-1} are achieved and a high product concentration accumulates within the capsule owing to the molecular cut-off of 100 000. However, the system is restricted to a batch operation, as at the end of the culture period (10–15 days) the capsules are allowed to settle out and the concentrate of capsules is lysed to release the product. To date, the method appears to be restricted to below the 100-litre scale and cell viability within the capsule is difficult to maintain. It is also a highly sophisticated and complicated procedure, and difficulties will be met with products that cause negative feedback problems to the cell. The other encapsulation technologies do offer the prospect of longer culture duration, as the product can diffuse out into the bulk culture fluid, and also since agarose is more stable than many other polysaccharide gels in culture media. However, the method is still considered at best a pilot-scale technology and not suitable for industrial scale-up, presumably because of the logistical and technical complexities.

A recent commercial development is the Bellco Bioreactor (*14*) which is a conical design and relies on airlift to circulate cells in, or on, dense slurries of hydrogel (alginate) beads. The beads are generated within the reactor and the 3.5-litre working volume is claimed to have 200 $cm^2\ ml^{-1}$ surface area.

Product Compartmentalization. The ability of some of these various immured culture systems to allow a product to become concentrated within a small compartment of the total medium volume is summarized in Table VII. Between 1.2% and 2% of the total volume (culture plus reservoir) can be isolated for product accumulation and this can be very advantageous for downstream processing.

TABLE VII Compartmentalization of Cells and Product Within the Bulk Medium

	Ratio		
Method	Product volume	Reactor medium volume	Total medium volume
Encapsulation	1	3	80
Membroferm	1	3	50
Hollow fibres	1	2	80

2.5.2. Entrapment (Fig. 9)

A methodology that is being increasingly used is entrapment. Inertial entrapment systems allow a fluid flow over a textured, or partitioned, surface which traps the cells and prevents them being washed through the unit. Examples are the tubular flow reactor (*46*) (Chapter 9) and the Opticell System (*4*) using the S-Opticore cartridges (Chapter 14). The claim is that over 95% of the cells are retained in the ceramic surface architecture, but some reliance is placed on the suspension cells "weakly" attaching to the surface. A

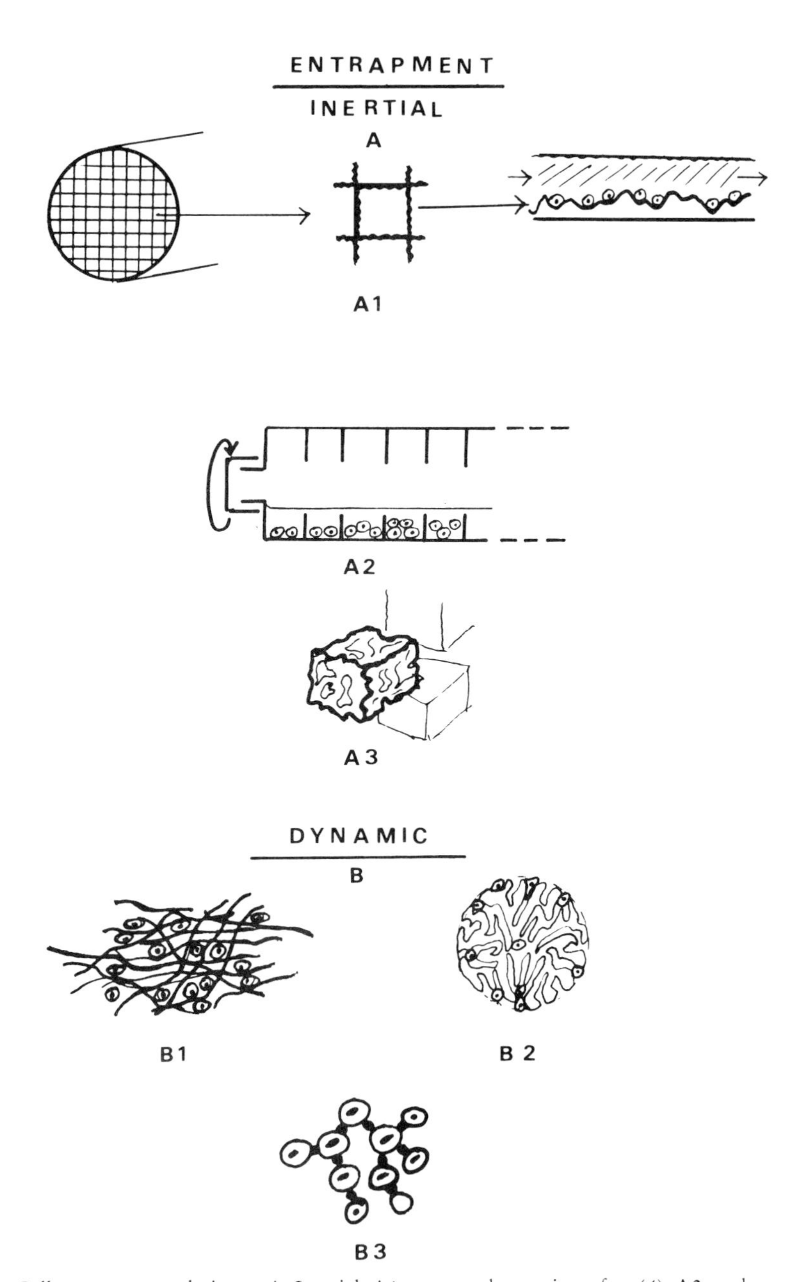

Fig. 9. Cell entrapment techniques. A, Inertial: A1—textured ceramic surface (*4*); A2—tubular flow reactor (*46*); A3—polyurethane sponge bed (*57*). B, Dynamic: B1—fibres (*62*); B2—matrix (macroporous) beads (*42, 72*); B3—electrostatic aggregates formed with Biocryl BPA particles (*122*).

recent development is the use of polyurethane sponges, in the form of 0.5–1 cm^3 cubes, in a packed column (*57*). The medium is free to circulate around the cubes and the porous matrix of the sponge entraps the cells as the medium is continuously circulated. Dynamic entrapment methods are also available based on clumps of (cellulose) fibres which enmesh cells in a stirred culture (*56, 62*) or a matrix-type microcarrier bead (*42*), or simply entrap them in a gell sphere (*72*).

An advantage of this type of culture is that volumetric scale-up possibilities are far greater than with immured systems, because of the lower pressure gradients across the bed. Thus, faster perfusion rates, with less sophistication of design, can be used which allow far larger volumes to be used before nutrient concentration gradients limit the scale-up. In addition, most of the substrates can be steam-sterilized.

2.6. Cell Substrates

Some cells need to be attached to a suitable surface either because they cannot grow in free suspension (anchorage-dependent cells, e.g. human diploid fibroblasts), or because expression of cell products are significantly lower, or altered (*85*), when cells are grown in suspension culture. Despite the unchallenged fact that cell production is more economical, and scale-up is more efficient, in suspension culture, substrate-attached cells still have a very important role in cell biotechnology.

Traditionally, scale-up of substrate-attached cells has been to increase the number of small culture units (flasks, roller bottles, etc.). This procedure is not only wasteful in labour but requires large volumes of working and incubation space. In addition, critical environmental control procedures which improve the productivity and reproducibility of cultures cannot be employed. The aim, therefore, is to scale-up by changing from a multiple to a unit process. This requires the provision of a large surface area in a compact volume in which media can flow through at a rate sufficient to keep the cells at optimum performance levels. A useful measure of the relative efficiency of various culture reactors in this respect is to compare them in terms of ratio of surface area to reactor medium volume. This parameter has been used to summarize anchorage-dependent culture systems in Fig. 10. Many of these methods have already been reviewed conceptually (Volume 1, Chapters 7–11) and only those currently being developed, or used commercially, are considered here. Basically, scale-up methods are based on either plate, immobilized beds, fluidized beds or membrane (cartridge) concepts.

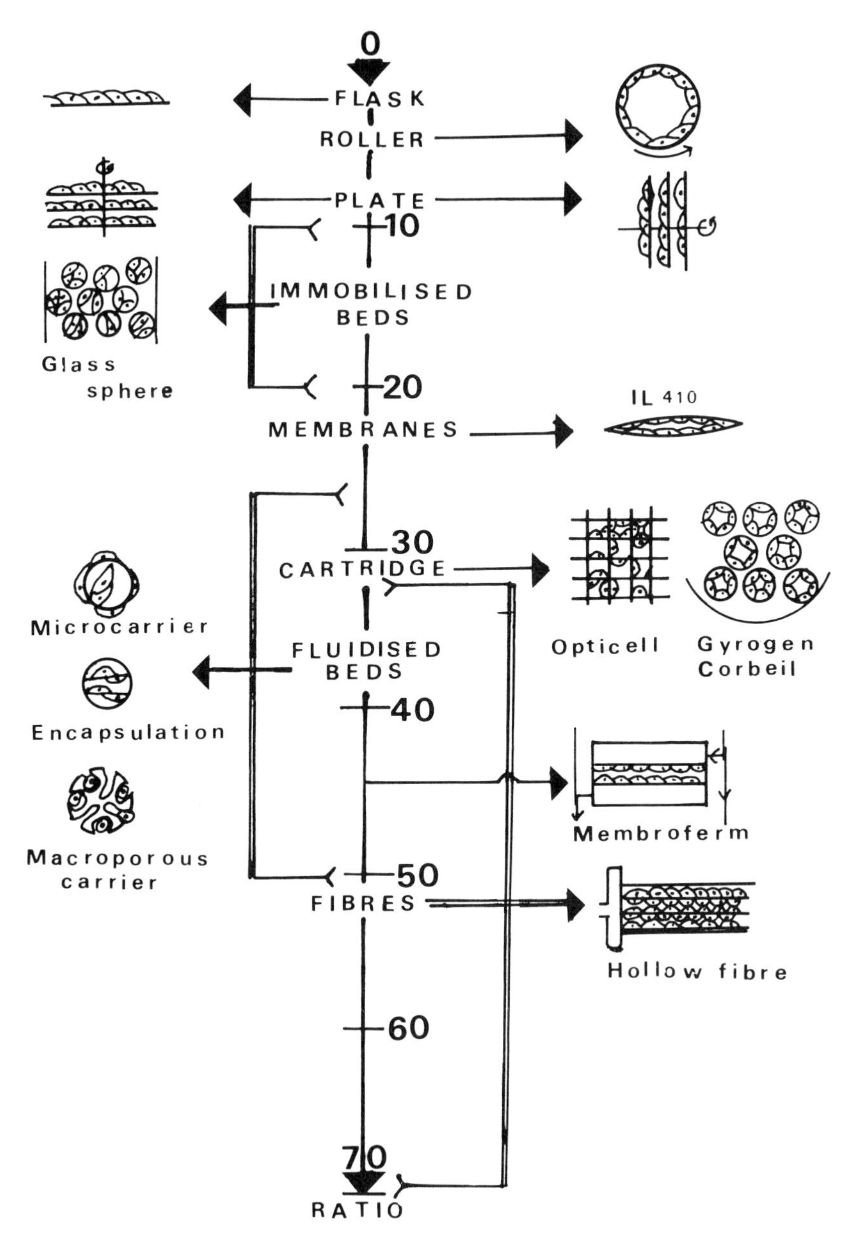

Fig. 10. Density scale-up of animal cells on surfaces showing the approximate surface-area-to-void-medium-volume ratios.

2.6.1. *Plate Reactors*

The inclusion of stacks of parallel plates in either a horizontal or vertical mode is a simplistic way of increasing the surface area within a reactor. Many systems have been developed and used at scales up to 200–300 litres (*67, 83*). However, they all have serious limitations. Vertical plates are difficult to inoculate evenly and, when confluent, the cell sheets have a tendency to slide off. Horizontal plates need a tilting mechanism to remove medium from between plates, only half the surface area can be used for cell growth, and there can be no variation in the surface-area-to-medium-volume ratio. However, high cell densities (1×10^6 cm^{-2}) have been obtained in glass plate units of 10 m^2 (*65*). This particular system has been used for the manufacture of measles vaccine.

One type of plate reactor which was thought to overcome many of these problems is the plate heat exchanger (Fig. 11) (*7, 8, 33, 35, 110*). This apparatus consists of a frame in which independent metal plates, supported on rails, are clamped between a head and follower. The stainless-steel plates (type 316) are sealed at their outer edges and around the ports by gaskets which are arranged so that the process fluid and the warming fluid are directed alternately into the passages formed by the plate. Although the plates are vertically stacked, they are ribbed in a herring-bone pattern, which means that most of the surface area is at 45°. However, cell attachment and growth in this apparatus is uneven except for cells which have a very rapid attachment time, or fibroblasts which are highly motile and can thus compensate for attachment gradients; i.e. the system is more suited to fibroblast than epithelial cells. Such cells grow to densities compatible with other substrates ($8–10 \times 10^4$ cm^{-2}). Nevertheless, the linear scale-up capacity of the system is enormous, as a rig can hold 100 plates available in surface areas from 850 cm^2 to 3.25 m^2. With suitable modifications it has the potential to be an efficient large-scale, although only low-intensity, culture unit.

2.6.2. *Immobilized Beds*

Many different substrates have been used to form an immobilized bed suitable for animal cells (for review see ref. (*87*)). The most-used substrate is the glass sphere, despite the fact that a sphere has the lowest surface area to volume ratio possible. An immobilized bed matrix has to be a compromise between maximizing the surface area and providing a passageway open enough for medium to be perfused at a rate sufficient to maintain homogeneity throughout the bed without subjecting the cells to damaging shear effects. Recent data (*63*) have demonstrated that a pore size of over 5 mm diameter is needed for maximum growth. The data in Table VIII show that using differently sized glass spheres in the diameter range 2–8 mm, the total cell yield

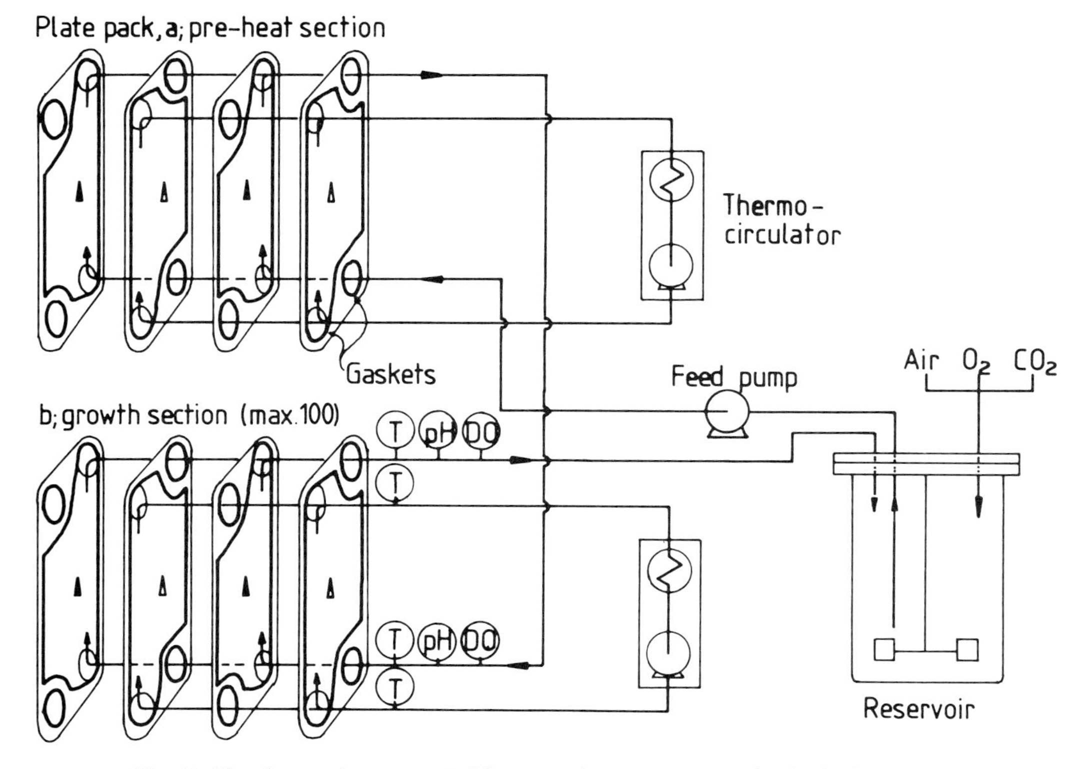

Fig. 11. Plate heat exchanger (*110*). Diagrammatic representation of individual plates in the culture and a flow diagram of the culture in operation (T, temperature sensors; pH, DO, pH and oxygen sensors).

TABLE VIII **Effect of Glass Sphere Diameter on Cell Yields**

Bead diameter (mm)	Cell number per		
	10^5 cm^2	10^5 kg beads	10^6 ml
2	0.36	6.7	2.7
3	0.78	4.0	1.6
4	1.25	5.0	2.0
5	2.50	8.0	3.0
6	2.20	6.0	2.4
7	2.40	5.0	2.0
8	2.30	4.3	1.7

was higher in beds of the larger spheres despite the reduction in total surface area. The interpretation is that the larger channel size overcame the problems of inhomogeneity and channelling (preferential flow through selected channels). To reduce inhomogeneity problems, the glass spheres could be coated with Fibronectin or the direction of medium flow could be alternated every 24 hr. In large beds, cells should be inoculated through a perforated tube running centrally the height of the bed to obtain an even distribution of cells (*63*).

Immobilized-bed cultures based on glass spheres have many advantages, in that the substrate is both inexpensive and re-usable, and scale-up to at least 100 litres is relatively problem-free (*104*). The disadvantages are that it is not a high-intensity system and can only support low cell densities per unit area; and it is limited to products that are secreted, i.e. it is ideal for long-term continuous cultures (see Chapter 10) but not for batch or cultures in which cells have to be harvested or the product extracted.

Several commercial systems based on packed cylinders of glass tubing have been developed. The simplest is the Corbeil which is used as a high-surface-area roller bottle (15 000 cm^2) for the manufacture of vaccines (*11*). The Gyrogen monolayer tissue culture fermenter (Chemap) was also developed for vaccine (FMDV) manufacture but is far more complex, with full process control capability (*21*). The largest unit provided 34 m^2 of surface area, but the system never became popular, mainly because it was complicated to use and to standardize, was not very versatile, and was expensive for the degree of process intensity and control sophistication it offered.

2.6.3. Fluidized Beds

Microcarrier Culture. Microcarrier culture, which can be considered as a dilute form of fluidized bed, is one of the few systems that has gained some acceptance as a commercially viable scale-up culture process. It is being used by several companies at scales exceeding 1000 litres for vaccine manufacture (*68*). The method has been greatly refined since van Wezel (*99*) published his preliminary results in 1967. Problems of unsuitable surfaces for attachment, even carrier toxicity, have been overcome and currently there are over 20 types of microcarrier commercially available (see Chapter 12). The advantage of being able to monitor and control the environment and achieve true homogeneity has allowed considerable scale-up in both volume and process intensity (to 5×10^7 cells ml^{-1}). Cell density has been increased by using high concentrations of microcarrier (e.g. 15 g l^{-1}) and maintaining the correct environment by medium perfusion. This has largely been made possible by the development of efficient spin filter systems for aeration (*88, 105*), and perfusion (*101, 28*), or both (*32, 34, 35, 117*). The ability to use fast perfusion rates (up to 2 volumes hr^{-1}) is extremely important for both nutrient supply and oxygenation (Table VI) (*32*).

Improvements in the mixing efficiency of specially designed flat-blade impellers which allow slow stirring rates to be used, and are thus potentially less damaging to cells, means that the possibility of microcarrier aggregation has to be monitored. This aggregation is usually a deleterious event, and the stirring speed should be increased above that necessary to maintain homogeneity in order to prevent it occurring.

Many efforts have been made to effect cell transfer from one bead to another within the culture (*19, 39, 61*). This is to facilitate the progressive scale-up of the cell seed. However, success is very limited and usually results in a very uneven distribution of cells between the beads. Although more costly, a good means of preparing a large cell inoculum in excellent physiological condition is to grow seed cultures on gelatin microcarriers. These are very easy and quick to dissolve, and leave the cells completely undamaged. This can even be done within the same reactor vessel in which the next higher volume of culture is to be grown. It also solves the perennial scale-up problem of how to prepare a large cell inoculum (e.g. for a 100-litre culture) quickly before trypsination damage occurs.

The flexible process capabilities of microcarrier culture is exemplified in Fig. 12 (*34*). This is a 100-litre volume culture vessel with a spin filter that allows a microcarrier (Cytodex 3) concentration of 12 g l^{-1} to be used. Cells are inoculated when the medium level is low (L2) and when attachment is complete this is raised to 100 litres (L1). The culture is perfused from the 30-litre reservoir vessel (medium at L3 mark), in which all environmental control procedures are carried out, at a rate of 2 volumes hr^{-1}. When the growth phase is complete, the medium levels can be reduced to the L2 and L4 marks (Fig. 12) so that perfusion can continue and a much higher product concentration can be achieved—in effect, the cell concentration is increased

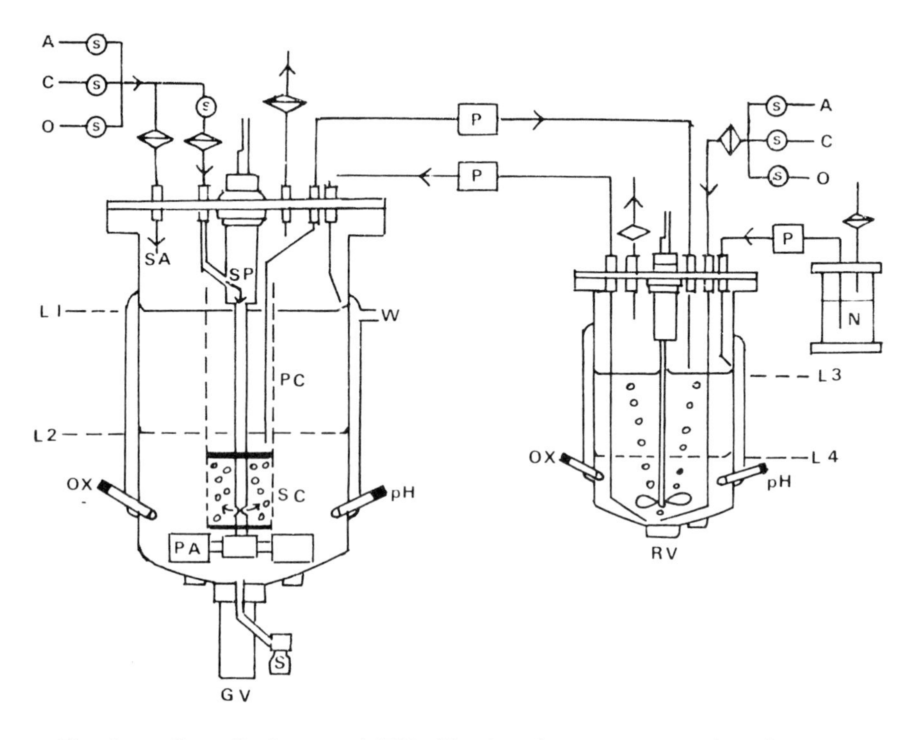

Fig. 12. 100-litre cell culture vessel (GV) with spin exchanger (*34, 117*) and a 30-litre reservoir (RV) in a complete perfusion culture system for high-density microcarrier culture. A, air; C, carbon dioxide; O, oxygen; N, NaOH; P, pump; SA, surface aeration; SP, sparging; L2, inoculation level; L1 and L3, growth levels; L4, product manufacture level; S, sampling; Pa, stirrer paddle; W, water jacket.

from approximately 2×10^7 to 4×10^7 ml^{-1}. A complete change of medium (e.g. to serum-free) can also be easily carried out between the growth and production stages by allowing the microcarriers to settle out.

Matrix (Macroporous) Carriers. It can be postulated that a microcarrier concentration of 15 g l^{-1} (27.5% of the total culture volume) may be the upper limit of a stirred system, owing to mechanical factors such as bead collision. There are two ways in which the method can be developed further, namely to provide a "matrix (porous) bead" which relies on protected surfaces within the bulk of the sphere, or to increase carrier density using fluid-lift, rather than stirred, mixing principles.

The use of matrix beads in a true fluidized bed in which 80% of the culture volume is the bead is described in Chapter 13 (*42*). In this system (CF-IMMO,

Verax Corp.) a density of 4×10^7 cells ml^{-1} of reactor volume, and $2–3 \times 10^8$ cells ml^{-1} inside the reinforced collagen sponge bead matrix, is run on a continuous basis. The microbeads are 200–600 μm in diameter with internal interconnecting pores of 30–100 μm, and are cultured as a thick slurry. Typically, about 6×10^6 beads per litre are used, giving a total surface area of 300 $m^2\ l^{-1}$. This system is suitable for both suspension and anchorage-dependent cells and the three-dimensional morphology of the matrix can be altered to meet the requirements of the different cell types (*42*).

A recent paper (*72*) has described the fabrication and potential use of macroporous gelatin beads. These gelatin spheres have large cavities and are capable of supporting double the number of similarly sized solid spheres. As well as affording mechanical protection, these cavities also allow the cells to develop and maintain a suitable microenvironment. The potential of fluidized-bed systems is being recognized by the appearance of several commercial reactors (*14*, *113*).

The many natural advantages of microcarriers over other competing systems (Table II), coupled with the developments discussed above, should make this technology one of the dominant commercial systems of the 1990s, coupling as it does volumetric and density scale-up. It is a method of great versatility with a huge range of surface configurations, the ability to be used in a wide range of reactor types from roller culture to airlift (if antifoam is added), in fibre cartridges (*90*), and in three-dimensional rotary-swivelling reactors (*114*). One disadvantage for some users is the high cost of the substrate if a reusable type is employed. However, the high cost of regenerating a substrate plus the high value of most cell products actually make substrate costs a fairly low proportion of the total costs.

2.6.4. Membrane Systems

Membranes were initially used for the growth of cells because they not only provided a large surface area but were a convenient means of aerating the culture. Early examples are the polymeric fluorocarbon bags (*69*) and the IL410 reactor (*41*) and the concept has now been utilized for the growth of suspension cells in specially developed plastic bags (1 litre) that allow rapid gas exchange (*50*). Cells preferentially absorb nutrients from their peripheral membranes and from the attachment surface; thus growth on permeable membranes would seem very advantageous. However, despite these cultures being based on such sound principles, they have not been used very extensively outside their laboratory of origin, possibly because they are over-elaborate in design. Owing to the present need to support high cell densities, membrane systems are now getting far more attention. The main examples, hollow-fibre units (Section 2.4.2) and the Membroferm (*48*), have been mainly used for suspension cells but have subsequently been modified for

TABLE IX Scale-up Factors—A Comparison of Different Culture Systems

(A) Anchorage-dependent cells

Factors	Roller bottle	Plate	Immobilized bed	Fluidized bed		Cartridge		Membrane
				Microcarrier	Matrix	Ceramic	Fibre	
Scale-up potential	0	2	2	3	3	2	1	1
Process simplicity	0	2	2	1	1	2	1	1
Mass transfer efficiency	3	1	2	3	2	2	3	3
Aseptic operation	1	3	2	3	3	3	1	1
Direct monitoring	3	0	0	3	2	0	0	0
High cell density	0	0	1	2	3	1	3	3
Steam sterilizablility	(Y)	Y	Y	Y	Y	Y	N	Y
Downstream compatibility	0	1	2	2	3	2	3	3
Re-utilizable substrate	(Y)	Y	(Y)	N	N	(Y)	N	N
Flexible area : volume ratio	3	0	1	3	3	1	3	2
Homogeneity/mixing	1	2	2	3	2	1	1	3
Critical control	0	1	2	2	2	2	3	2
Continuous process	0	1	3	3	3	3	3	3

(B) Suspension cells

Factors	Airlift	Stirred	Spin filter	Membrane	Ceramic	Fibre	Encapsulation	Gel Entrapment
Scale-up potential	3	3	3	2	1	1	1	2
Process simplicity	3	2	2	1	2	1	0	1
Mass transfer efficiency	3	2	3	3	2	3	1	2
Aseptic operation	3	3	3	1	3	1	1	1
Direct monitoring	3	3	3	1	0	0	1	0
High cell density	1	1	2	3	2	3	3	3
Steam sterilizablility	Y	Y	Y	(N)	Y	N	(Y)	(Y)
Downstream compatibility	1	1	2	3	2	3	3	1
Homogeneity/mixing	3	3	3	2	2	1	1	2
Critical control	2	2	2	2	2	3	1	2
Continuous process	1	1	2	3	2	3	0	2

Key: 0—not possible; 1–3—increasing ease or efficiency; Y—yes; N—no; ( )—with some alternatives.

substrate-attached cells. Of the range of fibre materials available, polysulfonate has been found to be most suitable. Attachment efficiency is increased by coating with poly-D-lysine (*92*).

If the membranes were totally covered by cells, a problem would arise of getting sufficient diffusion of nutrients into the cell compartment; thus in practice a lower-than-average cell yield is to be expected (possibly only 20% of the theoretical maximum). The cyclical flushing process between the lumen and extracapillary space used in one system (*115*) partially offsets this problem.

Membrane-type culture units have a great theoretical potential since they allow a maximization of surface area with the means of feeding the cells without having to rely on fast, shearing, medium flow rates. In practical terms such culture units only represent small-scale systems (below 1 m^2) at the moment, but there is a lot of activity in fabricating units which use various combinations of membranes, fibres and ceramics in order to scale up the size.

3. CONCLUSIONS

There is still a huge scope for improving reactors further to give greater versatility and scale-up capability. With scale-up the potential loss if a culture succumbs to contamination or system failure in terms of labour, materials, and time becomes increasingly costly. Thus, a keyword in design must be simplicity, since complexity of equipment design, media, support systems, and operational manipulations all lead to contamination risks. The parameters one looks for in a large unit process are listed in Table IX. Obviously, no one culture can feature all of these characteristics and a choice has to be made depending upon what is considered most important for a particular process. In order to compare the systems described in this review, each one is scored for each parameter in Table IX. These values should be taken purely as a guide and not as absolute values. There are several reasons for this, including personal preferences or interpretations, and also the fact that modifications which individuals have made (many of which have been described in this chapter) in order to reduce a problem, or to increase performance, have not been allowed for. The data are intended as a quick means of comparing different technologies and allowing a preliminary screening of which one might be the most suitable for a particular application.

Reactors should not, of course, be thought of in isolation, or as the key to improving productivity of animal cells. A process is as strong as its weakest component and therefore supporting equipment should not be ignored: items such as valves, gauges, sampling devices, attachment surfaces, etc. Probably the biggest increases in productivity will come as a result of studying cell physiology in more depth, and from media development. However, for this knowledge to be properly used we are largely dependent on the development of reliable and new biosensors (e.g. for glutamine, ammonia, lactate, and glucose). Computer technology is advanced enough to take advantage of such developments to give precise on-line monitoring and process control.

In conclusion, the ideal of a large-volume dense-culture system is tantalizingly close but needs technological developments in several related areas to bring it about. In the following chapters there is a chance not only to select a system which comes nearest to fulfilling one's present requirements, but also to try to decide which of the current culture systems or concepts will become the dominant technology.

REFERENCES

1. Arathoon, W. R., and Birch, J. (1986). Large-scale culture in biotechnology. *Science* **232**, 1390–1395.
2. Barteling, S. J. (1984). The skull fermenter. *Dev. Biol. Stand.* **55**, 143–147.
3. Bell, G. H., Davidson, J. N., and Scarborough, H. (1956). "Textbook of Physiology and Biochemistry" (3rd edn), p. 528. E. F. Livingstone Ltd., London.
4. Berg, G. J. (1985). An integrated system for large scale cell culture. *Dev. Biol. Stand.* **60**, 297–307.
5. Birch, J. R., Thompson, P. W., Lambert, K., and Boraston, R. (1985). The large-scale cultivation of hybridoma cells producing monoclonal antibodies. *In* "Large-Scale Mammalian Cell Culture" (J. Feder and W. R. Tolbert, eds.), pp. 1–18. Academic Press, Orlando.
6. Birch, J. R., Thompson, P. W., Boraston, R., Oliver, S., and Lambert, K. (1987). The large-scale production of monoclonal antibodies in airlift fermenters. *In* "Plant and Animal Cells: Process Possibilities" (C. Webb and F. Mavituna, eds.), pp. 162–171. Ellis Horwood, Chichester.
7. Burbidge, C. (1980). The mass culture of human diploid fibroblasts in packed beds of glass beads. *Dev. Biol. Stand.* **46**, 169–172.
8. Burbidge, C., and Dacey, I. K. (1984). The use of plate heat exchangers in growing human fibroblasts. *Dev. Biol. Stand.* **55**, 255–259.
9. Butler, M., Imamura, T., Thomas, J., and Thilly, W. G. (1983). High yields from microcarrier cultures by medium diffusion. *J. Cell Sci.* **61**, 351–363.
10. van Brunt, J. (1986). Immobilized mammalian cells: the gentle way to productivity. *Bio/Technology* **4**, 505–510.
11. Corbeil, M., Trundel, M., and Payment, P. (1979). Production of cells and tissues in a new multiple-tube tissue culture propagator. *J. Clin. Microbiol.* **10**, 91–95.
12. Duff, R. G. (1985). Microencapsulation technology: a novel method for monoclonal antibody production. *Trends Biotechnol.* **3**, 167–170.
13. Emery, A. N., Lavery, M., Williams, B., and Handa, A. (1987). Large-scale hybridoma culture. *In* "Plant and Animal Cells: Process Possibilities" (C. Webb and F. Mavituna, eds.), pp. 137–146. Ellis Horwood, Chichester.
14. Familletti, P. C., Smith, C. M., Cullen, B. R., Stremlo, D. L., and Fredericks, J. D. (1987). Mammalian cell culture production in an airlift bioreactor. *In Vitro* **23**(3), 21A (Abstract 37).
15. Fazekas de St.Groth, S. (1983). Automated production of monoclonal antibodies in a cytostat. *J. Immunol. Methods* **57**, 121–136.

16. Feder, J., and Tolbert, W. R. (1983). The large-scale cultivation of mammalian cells. *Sci. Am.* **248**, 24–31.
17. Feder, J., and Tolbert, W. R. (1985). Mass culture of mammalian cells in perfusion systems. *Int. Biotech. Lab.*, June, 40–53.
18. Fleishaker, R. J., and Sinksey, A. J. (1981). Oxygen demand and supply in cell culture. *Eur. J. Appl. Microbiol. Biotechnol.* **12**, 193–197.
19. Gebb, C., Lundgren, B., Clark, J., and Lindsog, U. (1984). Harvesting and subculturing cells growing on denatured-collagen coated microcarriers (Cytodex 3). *Dev. Biol. Stand.* **55**, 57–65.
20. Girard, H. C., Okay, G., and Kilvilcim, Y. (1973). Use of the vibrofermenter for multiplication of BHK cells in suspension and for replication of FMD virus. *Bull. Off. Int. Epiz.* **79**, 805–822.
21. Girard, H. C., Sutcu, M., Erdem, H., and Gurhan, I. (1980). Monolayer cultures of animal cells with the Gyrogen equipped with tubes. *Biotechnol. Bioeng.* **22**, 477–493.
22. Gori, G. B. (1965). Chemostatic concentrated cultures of heteroploid mammalian cell suspensions in dialyzing fermenters. *Appl. Microbiol.* **13**, 93–97.
23. Graff, S., and McCarty, K. S. (1957). Sustained cell culture. *Exp. Cell Res.* **13**, 348–357.
24. Graff, S., and McCarty, K. S. (1958). Energy costs of growth in continuous metazoan cell cultures. *Cancer Res.* **18**, 741–746.
25. Griffiths, J. B., Sargeant, K., and Whitaker, A. (1967). The use of the Graff cytogenerator to grow y BHK cells. *MRE Record* No. 17.
26. Griffiths, J. B., Thornton, B., and McEntee, I. (1982). The development and use of microcarrier and glass sphere culture techniques for the production of Herpes simplex virus, Type 2, in MRC-5 cells. *Dev. Biol. Stand.* **50**, 103–110.
27. Griffiths, J. B., and Thornton, B. (1982). Use of microcarrier culture for the production of Herpes simplex virus (type 2) in MRC-5 cells. *J. Chem. Technol. Biotechnol.* **32**, 324–329.
28. Griffiths, J. B., Atkinson, A., Electricwala, A., Later, A., McEntee, I., Riley, P. A., and Sutton, P. M. (1984). Production of a fibrinolytic enzyme from cultures of guinea pig keratocytes grown on microcarriers. *Dev. Biol. Stand.* **55**, 31–36.
29. Griffiths, J. B., and Riley, P. A. (1985). Cell biology: basic concepts. *In* "Animal Cell Biotechnology" (R. E. Spier and J. B. Griffiths, eds.), Vol. 1, pp. 17–48. Academic Press, London.
30. Griffiths, J. B. (1986). Scaling-up of animal cell cultures. *In* "Animal Cell Culture: A Practical Approach" (R. I. Freshney, ed.), Chapter 3, pp. 33–70. IRL Press, Oxford.
31. Griffiths, J. B. (1986). Can cell culture medium costs be reduced? Strategies and possibilities. *Trends Biotechnol.* **4**, 268–272.
32. Griffiths, J. B., and Looby, D. (1987). A comparison of oxygenation methods in a 40L stirred bioreactor. "Proceedings 8th. ESACT Meeting, Israel". Butterworths, Guildford.
33. Griffiths, J. B., Cameron, D. R., and Looby. D. (1987). Bulk production of anchorage-dependent cells—comparative studies. *In* "Plant and Animal Cells: Process Possibilities" (C. Webb and F. Mavituna, eds.), pp. 149–161. Ellis Horwood, Chichester.
34. Griffiths, J. B., and Electricwala, A. (1987). Production of tissue plasminogen activitors from animal cells. *Adv. Biochem. Eng. Biotechnol.* **34**, 147–166.
35. Griffiths, J. B., Cameron, D. R., and Looby, D. (1987). A comparison of unit process systems for anchorage dependent cells. *Dev. Biol. Stand.* **66**, 331–338.
36. Handa, A., Emery, A. N., and Spier, R. E. (1987). On the evaluation of gas–liquid interfacial effects on hybridoma viability in bubble column bioreactors. *Dev. Biol. Stand.* **66**, 241–253.
37. Harms, E., and Wendenburg, J. (1978). Large scale perfusion of cells growing on surfaces with automatic gas and medium control. *Cytobiologie* **18**, 67–75.

38. Himmelfarb, P., Thayer, P. S., and Martin, H. E. (1969). Spin filter culture: the propagation of mammalian cells in suspension. *Science* **164**, 555–557.
39. Hu, W-S., Giard, D. J., and Wang, I. C. (1985). Serial propagation of mammalian cells on microcarriers. *Biotechnol. Bioeng.* **27**, 1466–1476.
40. Hu, W-S., and Wang, D. I. C. (1986). Mammalian cell technology: a review from an engineering perspective. *In* "Mammalian Cell Technology" (W. G. Thilly, ed.), pp. 167–197. Butterworths, Boston.
41. Jensen, M. D. (1981). Production of anchorage-dependent cells—problems and their possible solution. *Biotechnol. Bioeng.* **23**, 2703–2716.
42. Karkare, S. B., Phillips, P. G., Burke, D. H., and Dean, R. C. (1985). Continuous production of monoclonal antibodies by chemostatic and immobilised hybridoma culture. *In* "Large-Scale Mammalian Cell Culture" (J. Feder and W. R. Tolbert, eds.), pp. 127–149. Academic Press, Orlando.
43. Katinger, H. W. D., Scheirer, W., and Kromer, E. (1979). Bubble column reactor for mass propagation of animal cells in suspension culture. *Ger. Chem. Eng.* **2**, 31–38.
44. Katinger, H. W. D., and Scheirer, W. (1982). Status and developments of animal cell technology using suspension culture techniques. *Acta Biotechnologica* **2**, 3–41.
45. Katinger, H. W. D., and Bleim, R. (1983). Production of enzymes and hormones by mammalian cell cultures. *Adv. Biotech. Proc.* **2**, 61–95.
46. Katinger, H. W. D. (1987). Principles of animal cell fermentation. *Dev. Biol. Stand.* **66**, 195–209.
47. Kitano, K., Shintani, Y., Ichimori, Y., Tsukamoto, K., Sasai, S., and Kida, M. (1986). Production of human monoclonal antibodies by heterohybridomas. *Appl. Microbiol. Biotechnol.* **24**, 282–286.
48. Klement, G., Scheirer, W., and Katinger, H. W. D. (1987). Construction of a large scale membrane reactor system with different compartments for cells, medium and product. *Dev. Biol. Stand.* **66**, 221–226.
49. Knazek, R. A., Guillino, P. M., Kohler, P. O., and Dedrick, R. L. (1972). Cell culture on artificial capillaries: an approach to tissue growth *in vitro*. *Science* **178**, 65–67.
50. Kolanko, W. (1987). Growth and antibody production in a flexible plastic culture vessel. *In Vitro* **23**(3), 22A (Abstract 44).
51. Kruse, P. F., Keen, L. N., and Whittle, W. L. (1970). Some distinctive characteristics of high density perfusion cultures of diverse cell types. *In Vitro* **6**, 75–88.
52. Kruse, P. F., and Miedema, E. (1965). Production and characterisation of multiple-layered populations of animal cells. *J. Cell. Biol.* **27**, 273–279.
53. Kruse, P. F., Whittle, W. L., and Miedema, E. (1969). Mitotic and non-mitotic multiple-layered perfusion cultures. *J. Cell. Biol.* **42**, 113–121.
54. Ku, K., Kuo, M. J., Delenti, J., Wildi, B. S., and Feder, J. (1981). Development of a hollow fibre system for large-scale culture of mammalian cells. *Biotechnol. Bioeng.* **23**, 79–95.
55. Kuhlmann, W. (1987). Optimization of a membrane oxygenation system for cell culture in stirred tank reactors. *Dev. Biol. Stand.* **66**, 263–268.
56. Larsson, B., and Litwin, J. (1987). The growth of polio virus in human diploid fibroblasts grown with cellulose microcarriers in suspension. *Dev. Biol. Stand.* **66**, 385–390.
57. Lazar, A., Reuveny, S., Mizrahi, A., Avtalion, M., Whiteside, J. P., and Spier, R. E. (1987). Production of biologicals by animal cells immobilised on polyurethane foam matrix. "Proceedings of 8th. ESACT meeting, Israel." Butterworths, Guildford.
58. Leist, C., Meyer, H. P., and Fiechter, A. (1986). Process control during the suspension culture of a human melanoma cell line in a mechanically stirred loop bioreactor. *J. Biotechnol.* **4**, 235–246.

59. Lehmann, J., Piehl, G. W., and Schulz, R. (1987). Bubble free cell culture aeration with porous moving membranes. *Dev. Biol. Stand.* **66**, 227–240.
60. Lim, F., and Sun, A. M. (1980). Microencapsulated islets as bioartificial endocrine pancreas. *Science* **210**, 908–910.
61. Lindner, E., Arvidsson, A. C., Wergeland, I., and Billig, D. (1987). Subpassaging cells on microcarriers: the importance for scaling up to production. *Dev. Biol. Stand.* **66**, 299–305.
62. Litwin, J. (1985). The growth of human diploid fibroblasts aggregates with cellulose fibres in suspension. *Dev. Biol. Stand.* **60**, 237–242.
63. Looby, D., and Griffiths, J. B. (1987). Optimisation of glass-sphere immobilised bed cultures. "Proceedings of 8th. ESACT meeting, Israel." Butterworths, Guildford.
64. McCoy, T. A., Whittle, W., and Conway, E. (1962). A glass helix perfusion chamber for massive growth of cells *in vitro*. *Proc. Soc. Exp. Biol. Med.* **109**, 235–237.
65. Mann, G. F. (1977). Development of a perfusion culture system for production of biologicals using contact dependent cells. *Dev. Biol. Stand.* **37**, 149–152.
66. Milenburger, H. G., and David, P. (1980). Mass production of insect cells in suspension. *Dev. Biol. Stand.* **46**, 183–186.
67. Molin, O., and Heden, C. G. (1969). Large scale cultivation of human diploid cells on titanium discs in a special apparatus. *Prog. Immunobiol. Stand.* **3**, 106–110.
68. Montagnon, B., Vincent-Falquet, J. C., and Fanget, B. (1984). Thousand litre scale microcarrier culture of Vero cells for killed Polio virus vaccine. *Dev. Biol. Stand.* **55**, 37–42.
69. Munder, P. G., Modolell, M., and Wallach, D. F. H. (1971). Cell propagation on films of polymer fluorocarbon as a means to regulate pericellular pH and pO_2 in cultured monolayers. *FEBS Lett.* **15**, 191–195.
70. Nilsson, K., and Mosbach, K. (1980). Preparation of immobilised animal cells. *FEBS Lett.* **188**(1), 145.
71. Nilsson, K., Scheirer, W., Merten, O.-W., Ostberg, L., Liehl, E., Katinger, H. W. D., and Mosbach, K. (1983). Entrapment of animal cells for production of monoclonal antibodies and other biomolecules. *Nature (London)* **66**, 183–193.
72. Nilsson, K., Buzsaky, F., and Mosbach, K. (1986). Growth of anchorage-dependent cells on macroporous microcarriers. *Bio/Technology* **4**, 989–990.
73. Nilsson, K., and Mosbach, K. (1987). Immobilised animal cells. *Dev. Biol. Stand.* **66**, 183–193.
74. Nilsson, K. (1987). Methods for immobilizing animal cells. *Trends Biotechnol.* **5**, 73–78.
75. Reuveny, S., Zheng, Z-B., and Eppstein, L. (1986). Evaluation of a cell culture fermenter. *Am. Biotechnol. Lab*, Feb, 28–36.
76. Reuveny, S., Velez, D., Miller, L., and MacMillan, J. D. (1986). Comparison of cell propagation methods for their effect on monoclonal antibody yield in fermenters. *J. Immunol. Methods* **86**, 61–69.
77. Robinson, J. H., Butlin, P. M., and Imrie, R. C. (1979). Growth characteristics of human diploid fibroblasts in packed beds of glass beads. *Dev. Biol. Stand.* **49**, 173–181.
78. Rose, C. G. (1967). The circumfusion system for multipurpose culture chambers. *J. Cell. Biol.* **32**, 89–112.
79. Rousseau, I., and Bu'Lock, J. D. (1980). Mixing characteristics of a simple air-lift. *Biotechnol. Lett.* **2**, 475–480.
80. Rupp, R. G. (1985). Use of cellular microencapsulation in large-scale production of monoclonal antibodies. *In* "Large-Scale Mammalian Cell Culture" (J. Feder and W. R. Tolbert, eds.), pp. 19–38. Academic Press, Orlando.
81. Santero, G. G. (1972). The rotary column method for growth of large-scale quantities of cell monolayers. *Biotechnol. Bioeng.* **14**, 753–775.
82. Sato, S., Kawamura, K., and Fujiyoshi, N. (1983). Animal cell cultivation for production of

biological substances with a novel perfusion culture apparatus. *J. Tissue Cult. Method.* **8**, 167–171.
83. Schleicher, J. B., and Weiss, R. E. (1968). Application of a multisurface tissue culture propagator for the production of cell monolayers, virus and biochemicals. *Biotechnol. Bioeng.* **10**, 617–624.
84. Schönherr, O. T., van Gelder, P. T. J. A., van Hees, P. J., van Os, A. M. J. M., and Roelofs, H. W. M. (1987). A hollow fibre dialysis system for *in vitro* production of monoclonal antibodies replacing *in vivo* production in mice. *Dev. Biol. Stand.* **66**, 211–220.
85. Spier, R. E., and Clarke, J. B. (1980). Variation in the susceptibility of BHK populations and cloned cell lines to three strains of FMD virus. *Arch. Virol.* **63**, 1–9.
86. Spier, R. E., and Griffiths, J. B. (1983). An examination of the data and concepts germane to the oxygenation of cultured animal cells. *Dev. Biol. Stand.* **55**, 81–92.
87. Spier, R. E. (1985). Monolayer growth systems: heterogeneous unit processes. *In* "Animal Cell Biotechnology" (R. E. Spier and J. B. Griffiths, eds.), Vol. 1, pp. 243–263. Academic Press, London.
88. Spier, R. E., and Whiteside, J. P. (1983). The description of a device which facilitates the oxygenation of microcarrier cultures. *Dev. Biol. Stand.* **55**, 151–152.
89. Spier, R. E., and McCullough, K. (1987). The large-scale production of monoclonal antibodies *in vitro*. Cambridge University Press. (In preparation.)
90. Strand, M., Quarles, J. M., and McConnell, S. (1984). A modified matrix perfusion-microcarrier bead cell culture system. *Biotechnol. Bioeng.* **26**, 503–507.
91. Tharakan, J. P., and Chau, P. C. (1986). Operation and pressure distribution of immobilized cell hollow fiber bioreactors. *Biotechnol. Bioeng.* **28**, 1064–1071.
92. Tharakan, J. P., and Chau, P. C. (1986). A radial flow hollow fiber reactor for the large-scale culture of mammalian cells. *Biotechnol. Bioeng.* **28**, 329–342.
93. Tharakan, J. P., and Chau, P. C. (1987). Modeling and analysis of radial flow cell culture. *Biotechnol. Bioeng.* **29**, 657–671.
94. Thayer, P. S. (1973). Spin filter device for suspension cultures. *In* "Tissue Culture Methods and Applications" (P. K. Kruse and M. K. Patterson, eds.), pp. 345–351. Academic Press, New York and London.
95. Tolbert, W. R., Feder, J., and Kimes, R. C. (1981). Large-scale rotating filter perfusion system for high density growth of mammalian suspension cultures. *In Vitro* **17**, 885–890.
96. Tolbert, W. R., White, P. J., and Feder, J. (1985). Perfusion culture systems for production of mammalian cell biomolecules. *In* "Large-Scale Mammalian Cell Culture" (J. Feder and W. R. Tolbert, eds.), pp. 1–18. Academic Press, Orlando.
97. Tovey, M. G. (1985). The cultivation of animal cells in continuous-flow culture. *In* "Animal Cell Biotechnology" (R. E. Spier and J. B. Griffiths, eds.), Vol. 1, pp. 195–210. Academic Press, London.
98. van Hemert, P. A., Kilburn, D. G., and van Wezel, A. L. (1969). Homogeneous cultivation of animal cells for the production of virus and virus products. *Biotechnol. Bioeng.* **11**, 875–881.
99. van Wezel, A. L. (1967). Growth of cell strains and primary cells on microcarriers in homogeneous culture. *Nature (London)* **216**, 64–65.
100. van Wezel, A. L. (1982). Cultivation of anchorage-dependent cells and their applications. *J. Chem. Tech. Biotechnol.* **32**, 318–323.
101. van Wezel, A. L. (1984). Microcarrier technology—present status and prospects. *Dev. Biol. Stand.* **55**, 3–9.
102. van Wezel, A. L., van der Velden-de Groot, C. A. M., de Haan, H. H., van den Heuval, N., and Schasfoort, R. (1985). Large scale animal cell cultivation for production of cellular biologicals. *Dev. Biol. Stand.* **60**, 229–236.

103. Varecka, R., and Scheirer, W. (1987). Use of a rotary wire cage for retention of animal cells in a perfusion fermenter. *Dev. Biol. Stand.* **66**, 269–272.
104. Whiteside, J. P., and Spier, R. E. (1981). The scale-up from 0.1 to 100 litre of a unit process system based on 3 mm diameter glass spheres for the production of four strains of FMDV from BHK monolayer cells. *Biotechnol. Bioeng.* **23**, 551–565.
105. Whiteside, J. P., Farmer, S., and Spier, R. E. (1985). The use of caged aeration for the growth of animal cells on microcarriers. *Dev. Biol. Stand.* **60**, 283–290.
106. Whiteside, J. P., and Spier, R. E. (1985). Factors affecting the productivity of glass sphere propagators. *Dev. Biol. Stand.* **60**, 305–311.
107. Wohler, W., Rudiger, H. W. & Passarge, E. (1972). Large scale culturing of normal diploid cells on glass beads using a novel type of culture vessel. *Exp. Cell Res.* **74**, 571–573.
108. WHO (1987). Acceptability of cell substrates for production of biologicals. Technical Report Series No. 747.
109. Amicon Corporation, Danvers, MA 01923, U.S.A. Technical publication (402).
110. APV International Ltd., Crawley RH10 2QB, U.K. Plate heat exchanger.
111. Bellco Glass Inc., Vineland, NJ 08360, U.S.A. U-carrier magnetic stirrer.
112. Bellco Glass Inc., Vineland, NJ 08360, U.S.A. Bellco-Corbeil Alternator Roller Culture Apparatus.
113. Bioengineering AG., CH 8636 Wald, Switzerland. Membrane Laboratory Fermenter and Fluidised Bed Reactor.
114. Edmund Buhler, D-7400 Tubingen, F.R. Germany. Buhler Cell Culture System ZKA.
115. Endotronics Inc., Coon Rapids, MN 55433, U.S.A. Acusyst hollow fibre reactors.
116. J. R. Scientific Inc., Woodland, CA 95695, U.S.A. Media supplied in plastic bags up to 500 litres volume.
117. LH Fermentation, Stoke Poges SL2 4EG, U.K. Spin exchanger filter.
118. New Brunswick Scientific, Edison NJ 08818, U.S.A. Perfusion swivel cap roller bottles.
119. SGI, 31100 Toulouse, France. Cellascenseur impeller.
120. Techne (Cambridge) Ltd., Duxford, U.K. MCS series of biological stirrers.
121. Techne (Cambridge) Ltd., Duxford, U.K. BR-O6 Bioreactor.
122. Rohm and Haas Co., Philadelphia, PA 19105, U.S.A. Biocryl BPA Bioprocessing aids.

8

Bubble-free Reactors and Their Development for Continuous Culture with Cell Recycle

J. LEHMANN,
J. VORLOP, and
H. BÜNTEMEYER
Gesellschaft für Biotechnologische Forschung mbH
Mascheroder Weg 1,
D-3300 Braunschweig

ANIMAL CELL BIOTECHNOLOGY VOL. 3
ISBN 0-12-657553 3

1. BUBBLE-FREE REACTORS—WHY?

In conventional cell-culture reactors, aeration is normally done by bubbling air or pure oxygen through the culture broth. This, of course, in protein-rich cell culture media, causes foaming; and with foam, cell flotation occurs. Phenomena such as wall growth, cell lysis, and uncontrolled release of proteases can be observed in relation to flotation. For these reasons, chemical antifoam agents are often used to prevent foaming. With culture broths containing antifoam difficulties arise in downstream processing procedures, because separation membranes and column chromatography materials are very sensitive to antifoam agents.

These problems cannot be neglected if one has to cultivate genetically manipulated mammalian cells or hybridoma for very long periods of time in continuous or perfusion cultures. Hence, there was a need for the development of a bubble-free reactor system for the continuous cultivation of such new cells both in small and large scale.

2. BUBBLE-FREE AERATION—HOW?

The simplest technique for bubble-free aeration is surface aeration, which is used in culture flasks or in multitray cultivation facilities. But for large-scale systems, surface is better supplied by immersing membranes into the culture broth, as proposed by Miltenburger 1979 (*1*) and others (*2, 3*). It was well known from blood oxygenation that silicone tubing could meet the demand for oxygen transfer capacity for artificial lungs, which have been produced as flat-plate-and-sheet membrane oxygenators or as compact units as spiral-coil membrane oxygenators (*4*). In Invitron's static maintenance reactor, silicone tubes are also used to supply oxygen and to remove CO_2 (*5*).

But since the wall of the silicone tube acts as a diffusion barrier, the surface-related oxygen transfer rate is still small and a tremendous amount of tubing is necessary to supply oxygen in such dense cell-culture systems. This disadvantage of a homogeneous membrane can partly be avoided by the use of porous membranes.

2.1. Aeration with Hydrophobic Microporous Membrane Fibres

Hydrophobic microporous membranes can be used for aeration. In particular, when the pores are filled with gas the oxygen transfer resistance across the membrane wall is less than that of a silicone wall of the same thickness. This is

Fig. 1. Electron micrograph of a cross-section of an Accurel membrane fibre wall. Wall thickness $S = 0.4$ mm.

due to the higher diffusivity in the pore-gas-phase and to the more convective mass transport pattern in the pores than in the homogeneous silicone material. There are two advantages due to the hydrophobic property of the membrane surface: water cannot soak into the pores, so that liquid is in direct contact with the gas phase; and cells have no contact with the membrane material. Since cells cannot stick to the membrane wall, there is no cell growth on the membrane.

Figure 1 shows an electron micrograph of a cross-section of a polypropylene Accurel (Enka, Wuppertal) membrane fibre. The wall thickness is $s = 0.4$ mm and the porosity is 75%. This membrane has the quality of a 0.2 μm sterilfilter, because the connections between the pores is of that magnitude. The pores themselves are relatively wide and they are at least one magnitude greater than the connections between them. The Accurel membrane is the best membrane for aeration purposes found so far. It can be steam-sterilized and it is rigid enough not to need mechanical support during sterilization. On the other hand, the membrane fibre itself is flexible and it is possible to coil it as a spiral coil. Hydrophilized with ethanol, Accurel fibre can be used as a microfiltration membrane (described in Section 3.2).

2.2. The Membrane Stirrer

In homogeneous suspension culture and microcarrier culture systems, gentle mixing is needed to distribute the cells and carriers homogeneously through the liquid volume. Cells and carrier should not be allowed to settle down onto the reactor bottom. On the other hand, the maximum tip speed of a stirrer plate should not exceed 14 cm s^{-1}, otherwise cell damage and cell disruption from the microcarrier can occur. It is difficult at such low stirring speeds to ensure that the reactor is well mixed and that the liquid velocity is not high enough to resuspend cells once they have settled down on the spiral coils. It is better therefore to use the spiral coil itself as stirrer. Such a stirrer is called a membrane stirrer.

Figure 2 shows schematically a reactor vessel with a membrane stirrer which is hanging on elastic silicone tubes on the top plate of the reactor. It is moved without rotation about its own axis. The membrane stirrer gently penetrates the culture broth, performing mixing and providing a uniform introduction of oxygen and desorption of CO_2 without mechanical damage to the cells.

For oxygen supply, the aeration gas streams through the fibre. To avoid bubbling, the reactor head-space has been connected to the gas input tube so that there is automatically a pressure equilibrium between the fibre input and

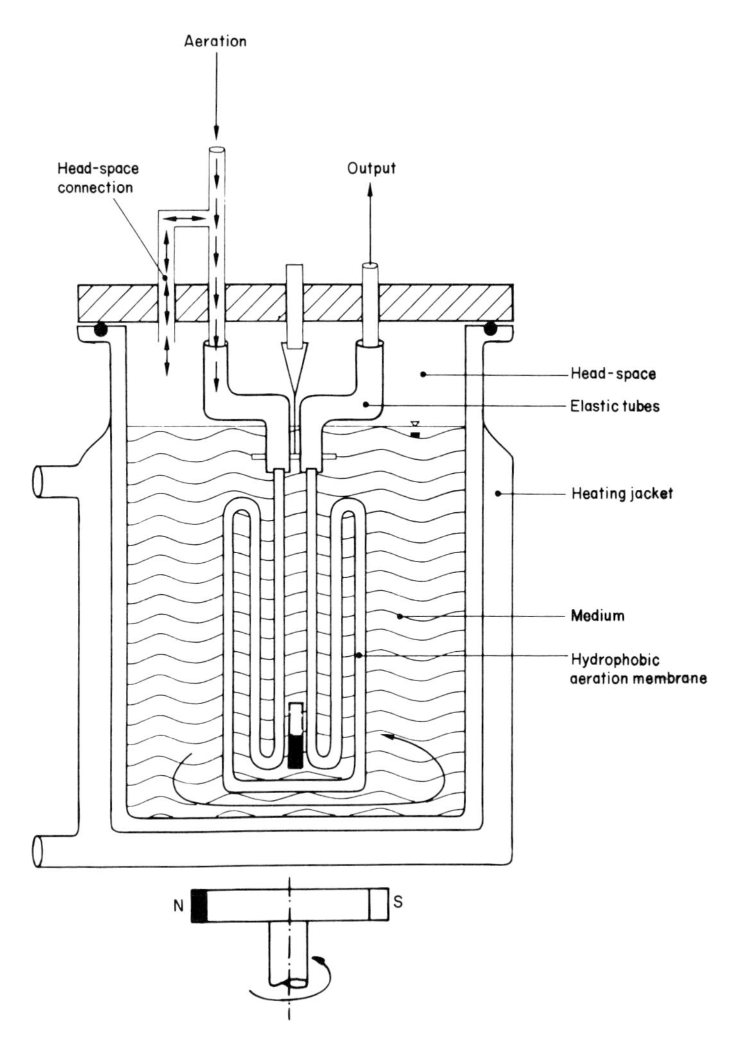

Fig. 2. Membrane stirrer reactor for bubble-free aeration with a porous membrane. The head-space is closed but connected to the gas input to achieve pressure equilibrium between the head-space and membrane. Bubbling cannot occur.

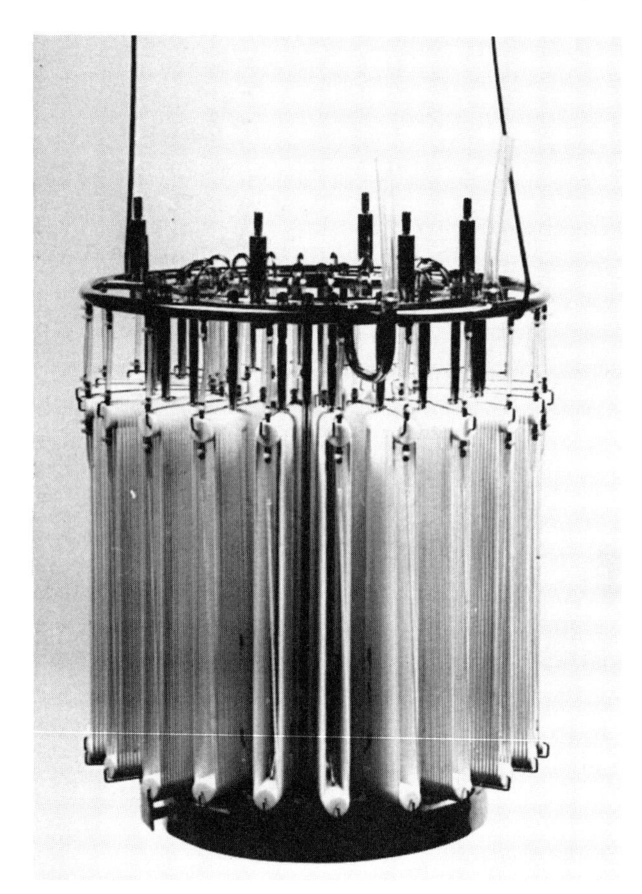

Fig. 3. Membrane-segment stirrer for a 150-litre microcarrier cell-culture reactor.

the head-space. That way the pressure in the liquid is everywhere equal to or greater than the static pressure inside the membrane fibre. Bubbles cannot come out of the pores under such conditions.

In large-scale reactors, the membrane stirrer consists not only of one membrane fibre but rather of several segments which are mounted on one support. The gas is fed parallel throughout the segments. Such a membrane segment stirrer is also moved without rotation about its own axis driven by an eccentric drive while hanging on the coverplate of the reactor. Figure 3 shows a membrane segment stirrer with 22 segments for a 150-litre microcarrier cell-culture reactor.

2.3. Oxygen Transfer Performance of Membrane Stirrers

It is obvious that the oxygen transfer rate is dependent on several parameters. The membrane area per unit volume corresponds to the length of the membrane fibre per unit volume. The gas flow through the segments can be parallel or in series and the speed of membrane movement relative to the surrounding fluid influences the boundary layer renewal.

Oxygen transfer rates have been determined for different membrane speeds, different segment connections, and different gas flow rates. The results for a 20-litre pilot segment membrane stirrer can be seen in Figs 4 and 5. From Fig. 4 one can clearly determine the relation between oxygen transfer rate (OTR) and membrane speed. The OTR is very strongly influenced by the membrane speed, so that it seems to be possible to use the agitation speed of the stirrer for mass transfer control purposes. On the other hand, Fig. 5 shows a saturation curve characteristic between the gas flow rate and OTR. Flow rates greater than 200 ml min^{-1} do not improve the OTR. This means that in the 20-litre

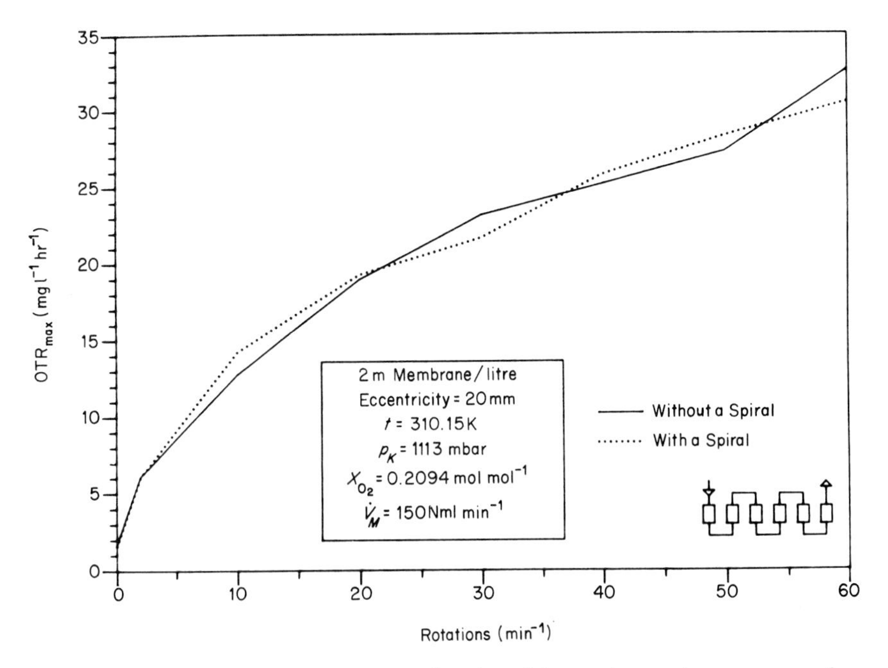

Fig. 4. Maximum oxygen transfer rate as a function of the membrane stirrer movement for a 20-litre pilot-scale reactor; p_k = head pressure; X_{O_2} = mole fraction of oxygen; V_M = aeration rate.

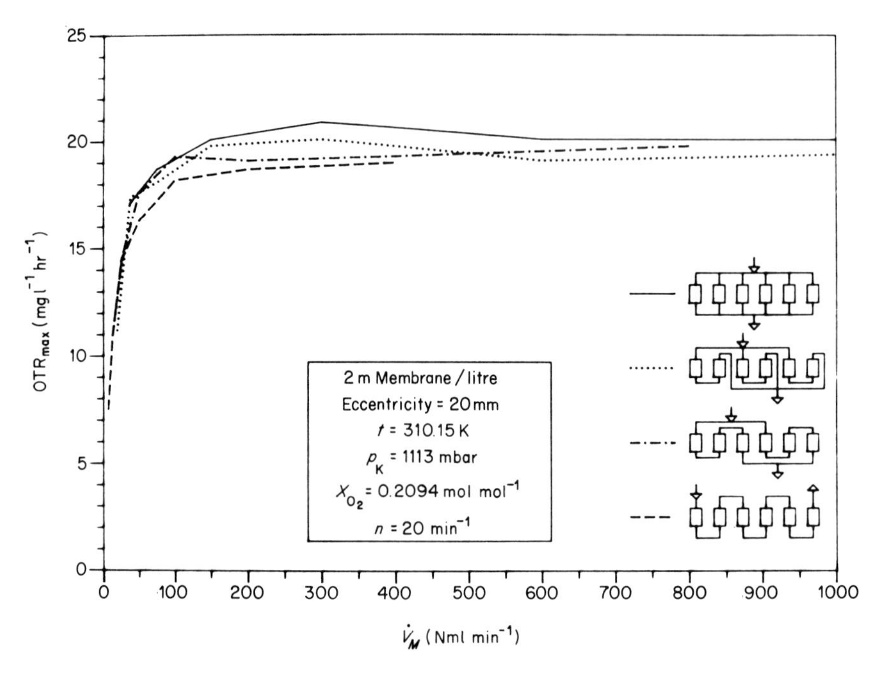

Fig. 5. Maximum oxygen transfer rate for different membrane connections as a function of the inlet gasflow into the membrane stirrer; p_K = head pressure; X_{O_2} = mole fraction of oxygen.

pilot reactor, gas flow rates above 200 ml min^{-1} are uneconomical. The gas flow rate is not a means for controlling the OTR.

The control strategy for DOT control via OTR should be to alter the oxygen concentration in the inlet gas stream.

2.4. Carrier Mixing in Membrane Stirrer Reactors

The membrane stirrer is used not only to aerate the culture broth, but also to mix it and to prevent carrier settling on the bottom of the reactor vessel. As the segment basket does not generate enough axial flow components in the liquid, a turbulence-generating spiral can be mounted beneath the segments (Fig. 6). It moves with the stirrer. The distance between the spiral and the bottom is only a few millimetres, so that strong turbulence vortexes are generated which guarantee adequate carrier mixing even at low stirrer speeds.

The microcarrier distribution has been determined for 20- and 150-litre scale reactors at different membrane stirrer speeds. It can be seen from Fig. 7

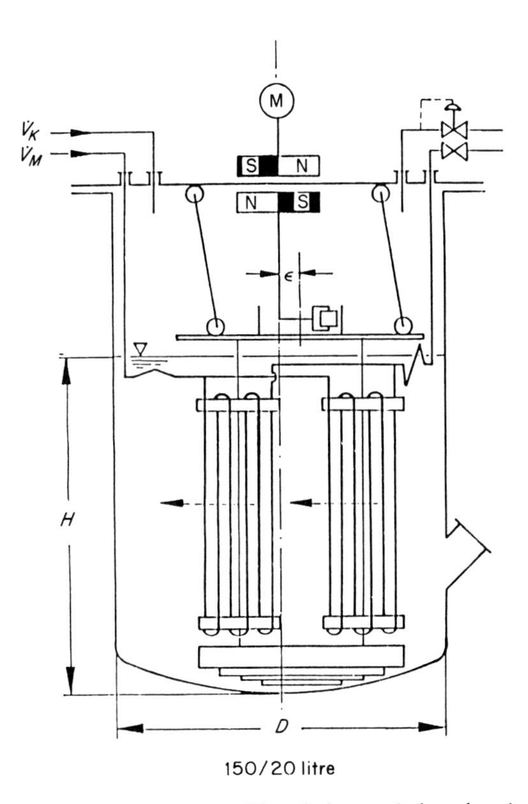

Fig. 6. Membrane stirrer arrangement with turbulence spiral on the stirrer; ε = eccentricity; $\dot{V}_K$ = head-space aeration rate; $\dot{V}_M$ = membrane aeration rate.

that at stirrer speed $u = 8.37 \text{ cm s}^{-1}$ the carrier distribution already behaves nearly uniformly.

3. PERFUSION IN BUBBLE-FREE REACTORS

A long-term continuous cultivation can be done with and without cell retention. There are several advantages when cells are retained. Firstly, there is the tendency that the higher the cell density, the higher the productivity which can be achieved; secondly, the self-conditioning ability of cells, i.e. secretion of growth factors and other proteins, seems to increase with the cell density. Therefore, perfusion culture seems to be preferable to chemostat culture, where the cell density cannot be freely extended by the dilution rate.

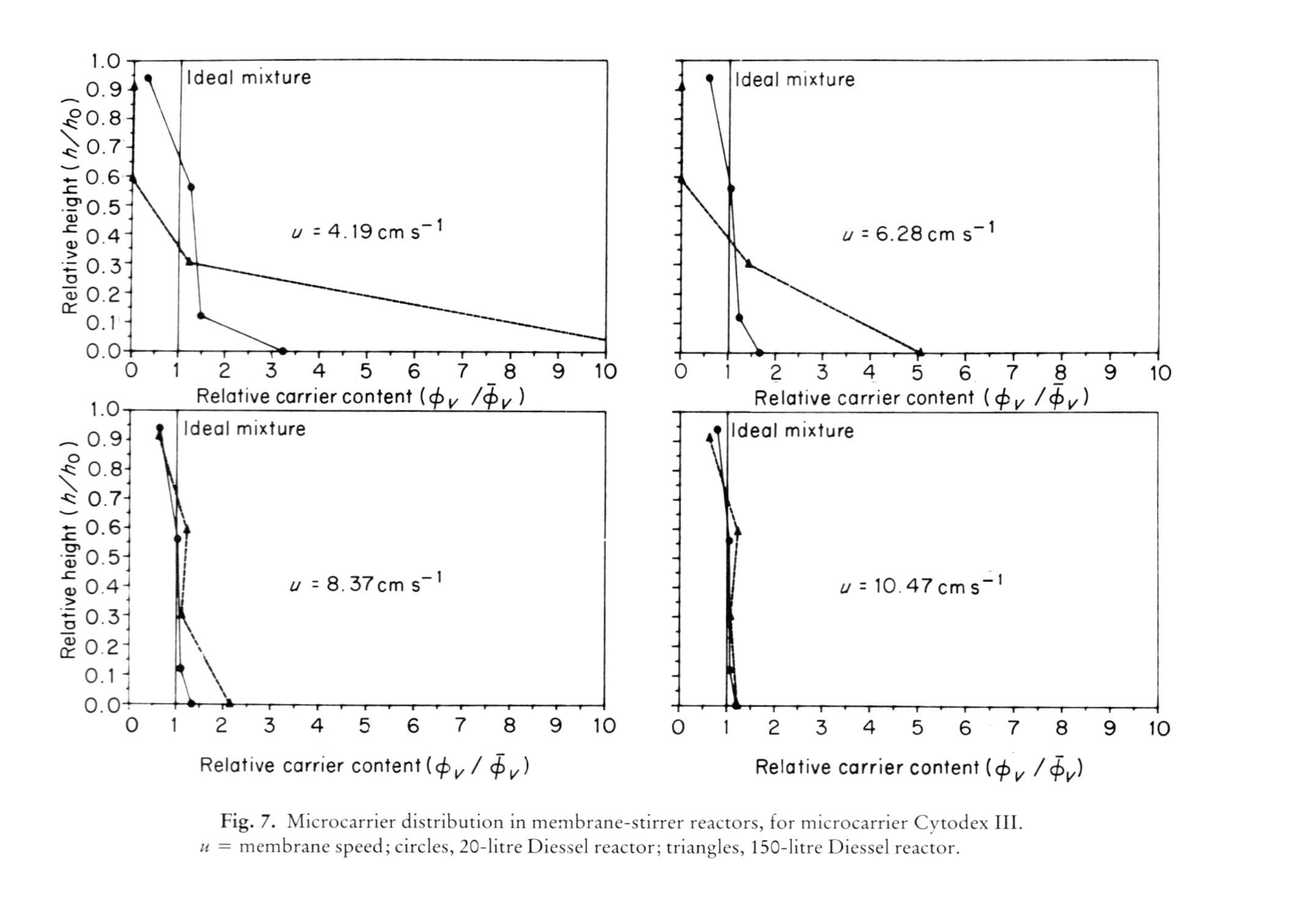

Fig. 7. Microcarrier distribution in membrane-stirrer reactors, for microcarrier Cytodex III. u = membrane speed; circles, 20-litre Diessel reactor; triangles, 150-litre Diessel reactor.

3.1. Microcarrier Retention with Fluttering Sieve Filter

In bubble-free aeration reactors, microcarriers can be retained using a very simple device—the fluttering sieve filter. Such a filter can be mounted in the centre of a membrane segment stirrer, and because of a very loose attachment of the textile screen on its support, it can flutter. This prevents clogging of the sieve by microcarriers.

As the head-space in large reactors requires a higher pressure than the surrounding, carrier-free but cell-containing broth can easily be withdrawn from the reactor by an aseptic overflow outside the reactor, as shown in Fig. 8. If the head-space pressure is carefully controlled by a back-pressure regulator, no harvesting pump is needed to bring the broth out of the reactor. Such a system has been employed for the successful long-term cultivation of various genetically manipulated cells, e.g. Mouse L, CHO, and BHK. For production of human β-interferon with Mouse L cells, the system has been used continuously for three weeks at 20-litre scale. The dilution rate was about 1 to 2 litres per litre per day (*6*). Figure 9 shows the time course for such a cultivation. Very high cell density, up to 8×10^6 ml^{-1} was reached, and interferon production could be maintained even without serum in the medium fed.

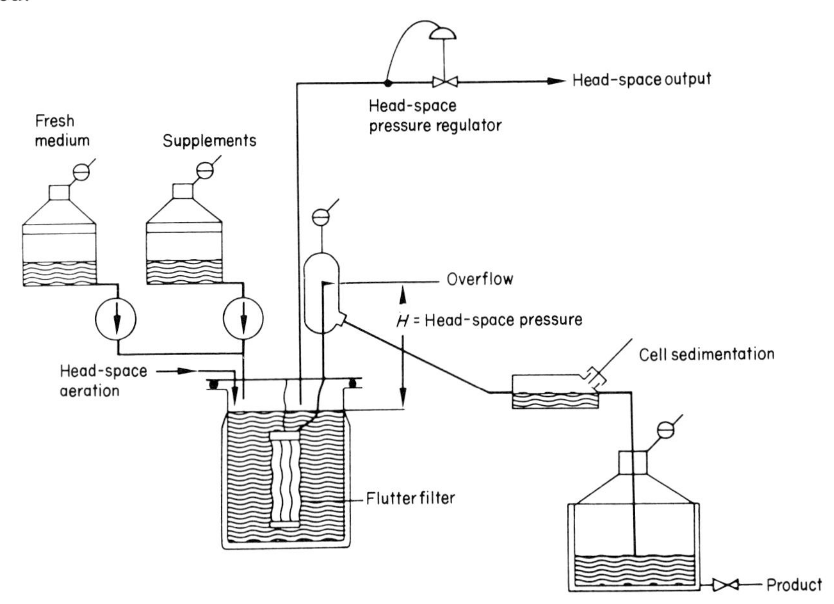

Fig. 8. Flow scheme of a perfusion reactor system with microcarrier retention using a fluttering sieve filter device.

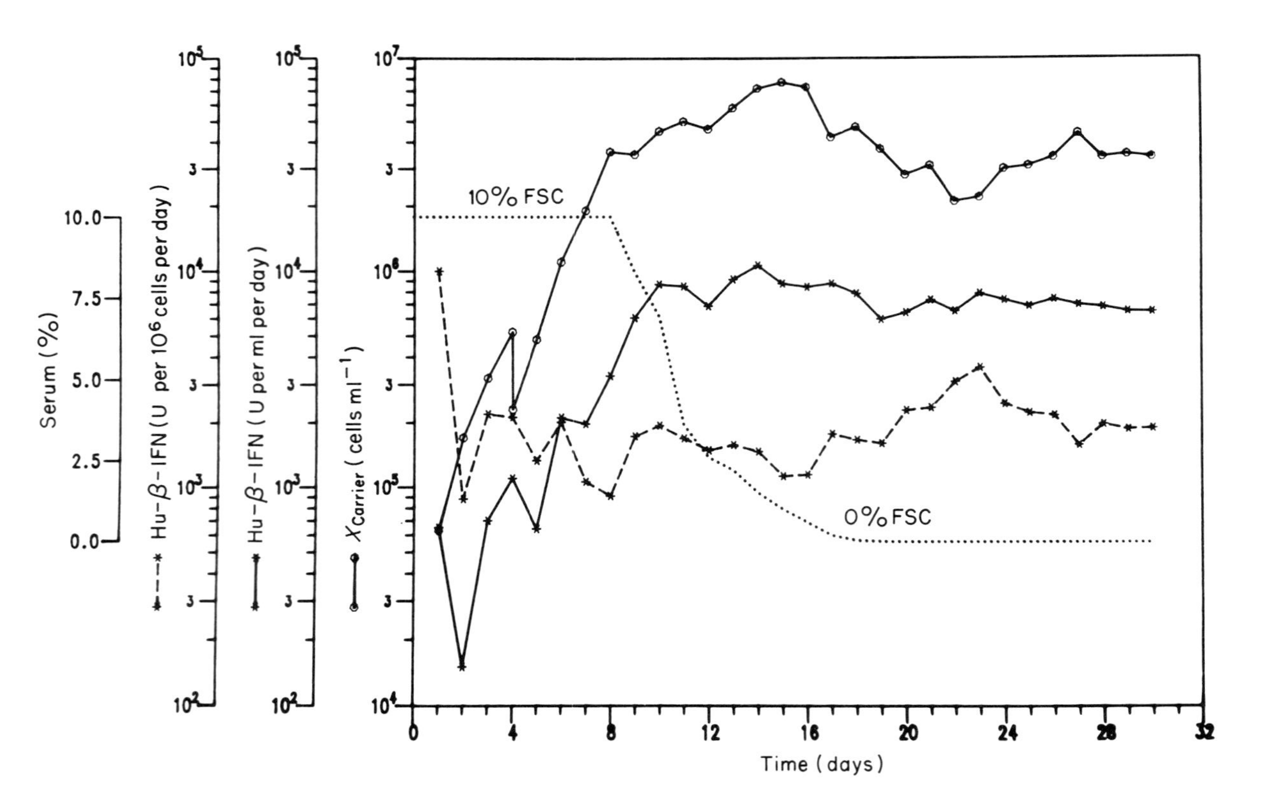

Fig. 9. Time course of a cultivation of human β-interferon producing Mouse L cells by carrier-free perfusion using a fluttering sieve filter (*6*).

3.2. Cell Retention with Microporous Membrane Filters

A new technique has been developed for retaining cells in suspension cultures in which a double-membrane stirrer has been employed.

3.2.1. The Double-membrane Stirrer

To the hydrophobic membrane which has been used for aeration only there is added another membrane filter with hydrophilic character, to form the double-membrane stirrer. Figure 10 shows such a double-membrane stirrer in

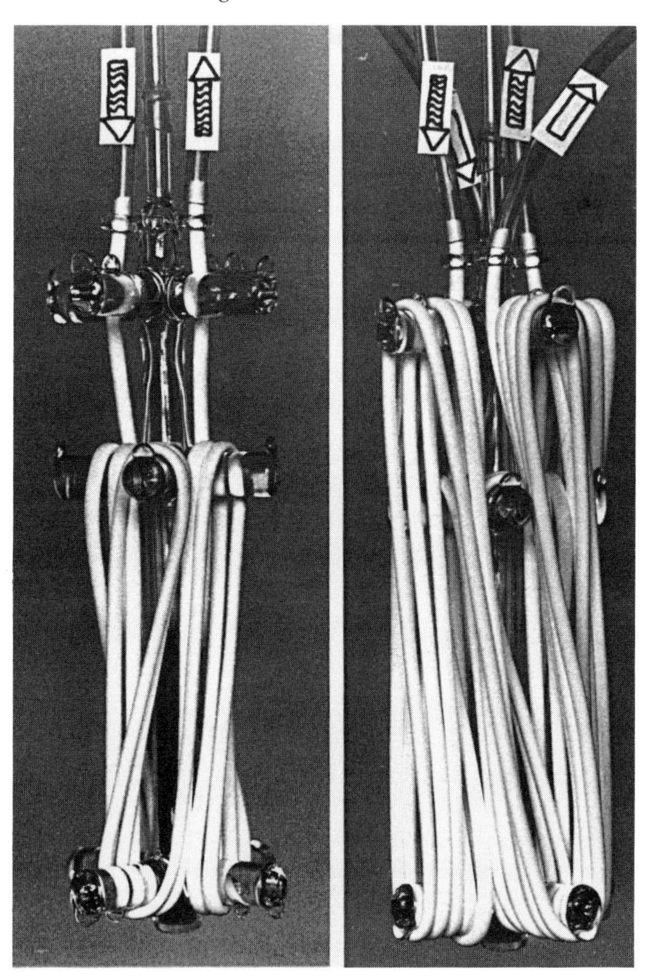

Fig. 10. The double-membrane stirrer for bubble-free aeration and cell retention for a 1-litre suspension culture reactor. Left: hydrophilized microfilter hollow fibre for perfusion. Right: hydrophobic and hydrophilic hollow fibres. medium gas

which both membranes are made from the polypropylene Accurel hollow fibre. The left-hand side shows the microfiltration membrane which is hydrophilized with ethanol, and the right-hand side shows the stirrer with both membranes. To prevent fouling and clogging of the microfilter membrane by cells, a special procedure for medium supply and harvest, involving the membrane in both steps, has been developed.

3.2.2. The Fresh Medium Backflush Procedure

The microfiltration membrane is used for both uniform medium supply and medium filtration for cell-free harvest (see Fig. 11). During one cycle period the flow direction through the fibre wall is alternated. During the first period, the feeding period, fresh medium is pumped through the membrane into the reactor while the harvesting pump stops the spent medium output. During the second period, the harvesting period, the cell broth is filtered cell-free and withdrawn with the harvesting pump while the feeding pump stops the fresh medium input line.

In this procedure the membrane fibre is filled with fresh medium when the harvesting period commences. To make sure that the portion of fresh medium

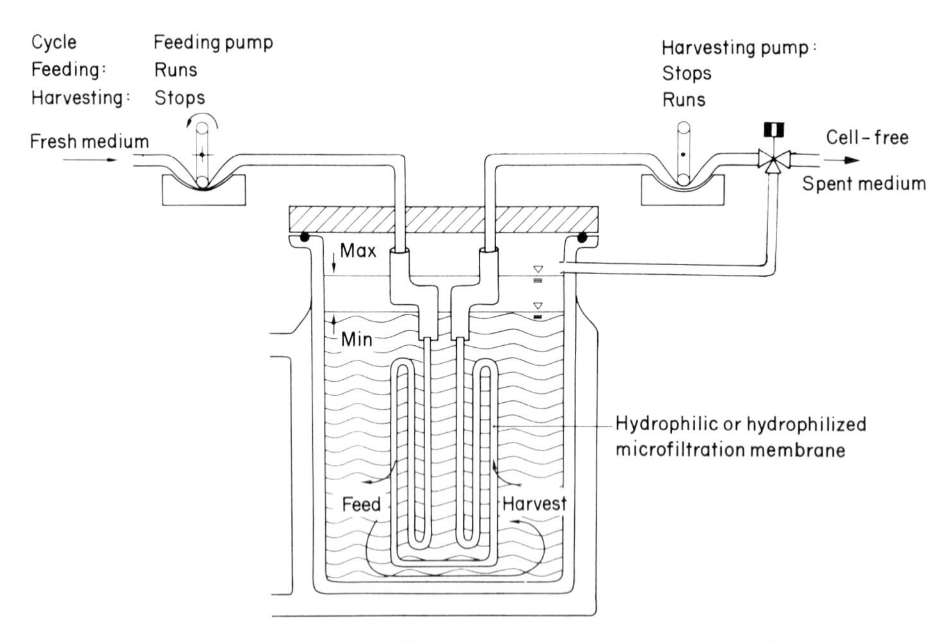

Fig. 11. Medium supply and microfiltration membrane-stirrer segments of a double-membrane stirrer. The pumps allow a two step procedure: feeding—medium is pumped into the reactor while the harvesting pump stops; harvesting—medium is pumped out of the reactor while the feed pump stops.

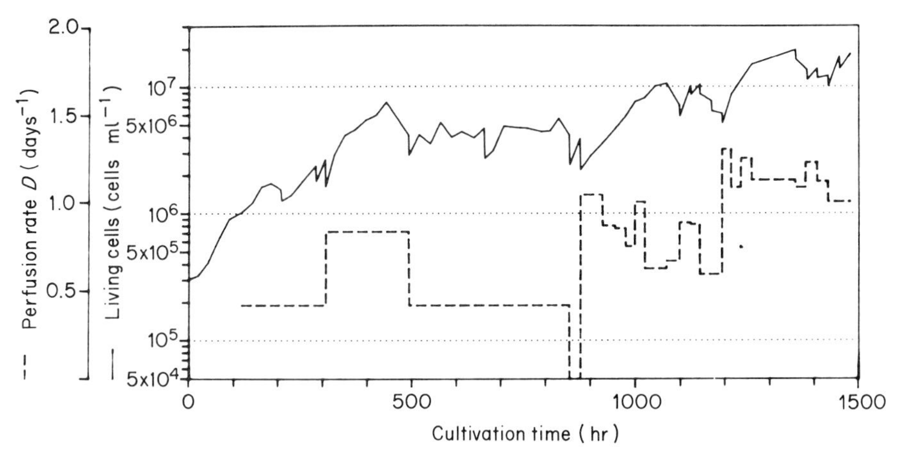

Fig. 12. Time course of a cultivation of a human B-cell line with the double-membrane stirrer for bubble-free aeration and cell-free perfusion of the medium.

is not wasted when harvesting starts, a three-way valve can be installed in the effluent line. This allows recycling of the first part of the spent medium back into the reactor until pure spent medium reaches the position of the valve. The valve is then reswitched and only spent medium will be harvested.

As the microfiltration fibre operates as a sterilization filter, the effluent is completely cell-free. It can be passed directly to any kind of purification device without further treatment.

One metre of Accurel membrane fibre per litre of volume allows a perfusion rate up to 1 litre of fresh medium per litre of broth per day. To guarantee stable long-term operation, that flux should not be extended.

A human B-cell line has been cultivated for a period of two months at cell densities up to 1.5×10^7 cells ml^{-1}, as shown in Fig. 12.

4. SCALING-UP

The new membrane stirrer reactor has been successfully tested for five months semi-continuously in repeated batch mode (*7*) before scaling-up to 30 litre and 150 litre volumes. For scale-up the height to diameter ratio was kept constant $H/D = 1$, and the amount of membrane surface per litre of broth was also kept constant in all scales.

The mass transfer characteristic as function of Sh (Sherwood number) over Re (Reynolds number) with d the membrane diameter and u the membrane speed has been estimated for 1- and 20-litre scale reactors. In both, we

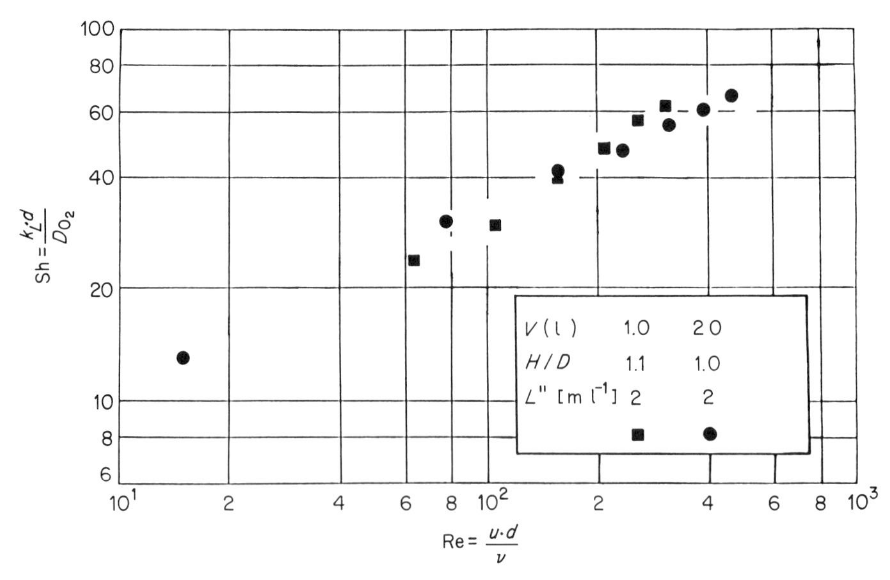

Fig. 13. Oxygen transfer characteristic for membrane-stirrer reactors of 1-litre and 20-litre scale. L^* is the length of membrane fibre per litre of broth.

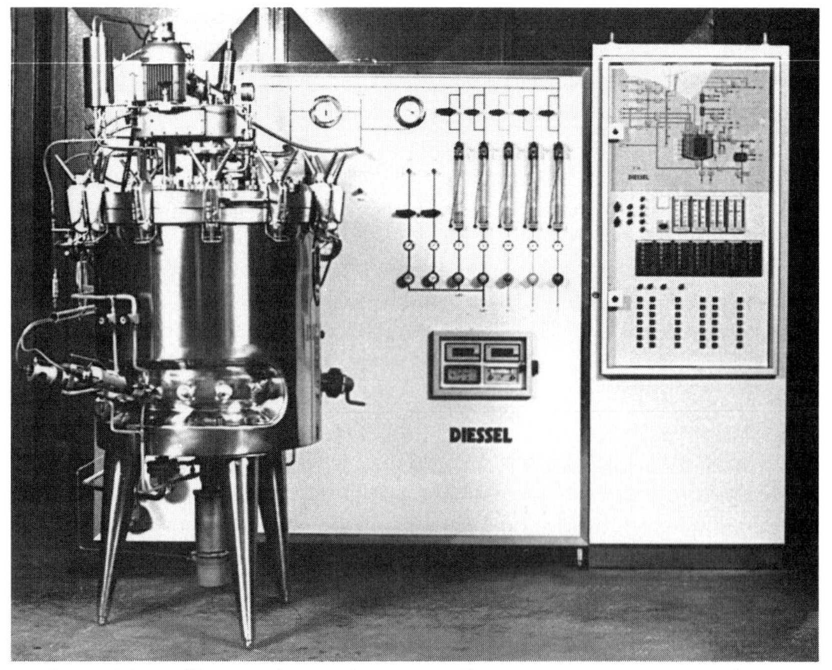

Fig. 14. The 150-litre membrane-stirrer reactor for continuous microcarrier culture.

obtained the same slope as shown in Fig. 13. The largest reactor built so far, a 150-litre reactor equipped with the 22-segment membrane stirrer, is shown in Fig. 14. The oxygen transfer rates obtained in the 150-litre reactor are slightly higher than in the smaller reactors. The reactor has not yet been fully investigated, so that total characteristic cannot be given.

5. CONCLUSIONS

The new membrane-stirrer reactor system seems to have promising qualities for long-term cultivation of genetically manipulated mammalian cells.

The trouble-free application of porous polypropylene Accurel fibres for both bubble-free aeration and perfusion of medium is a new example of successful utilization of membranes in animal cell culture technology.

REFERENCES

1. Miltenburger, H. G., and David, P. (1980). Mass production of insect cells in suspension. *Dev. Biol. Stand.* **46**, 183–186.
2. Tyo, H., and Wang, D. I. C. (1981). Engineering characterization of animal cell and virus production using controlled charge microcarriers. *In* "Advances in Biotechnology" (M. Moo-Young, ed.), pp. 141–146. Pergamon Press, Oxford.
3. Fleischaker, R. J., Giasch, D. J., Weaver, J., and Sinskey, A. J. (1981). *In* "Advances in Biotechnology" (M. Moo-Young, ed.), pp. 425–430. Pergamon Press, Oxford.
4. Hwang, S. T., and Kammermeyer (1975). "Membranes in Separations", pp. 251–253. Wiley, New York.
5. Tolbert, W. R., Lewis, C. Jr., White, P. J., and Feder, J. (1985). "Perfusion culture systems for production of mammalian cell biomolecules". *In* "Large-Scale Mammalian Cell Culture" (J. Feder and W. R. Tolbert, eds.), pp. 97–123. Academic Press, Orlando.
6. Fraune, E. (1987). Produktion und Aufarbeitung von humanem β-Interferon. PhD Thesis, University Hannover.
7. Lehmann, J., Piehl, G. W., and Schulz, R. (1985). Blasenfreie Zellkulturbegasung mit bewegten, porösen Membranen. *BTF-Biotech-Forum* **2**, 112–118.

9

A Tubular Biological Film Reactor Concept for the Cultivation and Treatment of Mammalian Cells

H. KATINGER
Institute of Applied Microbiology,
Faculty of Food- and Biotechnology,
University of Agriculture,
Vienna, Austria

The concept of a novel tubular film reactor for the cultivation, immobilization, and continuous perfusion of mammalian cells is described in this chapter. The reactor is capable of being used for both adhesive and non-adhesive cell types. The gas–liquid mass transfer is accomplished by gas transfer from a continuously gassed head-space to the continuously renewed liquid film in a manner which is comparable to the roller bottle technique. Alternatively, the reactor can be used as a continuous centrifuge. Thus, all unit operations

ANIMAL CELL BIOTECHNOLOGY VOL. 3
ISBN 0-12-657553 3

required for cell treatment such as cell washing, trypsinization, and changing nutrient media can be performed under aseptic operation within the reactor.

1. INTRODUCTION

The production of complex and structured glycoproteins such as viral vaccine antigens, various drugs, and monoclonal or chimeric antibodies has gained increasing importance. Those particular biomolecules are often not adequately synthesized in microbes; the use of mammalian cell culture expression systems and mammalian mass cultivation and production techniques become inevitable.

A wide array of potential production cell strains and cell lines has been derived either by direct isolation of cells from normal or transformed tissues or by the application of the various *in vitro* techniques of cell strain improvement. The resulting variety of the mammalian cells is enormous. Even if we limit our consideration only to those mammalian cell characteristics which might be relevant for the design of their respective *in vitro* mass culture systems, we still can make a long list of design criteria. Most of the important aspects have been worked out qualitatively at least (see Table I). Some unknown factors may still remain hidden in a "black box".

TABLE I Qualitative Mammalian Cell Characteristics Most Relevant for Bioreactor Design

Cell characteristic	Bioreactor and process design criteria
Non-adherent (suspension) cell	Static and/or submerged suspension culture as well as pseudo-adherent culture possible in single cells or mini-aggregates
Obligate-adherent cells	Culture adherent to matrices in monolayers, multilayers, aggregates, capsules
Growth	
Low cell densities	Improve nutrient media, go to open (continuous) culture, hold cells back in the bioreactor (i.e. immobilize) and perfuse
Low growth rate enrichment of toxic metabolites	Immobilize cells; perfuse (i.e. wash) the cells and improve the balance of nutrients in media, try to achieve C- limited and energy-source-limited culture
Toxic (negative) effects resulting from killed cells	Selectively immobilize viable cells only (use autopurifying immobilization techniques)
Sensitive to	
Shear stress	Reduce velocity gradients and use static mixing systems, protect cells by membranes or encapsulate cells, protect cells by macromolecules (e.g. by serum or reduce mass transport to diffusional forces)
Concentration gradients of reactants	Increase mixing or reduce distances of diffusional mass transport (reduce reactor size!)
Direct sparging of gas bubbles	Avoid direct contact between gas phase and cells (i.e. permeate, diffuse, or transport gaseous phases in dissolved fashion)

The compilation of the general design aspects given in Table I reflects only a very crude qualitative picture of the overall requirements. It is, for example, possible to cultivate a couple of cell lines under two circumstances; in submerged (suspension) culture as well as under strictly adherent conditions. There also exist phenomena which indicate that in parallel with the amended technique of *in vitro* culture maintenance from submerged to adherent state, some additional cell characteristics, such as the susceptibility to virus propagation, secretion of metabolites, etc., are also influenced. Hence it follows, that the cell characteristics as compiled in Table I are *per se* subjected to continuous changes; in other words, the technique of *in vitro* treatment of cells itself feeds back upon the cell characteristics expressed *in vitro*.

Until we can shed more light on those phenomena, animal cell technology will remain more a skilled craft based on qualified experience than a scientific discipline. I think that at present our knowledge lies somewhere in between these states.

Being aware of the phenomena described, I have chosen to develop the subject in this chapter, the description of the tubular film reactor concept, from a practitioner's point of view.

2. THE BIOLOGICAL FILM REACTOR—A CONCEPT

2.1. General Considerations

Biological film reactors are in widespread use in microbial technology and belong to the traditional reactor types used for aerobic waste-water treatment, the manufacture of vinegar, and so on. In the cultivation of mammalian cells, roller bottle culture most clearly evidences the principles of a biological film batch reactor (see Fig. 1). Both suspended and adherent cells can be cultivated very successfully by this technique. Adherent cells spread on the surfaces of the bottle and finally form a complete cell layer. As long as the bottle is kept in the rolling state, a liquid nutrient medium film permanently covers the inside of the bottle and the cell layer. The nutrient medium film is continuously renewed during the submerged phase in the liquid nutrient bulk and

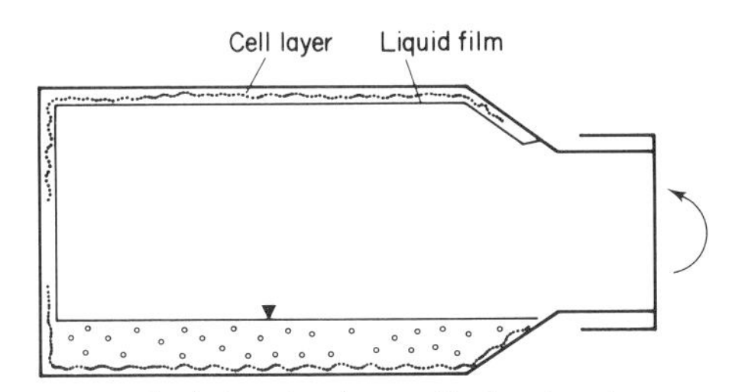

Fig. 1. Standard situation for a roller bottle culture.

supplied with gas during its residence time in the head-space. The roller bottle culture is a typical batch culture technique. Thus, as the mass transfer requirements change during the course of a batch culture, the ratios of surface film area to nutrient volume to head-space volume cannot be optimized. Such ratios could only be optimized if both the gas in the head-space and the nutrients in the liquid phase in the roller reactor were continuously renewed. On the other hand, the technique of treating cells in the roller bottle satisfactorily meets the mixing and mass transfer requirements as compiled in Table I:

1. Shear stress is minimized.
2. There is no direct contact between gas phase and cells (i.e. no gas bubble disengagement).
3. Obligate adherent cells can be cultivated.
4. The conditions for suspended non-adherent cells are somewhere between static and mixed.
5. There is reasonably homogeneous distribution of nutrients (i.e. no concentration gradients in the liquid bulk).

For these reasons roller bottle culture has, in practice, become a widespread and popular technique. In spite of the disadvantages that it cannot be used for continuous perfusion and of its expenditure of manual labour in the handling of multiple units, roller bottle culture is routinely used for even production scales in many laboratories. Therefore this generally acknowledged cultivation technique deserves further improvement, particularly with respect to aspects such as continuous perfusion and cell immobilization, continuous renewal of the gas head-space, scale-up, and the replacement of manual handling by automated unit operations. All of the approved mixing and mass transfer principles, as far as they are intrinsic to the roller bottle technique, should be retained.

It is possible to render a roller bottle reactor accessible to continuous perfusion by liquids and gases, but there are problems in the mechanical engineering of the apparatus. These aspects may be of great practical importance, but they are not within the scope of this theoretical analysis. However, if it is agreed that a roller bottle reactor can be converted from batchwise to continuous operation, the matter in hand must be reconsidered in quantitative and practical terms.

2.2. Quantitative Design Considerations

One very important requirement in the design of a bioreactor is reduction of all its relevant physical functions to a common qualitative and quantitative

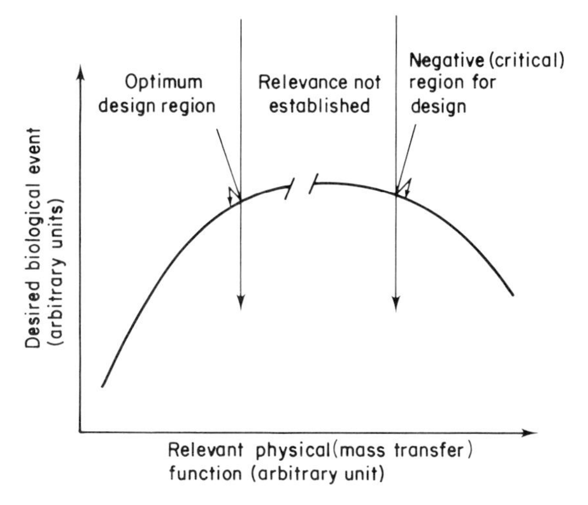

Fig. 2. Physiological response to physical input variables.

formula of performance. Another requirement for the designer is to achieve this object by the simplest, most practicable and safest configuration. Once the qualitative design criteria (as compiled in Table I) have been established, their quantitative dimensions have to be determined. This is usually very complicated. These quantitative dimensions have to be determined from an interphase in which complex biological effects (i.e. overall physiological reactions) are guided by complex physical functions. The problem here is that it is always difficult to determine the validity of a correlation between physical input variables and the physiological response observed. The problems starting are schematized in Fig. 2. Any physical function of a bioreactor, as far as it is relevant for the physiological response of a biological system, follows the generalized patterns indicated in Fig. 2. It is in the hand of the designer to find the optimum formula which meets the various physical functions quantitatively in the correct order of magnitude (i.e. within the design region). At least, this has to be achieved for those functions which dominate the overall performance of the reactor.

Often these quantitative relationships are difficult to establish, particularly when complex mixing functions have to be scaled up in mechanically agitated reactors. For a film reactor like the roller bottle, these correlations are relatively easy to establish. Shear stresses, detrimental to the cells, do not occur and mixing is generally adequate. All other functions of a film reactor can be achieved by harmonizing surface areas (i.e. liquid film surfaces for gas–liquid mass transfer and biological film surfaces for cell growth) and volumes (volumes for nutrient media and gaseous head-spaces) with each

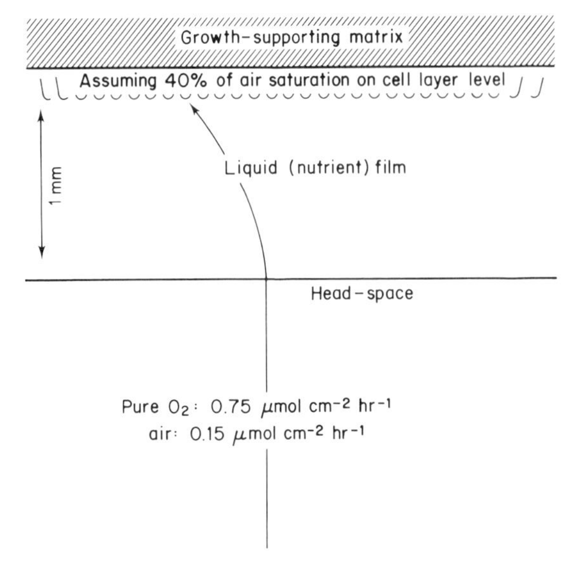

Fig. 3. Cells adhering to a growth-supporting matrix supplied with oxygen across a liquid film (calculated for a diffusivity $D \approx 4 \times 10^{-5}$ cm^2 s^{-1}).

other. The main problem for the reactor designer is to determine the correct geometrical proportions, which, once they have been properly established, can be scaled-up on the basis of geometric similarities without affecting the static and dynamic performance of the reactor.

For an *in vitro* system, in which gas–liquid mass transfer is a limiting step, and in which direct sparging of gas bubbles is not advisable, two main questions have to be answered: first, how much oxygen has to be supplied for the propagation and maintenance of a defined number of mammalian cells; second, how much gas–liquid film interphase is necessary for that oxygen supply? According to experiments done in our own laboratory with five different cell lines and applying four different cultivation methods, the specific oxygen demand of propagating cells correlates with the growth rate and is in the order of 0.12 μmol hr^{-1} for 10^6 propagating cells and roughly one-third for non-propagating (stationary) cells (exceptions not considered). The second question is that of oxygen transfer capacity across a liquid–gas interphase. Again, quantitative data have been published (*1*). The generalized situation is illustrated in Fig. 3, in which cells adherent to surface as a monolayer consume oxygen supplied by diffusion across liquid nutrient films.

A crude calculation shows that in a conventional film reactor, much more oxygen can diffuse across the liquid films than the cells below the surface can consume. The oxygen transfer capacity of 1 cm^2 of liquid film area would, in

TABLE II Relative Oxygen Transfer Capacities of Liquid Film Aeration for Adherent and Immobilized Propagating Cells

Cell type	Number of cells per cm^2 of growth supporting matrix	Dimensional surface growth matrix/interphase film for oxygen supply	Practical limits of immobilized cell density per cm^3	cm^2 of gas per litre interphase for oxygen supply of immobilized cells (cm^{-1})
Obligate adherent	low 2×10^4	100 (500)[a]	10^7	5 (1)[a]
	high 2×10^5	10 (100)[a]		
Pseudo-adherent (hybridomas)	10^6	5 (50)[a]	5×10^7	25 (5)[a]
Non-adherent (suspended)	—	—	10^8	50 (10)

[a] Pure oxygen in head-space assumed.

theory, meet the oxygen consumption of between 2.5×10^6 propagating cells (with air in the head-space) and 2.5×10^7 stationary cells (with pure oxygen in the head-space). Oxygenation is definitely not the limiting factor in a biological film reactor. This emerges clearly from Table II, in which the oxygen transfer capacity across films is related to the surface area necessary for the attachment of cells. The conclusion drawn from these data is simple. A one to two orders of magnitude larger area is required to keep these cells attached to surfaces than to supply those cells with sufficient oxygen. The disproportion between cell layer surface area and film area needs to be changed.

Cell immobilization to high cell densities is another important aspect of any animal cell reactor. The maximum cell densities reached by immobilization in any reactor are limited by its mixing parameters and also by the cell type, morphology, cell size, etc. The practical limits may be assumed to be somewhere between 10^7 and 10^8 cells per ml of reactor volume. Table II shows that the limits of the oxygen transfer capacity of a film reactor might be reached, at least theoretically, in the region of 5×10^7 cells ml^{-1}. We could imagine, as a first approximation, that a cubic volume containing cells surrounded by a gas–liquid interface leads to a quotient of 6 cm^{-1}. According to Table II, this would represent (theoretically) an oxygen transfer equivalent which meets the oxygen supply for 6×10^7 propagating cells contained in a cubic volume of 1 ml. It is probably not possible to translate the full potential of the film reactor principles into practical reality. There are, however, various approaches for optimization, as shown in Fig. 4. By simply modifying the surface structure of the cell carrier matrix, as visible in Fig. 4(*a*), the surface area for cell attachment might be easily increased 10-fold. As long as the nature of the surface is hydrophilic, it might hold a liquid film of roughly 0.1 cm^3 per

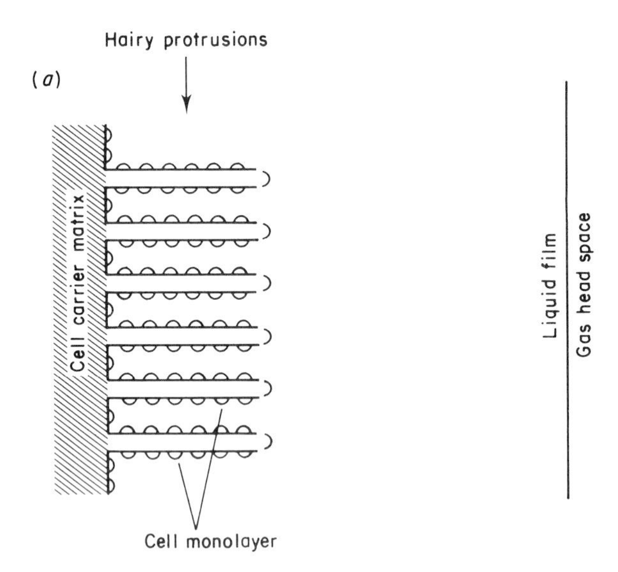

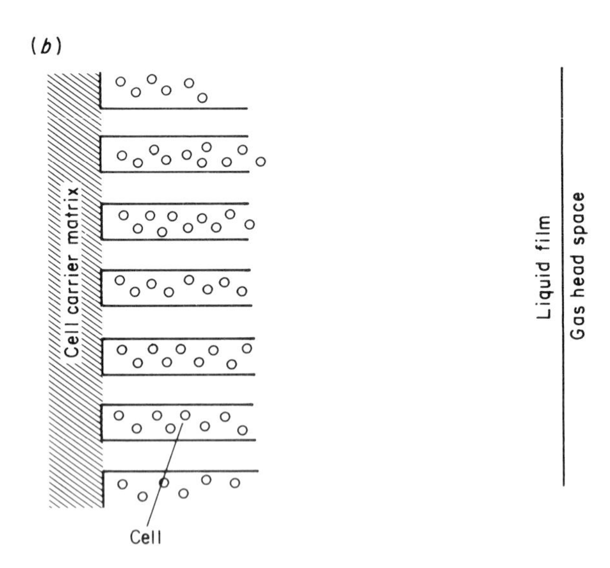

Fig. 4. Idealized situation in liquid biological film.

cm^2 as long as it is kept in gentle movement. This would lead to a completely changed relationship between the number of cells adhering (2×10^6 cm^{-3}) and the cell density in the liquid film (2×10^7 cm^{-3}). The oxygenation across the liquid film would still be far from becoming the limiting step. The situation sketched in Fig. 4(*b*), in which non-adherent cells are kept immobilized in a stagnant extracapillary liquid film, amounts to the same thing. Similar effects could be expected when a fibrous fleece is used as the material for the cell carrier matrix.

Thus far, only the detailed events at the film interphases have been considered. The next step is to coordinate the mixing detail with bulk mixing phenomena and other desired functions that the reactor should have. This can be achieved if the cell and film carrier parts are put together in a cylindrical vessel as shown schematically in Fig. 5. As long as the vessel is rotated with a speed comparable to that for a roller culture, the liquid films are permanently equilibrated with the gas head-space as well as with the liquid bulk during the submerged phase. If the distance between the single film carrier elements is large enough that liquid bridging does not occur, a permanent renewal of the head-space gas is also guaranteed. Continuous perfusion of the liquid phase and of gas mixture is possible.

The procedures for starting the culture are as follows. During steam-sterilization the outlet pipes are turned close to the wall of the vessel in order to drain off the condensate. Thereafter, the vessel is switched to rotate at a speed comparable to that of a spin-dryer. Most of the liquid adherent in film is then removed and the reactor is ready for inoculation. As determined by the amount of inoculum, either a part or the total volume of the reactor can be inoculated by feeding the cell suspension slowly through the inlet pipes. As long as the reactor is gently rolled, the cell suspension is sucked off into the

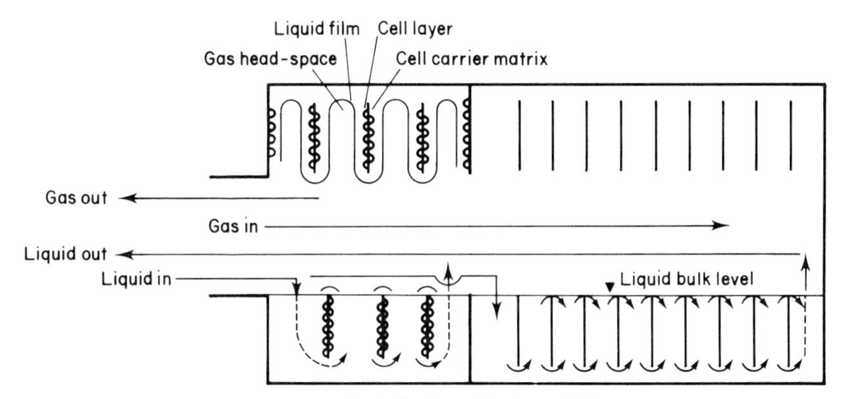

Fig. 5. Schematic of tubular liquid biological film reactor.

Fig. 6. Home-made prototype liquid biological film reactor.

film carrier elements. Thereafter, fresh medium is added. This procedure can be used for conditioning, trypsinization, cell-washing, and other purposes.

I will end these quantitative theoretical considerations with an outlook based on the film reactor concept developed recently. If the reactor configuration presented in Fig. 5 were expanded to a reactor size of 1 m in length and 1 m in diameter, more than 1000 m^2 of growth-promoting matrix (i.e. approximately 10 000 roller bottle equivalents), carrying roughly up to 10^{12} cells in approximately 300 litres of nutrient medium under continuous perfusion would be needed.

3. EXPERIMENTAL TESTING

All the experiments so far have been done in glass-blower-made pre-prototype versions and with workshop-manufactured cell-carrier matrices (see Fig. 6). Various materials for cell attachment have been tested according to their availability on the market. A mouse/mouse hybridoma cell line (U0208) was used in all experiments as reference. Preliminary results have been published. Several cheap fibrous matrices are available in bulk quantities,

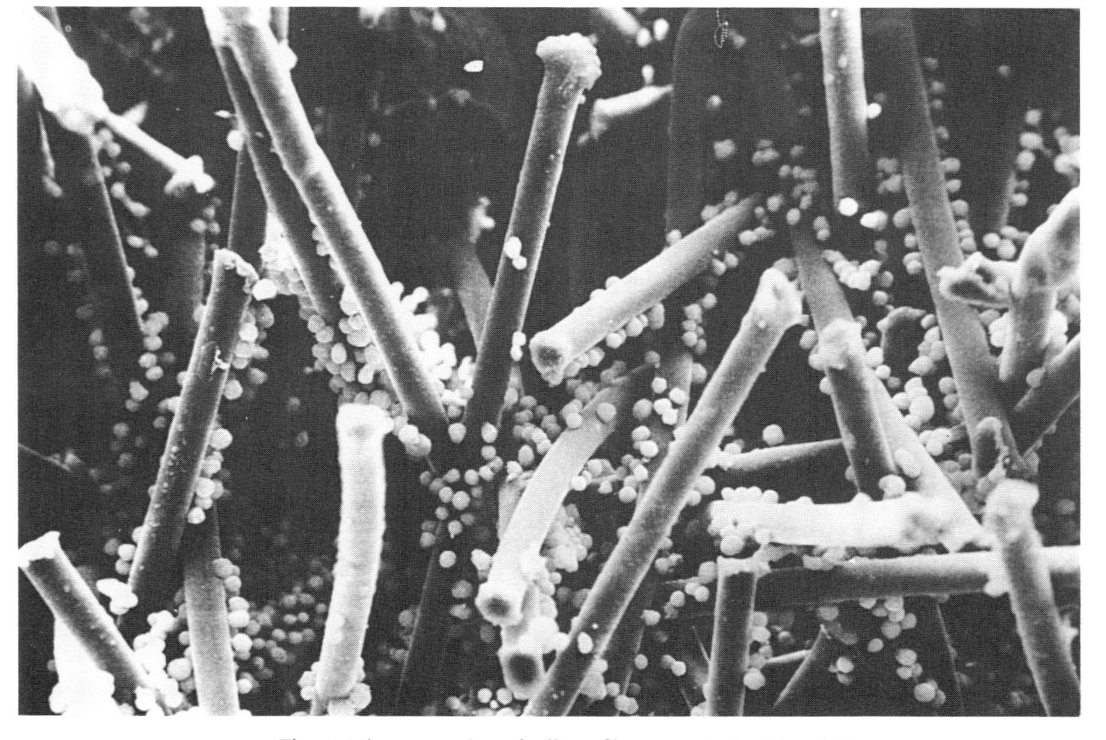

Fig. 7. Microscope view of cells on fibrous carrier ($\times 206 \times 0.8$).

such as glass, polyester, polystyrene, etc. (that can be used for the attachment of hybridomas). Up to 10^7 cells cm^2 can be attached on the fibrous surface (compare Figs 3 and 7).

In conclusion, it can be said that the concept of the biological film reactor as presented here could be fully achieved in experimental practice. It is now in the hands of the manufacturers to construct the first technically proficient prototype.

Acknowledgements

This work was supported in part by the Austrian Ministry of Science and Research and by the MBR Company, Wetzikon, Switzerland.

REFERENCE

1. Katinger, H. W. D. (1987). Principles of animal cell fermentation. *Dev. Biol. Stand.* **66**, 195–209.

10

Protein Production from Mammalian Cells Grown on Glass Beads

PETER C. BROWN, CARLOS FIGUEROA,
MAUREEN A. C. COSTELLO, ROBERT OAKLEY, and
SUSAN M. MACIUKAS
Bio-Response, Inc.
Hayward, California, U.S.A.

1. INTRODUCTION

At the most elementary level, mammalian cells of interest to biotechnologists may be classified according to their requirement for anchorage-

ANIMAL CELL BIOTECHNOLOGY VOL. 3
ISBN 0-12-657553 3

dependent growth or for the ability of the cells to grow freely in suspension. These two fundamentally different requirements most probably reflect the growth requirements of the tissue of origin of the cells: cells from solid tissues have evolved an intricate cell–cell and cell–substrate interdependence, whereas cells of the circulatory system such as lymphocytes have of necessity developed substrate-independent growth characteristics. The distinction between cells that require anchorage-dependent growth and cells that grow freely in suspension becomes clouded when one considers permanent cell lines.

For reasons that are both largely unknown and beyond the scope of this contribution, the establishment of permanent cell lines from anchorage-dependent primary cell isolates often confers on these cells the ability to grow either in an anchorage-dependent mode or, after a suitable adaptation period, in suspension. The Chinese hamster ovary (CHO) cell line is well known for this dual capability, and for this and a number of other reasons this cell line is a favourite host cell for the expression of heterologous genes introduced by recombinant DNA techniques. When grown in suspension, unlike cells of lymphoid origins, suspension-adapted cells usually grow as aggregates rather than as true single-cell suspensions. Conversely, the derivation of truly anchorage-dependent cell lines from cells of the lymphocyte series is, in our awareness, unprecedented.

For the large-scale production of secreted proteins and viruses from anchorage-dependent cell lines, there may be both biological and technical reasons for growing cells in the anchorage-dependent mode rather than as suspension aggregates. As has been alluded to above, some cell lines will not adapt to growth in suspension and one has no other choice as to the method of growth. In addition, it is known that with many cell lines, cellular productivity for proteins or viruses of interest is significantly reduced when these cell lines are grown as suspension aggregates and not in monolayer culture. The exact reasons for this decrease in productivity are obscure but it has been shown that the rearrangement of the cytoskeletal architecture and the loss of cell–cell contact and of a defined extracellular matrix influence both the patterns and levels of specific gene activity (*2, 3, 6, 9, 12*). In all likelihood, the intracellular processing and transport of newly synthesized proteins through the cytoplasm is affected as well, although we are unaware of data in support of this view.

From an engineering point of view, anchorage-dependent cells are ideally suited for immobilization within appropriately designed bioreactors. Cell attachment is mediated through specialized structures, called adhesion plaques, on the underside of the cell. Furthermore, unlike the majority of microorganisms which have been incorporated into immobilized cell systems, anchorage-dependent mammalian cells secrete at least a portion of the factors required for adherence to appropriate surfaces. For example, collagen and fibronectin facilitate cell attachment and are also produced by many anchorage-dependent cell lines (*1, 5*). Additional attachment factors are normally provided by serum supplements added to the basal growth medium. The primary advantages of immobilized cell systems are largely cell-free spent media (which simplifies downstream processing), extended production periods, high cell densities, and rapid media turnover rates within the bioreactor (*18*).

Numerous approaches to the large-scale growth of anchorage-dependent cells have been described. Outside of roller bottles, variations of microcarrier culture seem to be the most widely used technique (*17, 19, 20, 21*). Other techniques include growth on monolithic ceramic supports (*10*), semipermeable hollow fibres (*7, 16*) and flat sheets (*13*). Finally, a technique which holds great promise and which is the subject of this chapter is the growth of anchorage-dependent cells within packed beds of glass beads (*4, 15, 22*).

2. MAJOR COMPONENTS OF THE GLASS BEAD SYSTEM

Figure 1 shows a schematic of the packed-bed bioreactor system employed at Bio-Response. The essential components include the bioreactor chamber itself, the oxygenator, and the pump which circulates medium between the bioreactor and the oxygenator. Not depicted in this figure are the media delivery and accumulation systems which allow this system to be continuously perfused with fresh nutrients for periods in excess of 1 year.

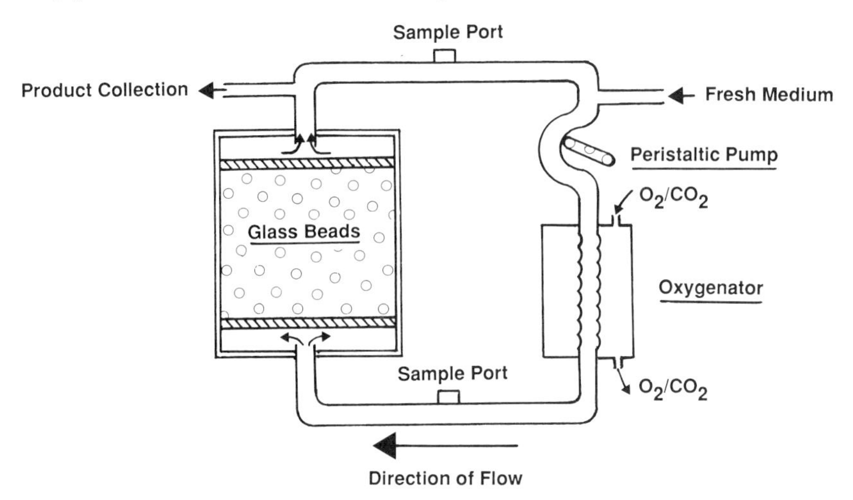

Fig. 1. Schematic diagram of the glass-bead system.

2.1. Bioreactor Chamber

Solid glass beads were chosen as the insoluble support matrix for the bioreactor because of the inertness of glass, the superiority of this surface for anchorage-dependent growth, and the regular packing geometry found in the bed formed by spherical glass beads. This latter characteristic makes possible the relatively unrestricted flow of media throughout the packed bed. Additionally, the beads are relatively inexpensive, are reusable, and readily conform to the container into which they are placed. While a range of sizes is available, 3-millimeter diameter beads were chosen because this size represents the best compromise between available surface area and openness of the channels formed within the packing array. These channels, in the range of 200–400 μm in diameter, minimize the possibility of channelling resulting from occlusion by cellular debris which accumulates during extended culture periods. The calculated surface area of 3-mm glass beads is approximately 1 m^2 per litre of packed beads.

The chamber of the bioreactor is basically modelled after a chromatography column, with dispersion plates included top and bottom. The geometry of the chamber is critical and, while rigorous studies have not been attempted, aspect ratios in the range of 1 : 1–2 : 1 (height : diameter) appear to be appropriate for a number of cell lines. Similar conclusions were reached by others (*22*). Higher aspect ratios may result in uneven distribution of cells with higher localized densities accumulating proximal to the inlet of the chamber. More will be said of this important observation later. While not shown in the schematic, an optional medium dialysis module may be interposed within the circulating loop downstream of the bioreactor chamber. These modules consist of hollow-fibre cartridges. Because of the low-molecular-weight cut-off of these hollow-fibre membranes, components in the basal media may be exchanged continuously with an external media source while higher-molecular-weight components such as costly media supplements or serum, as well as most relevant products, remain within the bioreactor circuit. This dialysis option, while adding to the complexity of the system, has proved effective in reducing the costs associated with production in complex medium and has resulted in higher concentrations of relevant product in the conditioned-medium effluent stream.

2.2. Oxygenator

A variety of gas-permeable membranes has been developed for blood oxygenation during thoracic surgery. Having evaluated a number of the materials used in these devices, we have found that solid silicone-rubber-based membranes are superior to hydrophobic semi-permeable materials, in that the silicone materials do not permit the passage of medium as do hydrophobic materials after 1–2 weeks of service. Further, silicone may be cast in either flat sheets or hollow fibres and both configurations have been successfully used in the packed-glass-bead system.

While a number of designs are possible, turbulence and fluid shear forces within the oxygenator should be sufficient to minimize boundary layer effects and simultaneously prevent the accumulation of cells and cellular debris which will eventually compromise the gas transfer properties of the silicone membrane.

Finally, sparging alone within an oxygenator chamber has been successfully used on small systems under 0.2 litre. Sparging in larger systems has not been tried, although we believe that a bubble-free, oxygen-saturated effluent stream is achievable from such a sparged system. However, foaming would be a serious problem, especially in media with substantial proportions of serum (>3%).

2.3. Pumps

Peristaltic pumps of various sizes are available and have been used with bioreactors up to 60 litres in volume. These pumps are especially well suited to continuous perfusion because of simplicity and reliability, ease of sterilization, and low shear forces. As systems are scaled upwards beyond 60 litres, pumping demands may not be met by peristaltic pumps that are currently available.

3. OPERATIONAL CONSIDERATIONS

Assembled systems are autoclaved, installed within an incubator chamber, and connected to a media source in an adjacent cold room as well as to a reservoir for spent medium. A cell inoculum is prepared, usually from roller bottles, and introduced rapidly into the bioreactor chamber. To allow for attachment, fluid circulation is reduced for the first 2–4 hours. Thereafter, circulation is gradually increased.

3.1. Process Control

Within a matter of weeks, continuous perfusion systems such as the packed-glass-bead systems can achieve near steady-state conditions with

TABLE I Process Control

Parameter group	Response
1. O_2, CO_2, pH	Alter gas composition, medium perfusion rate, circulation rate
2. Glucose, lactic acid ammonia	Alter medium perfusion rate
3. Relevant product, lactate dehydrogenase	Alter medium perfusion rate

respect to pH, dissolved oxygen, glucose and lactate levels. As a result, the benefits of on-line monitoring are not as obvious as with the continually changing conditions found in discontinuous batch processes. Furthermore, because of the duration of production runs achievable (greater than a year), the reliability of even the most durable pH and dissolved oxygen probes is questionable. Since off-line measurements are required for parameters that cannot currently be measured by indwelling probes, we perform all process control analyses on samples taken from sample ports within the fluid circulation loop. A summary of process control parameters that we currently monitor is shown in Table I. Data taken at 48–96 hr intervals in addition to equivalent data acquired at earlier times dictate decisions related to changes in gas composition (for pH control), media perfusion, and circulation rate.

3.2. Circulation Rate

The components of the glass-bead system are united through a circulating fluid loop of constant volume to which fresh medium is added and from which conditioned medium is removed. The primary function of continuous medium circulation is to shuttle oxygen from the oxygenator to the cells within the packed bed and to remove excess CO_2 produced through cellular respiration. A secondary and less well recognized function of active medium circulation is to prevent the overgrowth of cells and the resulting channelling of fluids within the packed bed.

The relationship between circulation rate and bioreactor shape is complex and should be optimized for each cell line. For most cell lines, high lineal velocities appear to give the best production of secreted products. With lower velocities, dissolved oxygen differences across a mature 12-litre bioreactor in the range of 80 mmHg have been observed. Further, we suspect that other significant nutrient gradients exist as well, especially at these lower flow rates, and that these gradients are responsible for the higher localized cell densities observed proximally to the medium inlet ports, regardless of whether the inlet is at the top or the bottom of the reactor. With higher circulation rates, these cell density gradients are diminished and total bioreactor productivity over time is increased.

None of the cells that we have grown in the packed-glass-bead system have been maintained in a non-dividing state. Accordingly, both viable and non-viable cells may be found in the circulating loop. To facilitate removal of these cells from the system along with the conditioned media effluent, no stagnant flow regions should be accessible to the fluid stream unless these regions are specifically designed to accumulate and remove these cells. Indeed, cell traps of proprietary design have been designed around this principle.

4. EXPERIENCE

4.1. Cell Lines

In Table II are listed the number and variety of anchorage-dependent cells grown in the packed-glass-bead system. Where possible, the identity of the protein of interest produced by each cell line is also indicated. Noteworthy among the cell lines listed is the human Bowes melanoma (*8*), which, after a brief expansion in serum containing medium, was switched to totally protein-free medium and maintained for periods in excess of a year. While the cell line has been uniquely selected for protein-free growth (*23*), we found that two CHO cell lines and the human kidney carcinoma could also be maintained for periods of at least 4 weeks in protein-free medium in the packed-glass-bead system. We believe that the combination of high cell density and continuous perfusion enables cell populations to condition basal media adequately without total depletion of nutrients, as would be found in most discontinuous processes.

TABLE II Cell Lines Grown in Glass-bead Bioreactors

Cell line	Relevant product
Mouse mammary fibroblast	Heterologous protein (via rDNA)
Chinese hamster ovary (5 lines)	Heterologous protein (via rDNA)
Human kidney carcinoma	Constitutive product
Human melanoma	Tissue plasminogen activator
Hepatocyte/hepatoma hybridoma	Acetylcholinesterase

4.2. Cell Density

An unexpected advantage of the packed-glass-bead system is that markedly higher cell densities may be achieved than would be predicted on the basis of surface area alone. For example, a genetically engineered CHO cell line was grown in a 1-litre glass-bead bioreactor in protein-free medium for the last 41 days of the 61 day project. Cells were trypsinized within the culture chamber and a total of 1.3×10^{10} viable cells were recovered. If this number of cells were cultured on monolayers, approximately 2–2.5 m^2 of surface would have been required rather than the 1 m^2 surface area available on 1 litre of glass beads.

When viewed through the transparent wall of some bioreactors, aggregates of cells are seen to accumulate between the packed beads. These loose aggregates may explain the unexpectedly high densities achieved in the CHO cell project mentioned. Furthermore, approximately 33% of the total viable cells recovered at the end of this project were found in the initial drain of the bioreactor prior to trypsinization. Therefore, these aggregates were passively restrained within the packed bed under the low lineal flow of medium employed in the bioreactor. Finally, it is unlikely that the accumulation of aggregates would occur with cells that are truly anchorage-dependent. The CHO cells, as has been mentioned, will readily adapt to suspension aggregate growth and this characteristic undoubtedly explains the presence of loosely adherent sub-populations within the packed-glass-bead system.

4.3. Production of Acetylcholinesterase

The large-scale production of acetylcholinesterase (AChE) from a hepatocyte/hepatoma hybridoma (*14*) was solved in a somewhat unique manner utilizing the packed-glass-bead system. Because of the extremely low level of secretion (0.3 ng/10^6 cells/day) and the presence of the enzyme in serum, production under serum-free conditions was required. Unfortunately, none of the serum-free media with which we were working at the time would sustain growth upon insoluble supports such as glass beads; 6–7 days after removal of serum, the cells detached. These detached cells remained viable and we attempted growth as suspension aggregates. This proved untenable because of our inability to grow these cells with agitation in less than 1–2% serum and the lower levels of productivity observed with suspension aggregates (0.1 ng/10^6 cells/day).

Accordingly, we developed a protocol whereby packed-glass-bead bioreactors, once seeded with E-2 cells, were pulsed with calf serum on a weekly

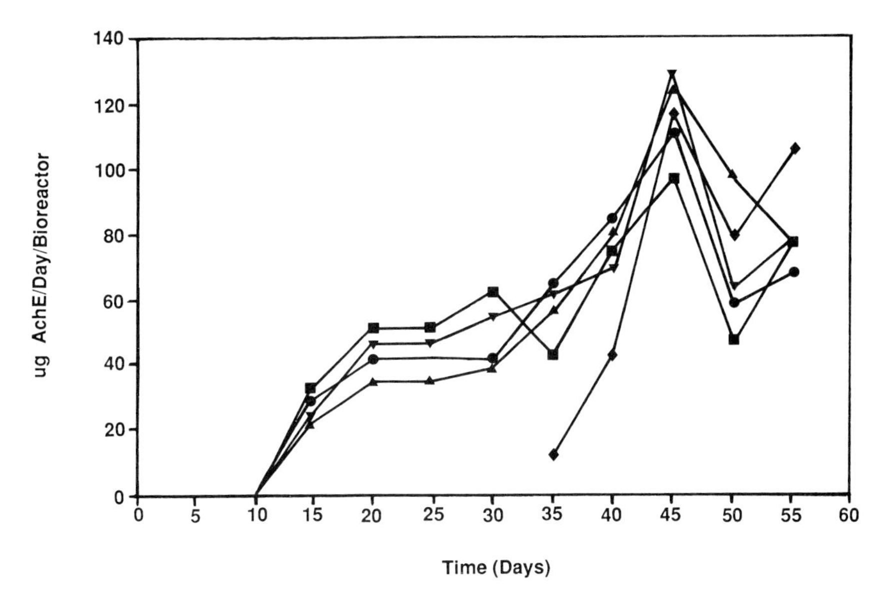

Fig. 2. Production of acetylcholinesterase. A subclone of the E-2 hepatocyte × hepatoma cell line (*14*) was inoculated (2×10^9 cells) into each of 5–12-litre glass-bead bioreactors and cultivated for 57 days in a serum-free, defined medium without antibiotics except for pulses of heat-inactivated calf serum at 7-day intervals (see text). Process control was as indicated in Table 1.

basis for the duration of production. During the 6–8 hr pulse, serum content rose to approximately 2.5%. By the end of a 24 hr wash-out period, with serum-free, defined medium, serum levels were well below 1%. Perfusion with serum-free, defined medium was continued for the next 6 days until the process was repeated. Using this procedure, negligible quantities of serum were carried over into the pooled conditioned medium. As a result, this medium could be concentrated approximately 100-fold using conventional membrane concentrators prior to more selective protein recovery processes.

Five 12-litre glass-bead bioreactors seeded with E-2 cells were established and maintained for 59 days. Productivity for the five systems is shown in Fig. 2. Over the course of the production, approximately 2400 litres of conditioned media were collected which contained about 15 mg of AChE. For comparative purposes, we estimate that to produce an equivalent amount of material in static or roller bottle culture, the combined 24 hr harvests from about 4×10^{13} cells would have been required. Production data aside, these data show the reproducibility of production for four out of the five bioreactors represented in this figure.

5. DISCUSSION

The growth of anchorage-dependent cells in large-scale culture introduces a variable not encountered in the routine culture of cells in suspension. In particular, cells must be provided with a compatible matrix possessing high surface-to-volume ratios and a structure which allows for the unrestricted delivery of oxygen and non-volatile components of growth medium. The packed-glass-bead system described herein and elsewhere (*4, 15, 22*) satisfies this additional requirement.

The feature that has most impressed us about the glass-bead system is that, with very little direct monitoring or hands-on management, production runs routinely exceed 4 months and occasionally may exceed 1 year. Naturally, for such extended production, the cell line chosen for production must be stable with respect to production of the relevant protein. A second characteristic which not all cell lines share, and which is highly desirable for the glass-bead system, is that the synthesis and secretion of the protein of interest should not be strongly coupled with the cell division cycle. Regardless of cell line chosen or the media in which it is grown, cells do not divide as rapidly in a mature glass-bead system as they would in monolayer culture. In this sense, the cultures assume the characteristics of an extended stationary phase seen in re-fed static cultures. If the protein of interest is not expressed well in healthy stationary phase cultures, then it probably will not be expressed well in mature glass bead cultures, and other culture methods should be pursued.

A question which arises is the scale to which the glass-bead system could be developed. While packed beds up to 3300 litres have been proposed (*15*), we believe that multiple glass-bead bioreactor systems in the order of 200 litres represent a workable compromise between redundancy and individual bioreactor capacity. With 20 such systems operational for a full year, and with cells that produce an average 5 mg of relevant protein per litre of medium, upwards of 20 kg could be produced. The major problems to be overcome with scale-up include adequate pumping, high capacity oxygenators, sterilizability, and inoculation strategies. Currently, our direct experience stops at bioreactors larger than 60 litres, but scale-up to 200 litres is being pursued.

In summary, the packed-glass-bead bioreactor is a superbly serviceable design for the growth of anchorage-dependent cells. Cultures may be maintained for very long periods; the productivity of the cells maintained for these periods is largely dependent upon how well the fundamental needs of the cells within the bioreactor are met by components of the system outside the bioreactor itself.

REFERENCES

1. Alitalo, K., Keski-Oja, J., and Vakeri, A. (1981). Extracellular matrix protein characterize human tumor cell lines. *Int. J. Cancer* **27**, 755–761.
2. Ben-Ze'ev, A. (1984). Cell/cell interaction and cell shape related control of intermediate filament protein synthesis. *In* "Molecular Biology of the Cytoskeleton" (G. Borisy, D. Cleveland, and D. Murthy, eds.), pp. 435–444. Cold Spring Harbor Laboratory, New York.
3. Bissel, M. J., Hall, G. H., and Perry, G. (1982). How does the extracellular matrix direct gene expression? *J. Theor. Biol.* **39**, 31–68.
4. Brown, P. C., Costello, M. A. C., Oakley, R., and Lewis, J. L. (1985). Applications of the mass culturing technique (MCT) in the large scale growth of mammalian cells. *In* "Large Scale Mammalian Cell Culture" (J. Feder and W. Tolbert, eds.), pp. 59–71. Academic Press, New York.
5. Castellani, P., Siri, A., Rossellini, C., Infusini, E., Borsi, L., and Zardi, L. (1986). Transformed human cells release different fibronectin variants than do normal cells. *J. Cell Biol.* **103**, 1671–1677.
6. Farmer, S. R., Wan, K. M., Ben-Ze'ev, A., and Penman, S. (1983). Regulation of actin mRNA levels and translation responds to changes in cell configuration. *Mol. Cell Biol.* **3**, 182–189.
7. Knazek, R. A., Gullino, P. M., Kohler, P. O., and Dedrick, R. L. (1972). Cell culture on artificial capillaries: An approach to tissue growth *in vitro*. *Science* **178**, 65–69.
8. Leist, C., Meyer, H. P., and Feichter, A. (1986). Process control during the suspension culture of a human melanoma cell line in a mechanically stirred loop bioreactor, *J. Biotechnol.* **4**, 235–246.
9. Liang Li, M., Aggeler, J., Farson, D. A., Hatier, C., Hassell, J., and Bissel, M. J. (1987). Influence of a reconstituted basement membrane and its components on casein gene expression and secretion in mouse mammary epithelial cells. *Proc. Natl. Acad. Sci. U.S.A.* **84**, 136–140.
10. Lydersen, B. K., Pugh, G. G., Paris, M. S., Sharma, B. P., and Noll, L. A. (1985). Ceramic matrix for large scale animal cell culture. *Bio/Technology* **3**, 63–67.
12. Saadat, S., and Thoenen, H. (1986). Selective induction of tyrosine hydroxylase by cell–cell contact in bovine adrenal chromaffin cells is mimicked by plasma membranes. *J. Cell Biol.* **103**, 1991–1997.
13. Scheirer, W. (1987). High density growth of animal cells within cell retention fermenters equipped with membranes. *In* "Animal Cell Biotechnology," Vol. 3. (R. Spier and J. Griffiths, eds.), pp. 263–281. Academic Press, London.
14. Schuman, R. F., and Hunter, K. W., Jr. (1986). Secretion of acetylcholinesterase by a mouse hepatocyte × rat liver cell hybrid culture. *In Vitro Cell Dev. Biol.* **22**, 670–676.
15. Spier, R. E. (1985). Monolayer growth systems: hetergeneous unit processes. *In* "Animal Cell Biotechnology," Vol. 1. (R. Spier and J. Griffiths, eds.), pp. 243–263. Academic Press, New York.
16. Tharakan, J. P., and Chau, P. C. (1986). The radial flow hollow fibre bioreactor for the large scale culture of mammalian cells. *Biotechnol. Bioeng.* **18**, 329–342.
17. Tolbert, W. R., Lewis, C., White, P. J., and Feder, J. (1985). Perfusion culture systems for production of mammalian cell biomolecules. *In* "Large Scale Mammalian Cell Culture" (J. Feder and W. Tolbert, eds.), pp. 97–128. Academic Press, New York.
18. Van Brunt, J. (1986). Immobilized mammalian cells: The gentle way to productivity. *Bio/Technology* **4**, 505–510.

19. Van Wezel, A. L. (1985). Monolayer growth systems: homogeneous unit processes. *In* "Animal Cell Biotechnology," Vol. 1. (R. Spier and J. Griffiths, eds.), pp. 266–282. Academic Press, New York.
20. Varani, J., Hasday, J. D., Sitriu, R. G., Brabaker, P. G., and Hillegas, W. (1986). Proteolytic enzymes and arachidonic acid metabolites produced by MRC-5 cells on various microcarrier substrates. *In Vitro Cell Dev. Biol.* **22**, 575–582.
21. Verax, Inc., technical brochure.
22. Whiteside, J. P., and Spier, R. E. (1985). Factors affecting the productivity of glass sphere propagators. *Dev. Biol. Stand.* **60**, 305–311.
23. Wilson, E. L. (23.12.1983). European Patent Application 0113319.

11

High-density Growth of Animal Cells Within Cell Retention Fermenters Equipped with Membranes

W. SCHEIRER
Sandoz Forschungsinstitut GmbH,
Vienna, Austria

1. INTRODUCTION

The growth of animal cells within suspension fermenters can have some shortcomings owing to the relatively low cell densities which are normally

ANIMAL CELL BIOTECHNOLOGY VOL. 3
ISBN 0-12-657553 3

reached in classical fermenter designs. The reason for limited cell growth may be found in the components of the culture medium. The possible concentration range is quite narrow and some of the essential components are unstable under cultivation conditions. Animal cells, which are not adapted for extracorporeal survival, are used to re-feeding and conditioning by the somatic systems. Under such tissue conditions, animal cells show about 100- to 1000-fold higher cell densities compared to batch cultivation *in vitro*. Furthermore, there is some evidence of beneficial cellular interaction at high cell densities. One may see this in two phenomena; one is the better growth of most cell lines in soft agar cultures, where the convection of medium is prevented and, therefore, the build-up of a microenvironment by the cells is favoured; the other is that researchers working on high-density cultures have frequently found a reduced serum demand in their cultures, leading to the assumption that growth factors from the serum have been partially replaced by cellular products (personal communications).

These considerations might suggest a high-density cultivation system to be superior to a conventional one.

There are other differences between high-density and low-density systems, which should be considered. As one must use some kind of perfusion of nutrient medium within such systems, the mean residence time of medium and product respectively are drastically reduced in high-density cultures. The factor of reduction is inversely correlated with the cell concentration.

The oxygen demand of the culture increases proportionally to the cell concentration and may become critical because of the low solubility of oxygen in aqueous solutions. Another very important aspect is the relation of product generation to the growth rate of the culture. Many of the high-density culture systems are designed for perfusion operation at a growth rate approaching zero (e.g. with hollow-fibre systems). Since there are some cell lines which have a more or less strong growth association of product generation, one has to consider carefully the correspondence of product generation characteristics with the production system (*2, 12*).

Looking at the currently available systems and considering the advantages, shortcomings, and current technical possibilities, one might be able to find advanced systems. However, since an "optimal" system for all applications is still a fiction, the specifications of the process must be identified before making a choice.

2. SELECTION OF PROCESS

After identification of a cell line for production purposes, it is important to analyse the characteristics of the particular cell line in order to design the process. Many of the variations may have an impact on the process, and, therefore, the identification of relevant characteristics must be carried out.

2.1. Surface Adherence

The most important point for the selection of a production system is the question of whether surface cultivation is necessary, or whether the cells can be grown in any kind of suspension system. Many attempts have been made to use surface-dependent cells within suspension systems. The best known of these is the microcarrier-approach (*24*), but other possibilities, such as fibre culture (*18*) or aggregate culture (*22*) have been used successfully. However, if there is a preference for a classical surface-type cultivation, it should be noted that there are some fermenter types which have been reported as successful units (*17, 21*). These include some of the filled-column-type (*11, 25*), others of the membrane type, such as hollow-fibres or flat membranes (*1, 10, 20*). The difference between them is that the cells are protected from hydraulic and mechanical shear within the membrane devices, and the separation effect by the membrane. This separation effect may be used for retaining cells, for retaining the product, or for the exclusion of distinct substances, e.g. immunoglobulins, from the medium. Oxygenation of the cells through membranes may also be possible (*23*). These devices duplicate in principle the environment of cells within organs and may be used for suspension-type cells as well as for surface-dependent ones.

2.2. Fragility of Cells

Another point to consider is the extreme fragility of most animal cells to mechanical and hydraulic shear forces. This is a particular problem when using cultivation vessels on a large scale, because of the exponential growth of peak energy within turbulent systems (*8*). In some cases, the use of low-shear mixing systems like air-lift or draft-tube reactors might solve the problem.

To reach a large scale with a sensitive production cell usually requires skill. For example, one way might be the use of low-shear systems mentioned above and replacing increased vessel volume by increased cell concentration. This is done by some kind of cell retention system (*5, 23*). With these systems, a cell concentration of more than $30 \times 10^6\ ml^{-1}$ may be reached, allowing the reduction of the vessel volume to 5–10% compared to a normal suspension type system, where cell concentrations of $1–3 \times 10^6\ ml^{-1}$ are common.

The other way of protecting cells from shear forces is shielding. This is done with systems in which some kind of "encapsulation" is used. The capsules may be made of semi-permeable membranes like polylysine (*19*), or of porous, spongy materials (*7, 15*). In principle there is no difference between microcapsules over one or more cells or "macrocapsules" like hollow-fibre or flat-membrane systems. From a practical point of view, one has to evaluate such systems very carefully for their technological differences, such as uniformity of system (geometry), scale-up, handling time and steps, saving of cell mass, exclusion limit of capsule material, economics, and so on.

2.3. Nutritive Demand

Considerations of the nutritive demand of cell cultures in production processes may also influence the selection of a production system. The most critical feeding substance, particularly with high-cell-density cultures, is oxygen, because of its low solubility in aqueous solutions. At oxygen concentrations below approximately 1 mg l^{-1}, the metabolism is altered drastically (*4*); one has to ensure that no portion of the cell suspension is exposed to such low concentrations of oxygen. This is particularly important with encapsulation systems in which the oxygen supply is governed by diffusion only. In this case, at the maximum cell density (packed cells, 10^8–10^9 cell ml^{-1}), the possible maximum cell layer thickness above or around a membrane is approximately 1 mm (see Fig. 1).

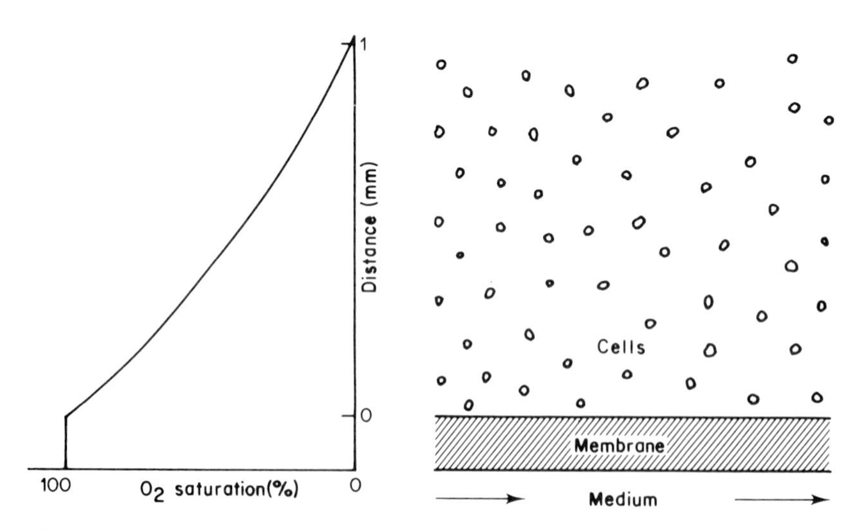

Fig. 1. Schematic situation of oxygen tension within a cell layer which is oxygenated by perfusion.

There may be similar problems with other nutrients. Even when the solubility of these is sufficient, high concentrations are very often dictated against by toxicity problems. For this reason, systems which offer the ability for re-feeding, medium regeneration, etc., have many advantages. A nutritional study in a representative system is, therefore, a very important prerequisite for process design.

2.4. Genetic Stability

When dealing with a cell line whose genetic stability is limited, there is risk of loss of productivity on scale-up; due consideration must be given to this possibility when selecting a culture system. Examples of such cell lines may be found among hybridomas, particularly interspecies hybridomas, and genetically engineered cell lines. First of all, one should try to stabilize the cell line by cell-biological means. Then one might try to slow down the loss of genes by reducing the overall number of cell duplications and by decreasing the population growth rate. Both can be achieved by appropriate cultivation systems and methods. Total cell duplications are minimized when the loss of cell mass is prevented with the help of retention systems. Such systems also enable the control of growth rate simply by using a temperature shift. With a capsule system, there is an inherent growth rate reduction by space limitation. Other growth rate reductions are also possible.

2.5. Stability of Product

The stability of the product may be another problem. With a product which is stable under cultivation conditions, there is a free choice between system and methods. With a less-stable product, there is probably the possibility of using chemical stabilizers. During the fermentation process, there is the possibility of shortening the product residence time within the system. This can be accomplished by a high-cell-density system in which the perfusion rate for the medium is high and, therefore, the product is removed earlier. Another way might be the separation of the product from the feeding medium by membranes and the application of an optimized, feeding-independent harvesting scheme.

2.6. Product-generation Characteristics

The most important and complex point to consider is the product-generation characteristics. There are different influences to analyse in deciding on methods which can be used to improve the process design: excretion rate, type of excretion, and, eventually, feedback regulation must be considered together with the relation of the production rate to culture growth (*6, 12*). If there is a feedback requirement, a specific optimization strategy must be applied. If there is no possibility of increasing product generation per volume of medium, the medium volume must be increased and medium cost and protein load have to be minimized. However, the product concentration should be programmed to be as high as possible to facilitate the downstream operation (purification).

The excretion rate is usually related to the growth rate of the culture. In the case of a growth-dependent type of secretion, there is no way of using perfusion-operated capsule systems because there is almost no growth after confluence within the capsules. In such cases, any system with logarithmic growth, like a continuous culture or a batch culture, must be used. If there is a good production rate during the stationary culture phase, one should avoid logarithmic growth systems and prefer either retention systems or long-lasting batch cultivations. As there are many variations in productivity patterns, it is necessary to investigate this very carefully before selecting a cultivation system (*12*).

Regarding the excretion level, there may be the problem of too low a productivity. In such cases, the so-called three-chamber membrane systems (see Section 4.1) may open a way of collecting the product and using inexpensive media.

If the excretion characteristic is not uniform, but either cell-cycle-associated or an occasional release from the cell internal pool (e.g. during cell disintegration at death phase), the chosen system will obviously have to meet those requirements also.

3. CULTURE PRINCIPLES

For the selection of a relevant cultivation and production system, it is necessary to be aware of the principal culture patterns of the systems in order to find the optimum combination for the respective process (*8, 4*).

A batch culture is characterized by decreasing substrate concentration. Cell growth is zero at the onset and passes through a maximum before reaching the highest cell concentration. It decreases at the end of the batch and may exhibit negative values, caused by cell death and disintegration (Fig. 2(*a*)).

A continuous culture is characterized by steady states which are reached after the initial growth phase (Fig. 2(*b*)). The relation of cell concentration, substrate concentration, and growth rate is determined by the perfusion rate and by operation as a chemostat or a turbidostat, respectively. The growth

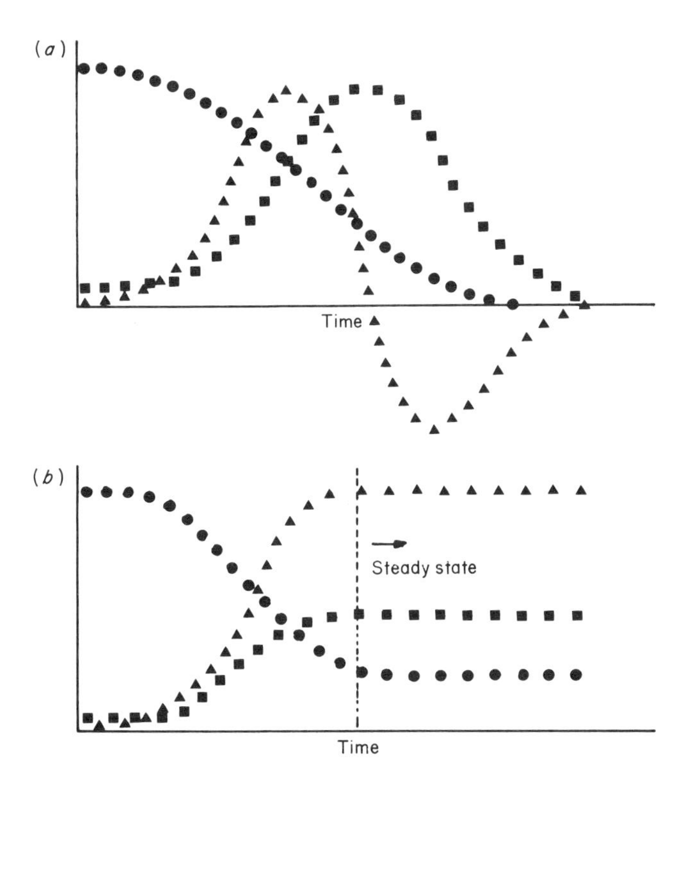

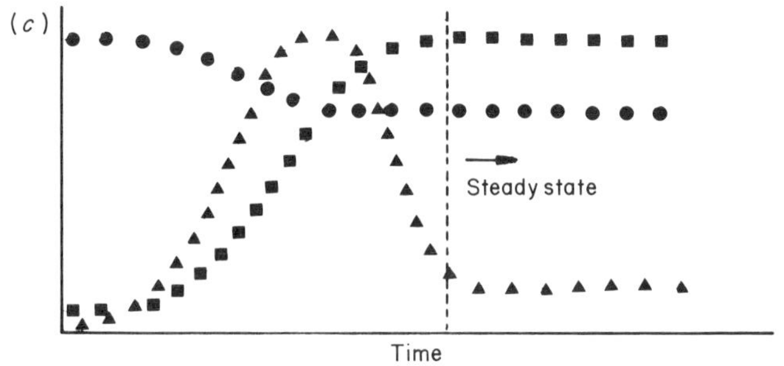

Fig. 2. Principal culture characteristics of (*a*) batch, (*b*) continuous and (*c*) perfusion culture with respect to substrate concentration (circles), growth rate (triangles), and cell density (squares).

rate can be varied from the maximum in the turbidostatic mode to about 25% of this in a strong chemostatic mode. Though the limiting nutrient for the chemostatic mode is usually not defined, it must not interfere with productivity.

A perfusion system with cell retention is also characterized by steady states which are reached after the initial growth phase (Fig. 2(*c*)). The cell concentration reaches a maximum, which is determined by space or other limitations but not usually by chemical ones. The growth rate decreases to almost zero after confluence of the system. Some residual growth is always detected in practice, because of the replacement of dead cells. The substrate concentration can be regulated by altering the perfusion rate.

For an optimized system, it may be necessary to use a combination of continuous and perfusion systems to enable flexible choice of growth rates between maximum and almost zero. Flexible choice of cell density and nutrient concentration are also necessary to determine the optimum production conditions. Owing to modern membrane and separation technology, such flexible systems exist, two of which I shall introduce in the following sections.

4. HIGH-DENSITY FERMENTATION SYSTEMS

4.1. Membroferm

The Membroferm is a modular fermentation system in which the cells can be grown between flat sheets of membranes. As the membranes are separated by a fluorocarbon matrix, cells can be grown either on the surface of the matrix fibres or, for suspension cells, in the openings of the fabrics. The goals for the functional design have been:

1. Free choice of any commercially available flat membrane.
2. Good geometry for diffusion and circulation of medium.
3. Scale-up potential.
4. *In situ* sterilization.
5. Accumulation of product within the system.
6. Harvesting of product independently from cells and medium.
7. Flexibility in operation mode and process control.

The system is composed of three different types of fabric spacers, which are simultaneously the reactor chambers (Fig. 3). We call them medium, cell, and product chamber. They are made of fluorocarbon fabrics with silicon gaskets and are the supports for the membranes. The size of the chambers is 28 × 32 cm, giving about 500 cm^2 of membrane working area. For growth of surface-dependent cells, the fabrics are treated chemically. The surface pro-

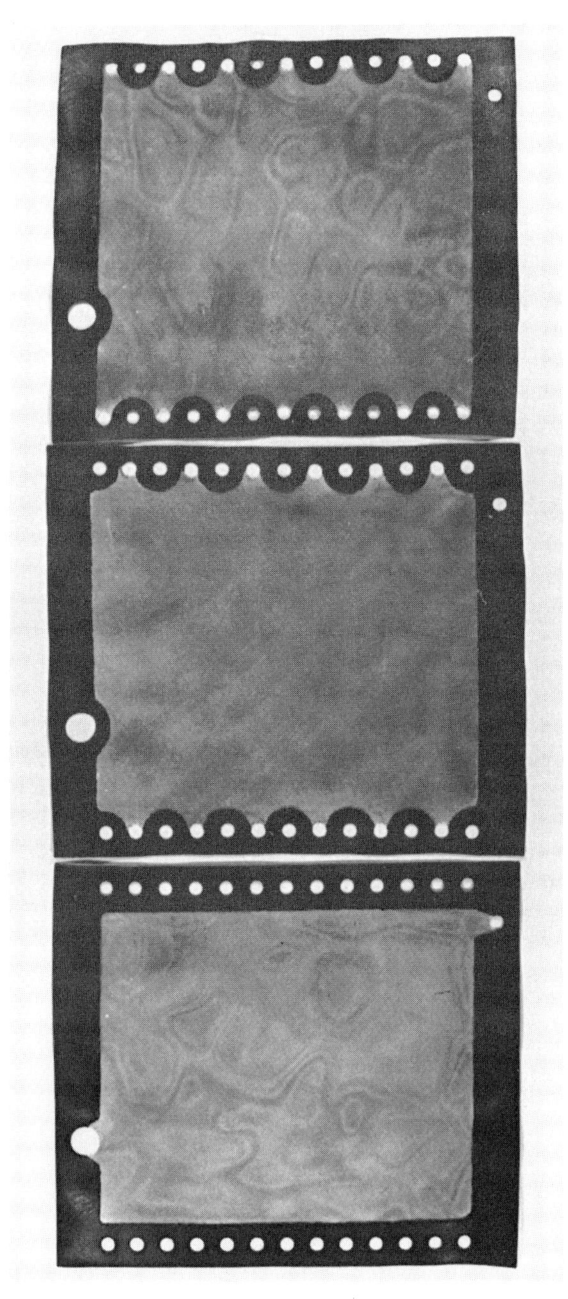

Fig. 3. Fabric spacers, which are simultaneously the reactor chambers. From top to bottom: medium, product, and cell chamber. The large hole open to the cell chamber is for inserting an oxygen probe.

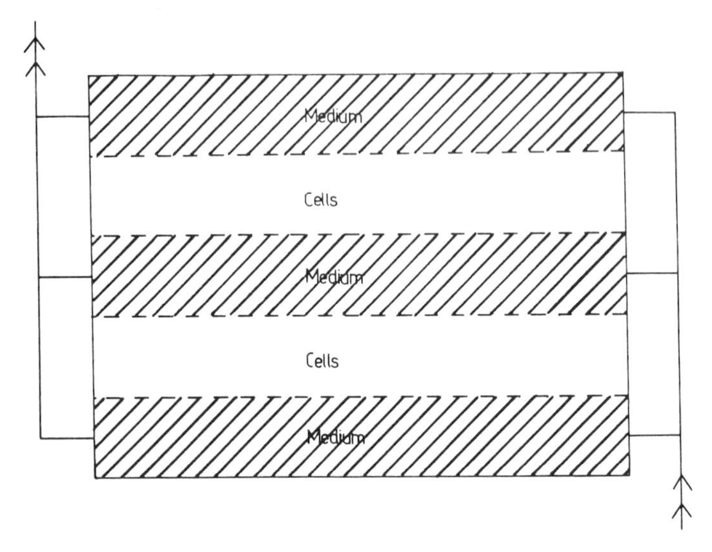

Fig. 4. Two-chamber system.

vided per chamber is, dependent on the type of fabric, between 1400 and 7000 cm^2 at a volume of about 20 ml (0.6 mm chamber height). Stacking of 30–400 layers is possible with the construction, corresponding to a maximum membrane area of 20 m^2, a cell chamber column of 4 litres and a surface area of fabric between 7 and 35 m^2. There is the possibility of mounting electrodes directly within the cell chamber through the filling channels. A fourth chamber, for example an oxygenation chamber, can be employed by using a hollow-fibre matrix as spacer material. On the other hand, there is the possibility of using the product chamber as an oxygen chamber when using the equipment for two-chamber runs (medium plus cells plus oxygen).

By different combinations of the spacers, several different operation modes become possible. The simplest is the alternating stacking of medium and cell chambers to give a two-chamber system (Fig. 4). Any microfiltration membrane which retains the cells can be used. For retaining the product, an appropriate ultrafiltration membrane can also be used. In principle, there is no difference from a hollow-fibre cartridge, but there is free choice of membranes, a better geometry for diffusion and a volume of up to 4 litres of cell mass.

With the three chambers set up, the cells are immobilized between two different flat membranes (Fig. 5). The membrane between the nutrient medium and the cells is an ultrafilter with a molecular weight cut-off

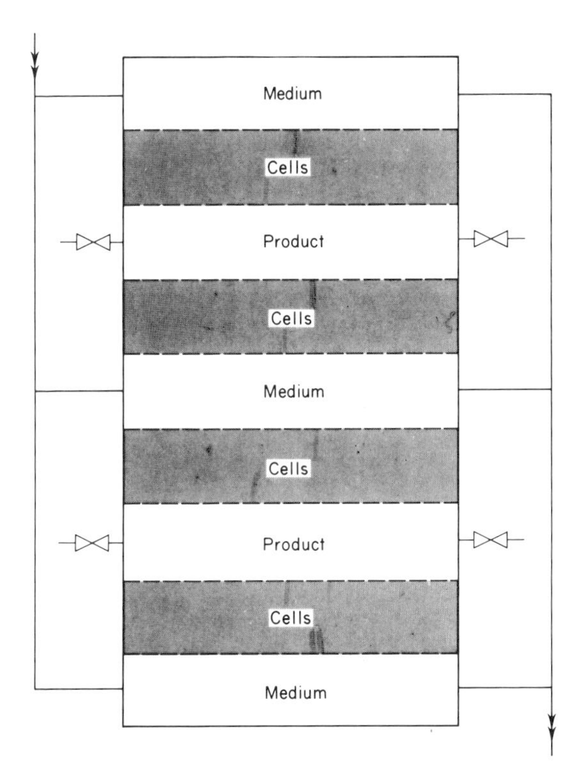

Fig. 5. Three-chamber system.

appropriate for the particular product. It must be open enough to feed the cells with nutrients and provide them with oxygen by diffusion. The antibodies or another product must not pass this barrier and will remain within the cell compartment. The other side of the cell chamber is a microfiltration membrane of a pore size suitable for retaining the cells from the product chamber, e.g. a 0.2 μm sterile filtration membrane. The product traverses this membrane by diffusion into the product chamber.

With this configuration it is possible to feed the cells continuously and wait for a suitable product concentration. This closely resembles the situation for mouse ascites, i.e. product harvest can be operated repeatedly at suitable time intervals by replacing the contents of the product chamber with fresh medium without removing cells from the reactor. There may be an advantage over mouse ascites, because of the exclusion of large serum proteins from the product by the feeding membrane barrier. As mentioned above, there is the possibility of using the third chamber for other purposes, e.g. oxygenation or induction. Stacking of the system is possible, as shown in Fig. 5.

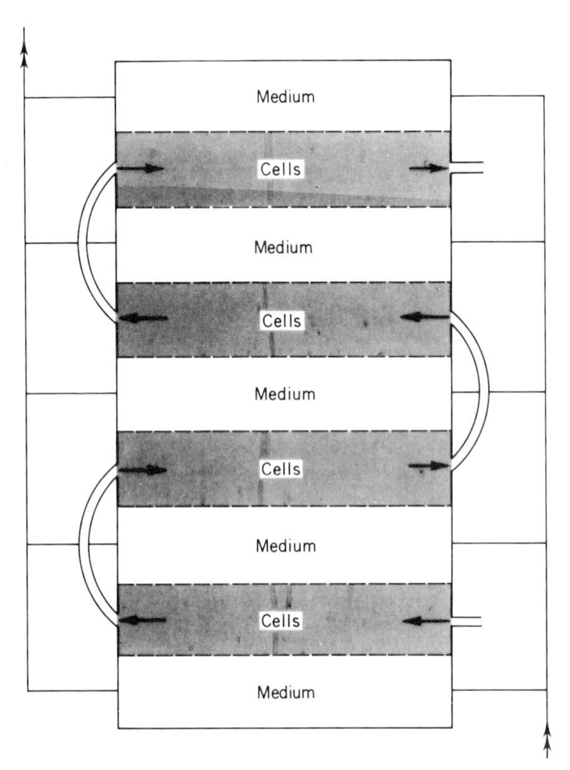

Fig. 6. Operation of Membroferm as a tube reactor.

Another operational mode with this system is its use as a tube reactor (Fig. 6). This operational mode is more suitable for microorganisms than for cells and is possible with slight modification of the gaskets. The new principle of this construction compared to a conventional tube reactor is that the tube is mantled by membranes, allowing, for example, re-feeding of cells or enzyme treatment of material for bioconversion all the way through the tube, which is up to 50 m long.

To achieve the proper environmental conditions for the cells, a peripheral system is installed (Fig. 7). In principle, it is the same as for any other perfused immobilization system. The conditioning vessel must be as small as possible for short mean residence time of the medium in order to minimize inactivation and denaturation of nutrients. This is particularly important for media with low protein content. From this vessel, the medium is circulated through the system by a rotating pump; the circulation speed is controlled in such a way that the oxygen tension at the outlet of the medium chamber is high enough to avoid starving the cells. This can be measured by a probe mounted directly

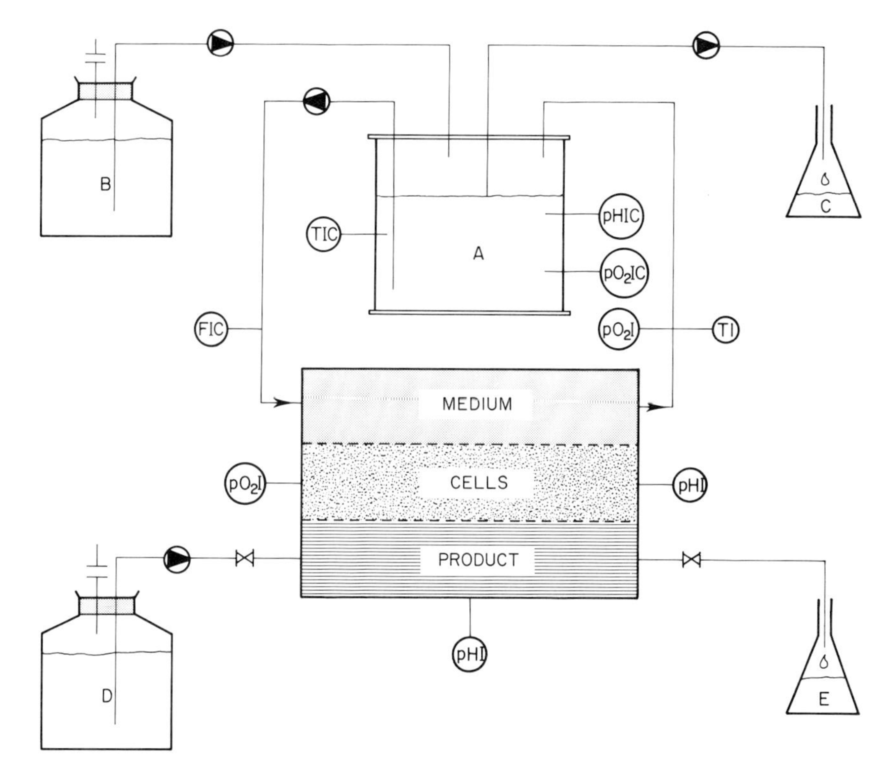

Fig. 7. Periphery installation of a Membroferm with conditioning vessel (A), reservoir of growth medium (B), spent medium (C), product-collecting medium (D), and product (E). Probes are indicated as follows: TI, temperature indicator; FI, flow rate indicator; pO_2I, oxygen tension indicator; pHI, pH indicator. (C) indicates an automated control circuit.

within the cell compartment. The perfusion rate of medium through the conditioning system depends on the consumption of nutrients by the cells. It is controlled by appropriate guide parameters, e.g. glucose concentration.

The commercial form of the system (Fig. 8) is fully sterilizable *in situ*. All connections can be resterilized during operation as with a conventional cell fermentation system.

4.2. Rotating Sieve Fermenter

A retention system with freely suspended cells is the rotating sieve system, and its operation is very flexible. The idea of the spin filter system was first mooted by Himmelfarb and Thayer (*5*) and described later by other authors

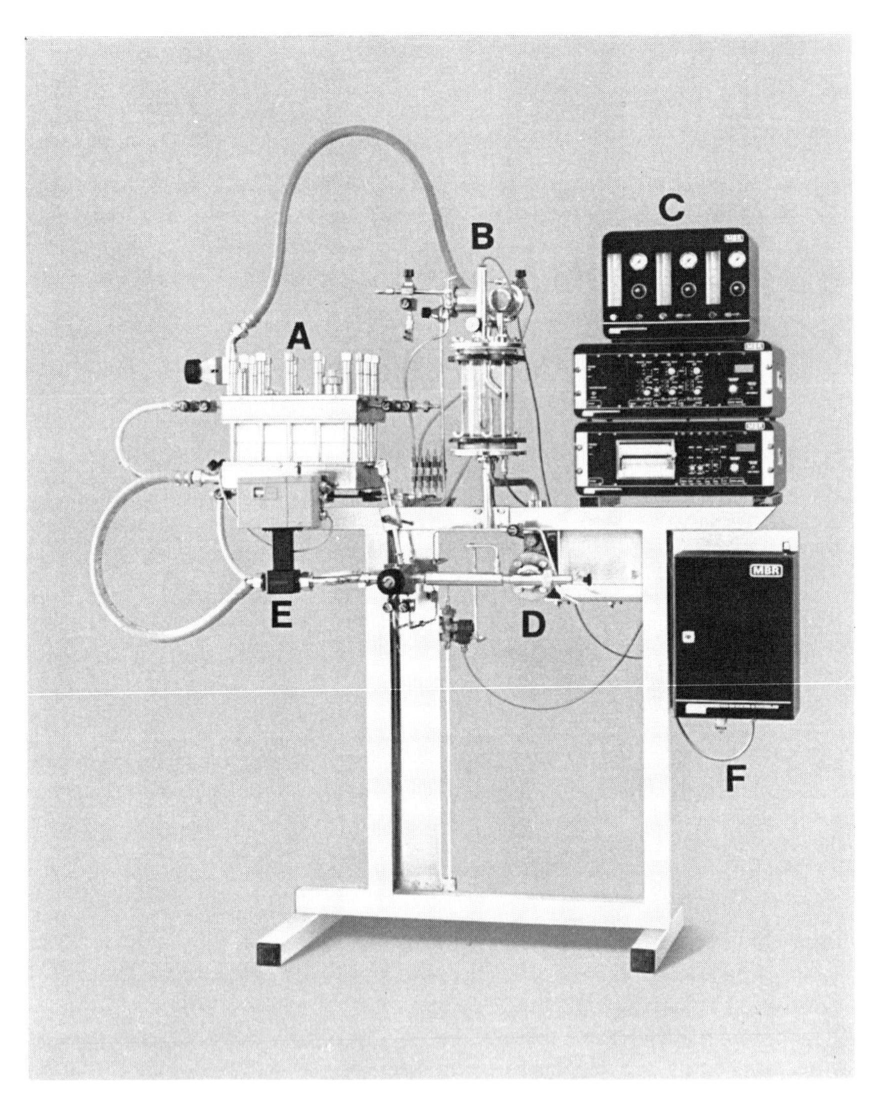

Fig. 8. Commercial form of the Membroferm: (A) the reactor; (B) the conditioning vessel; (C) the control unit; (D) circulation pump; (E) flow indicator; (F) control for pump speed.

(*23, 26*). The disadvantage of all these systems was the fouling of the separation screen by cells and cellular particles after a relatively short time of operation. This made it necessary to clean or to replace the spin filter periodically, which was inconvenient because of the resterilization process. We developed a design in which the operation period could be prolonged and allowed removal of more than 100 reactor volumes of spent medium with no sign of screen fouling. The reason was the use of a special screen whose pores are larger than the medium cell diameter. Surprisingly, there is excellent retention, reaching a separation efficiency of 96–99%. The effect is not due to centrifugal forces, which have been estimated to be below 0.2*g*, but to other, unknown, influences.

The system has a special design, based on a low-turbulence fermenter, which has a geometry similar to that of an air-lift fermenter. The unit is driven by a slowly rotating marine impeller and has a spin filter insert which operates with a hydrostatic pressure differential between outside and inside of the filter alone (Fig. 9). The system is equipped with a silicone oxygenation system and

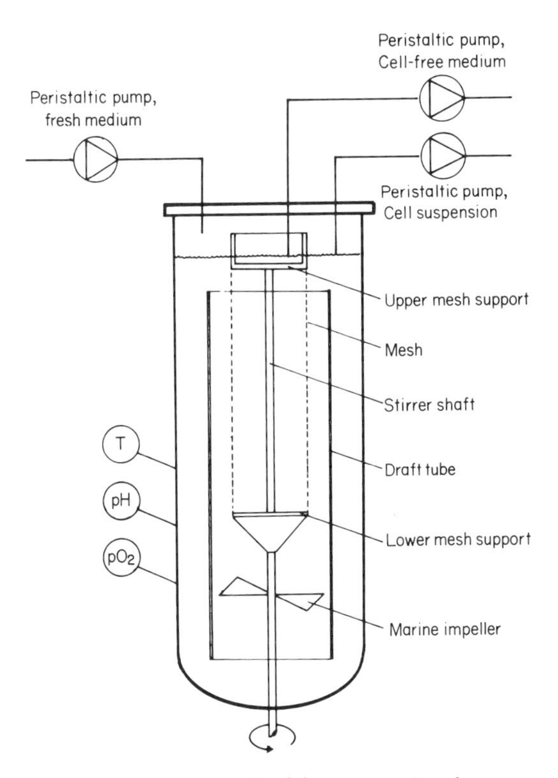

Fig. 9. Schematic principle of the rotation sieve fermenter.

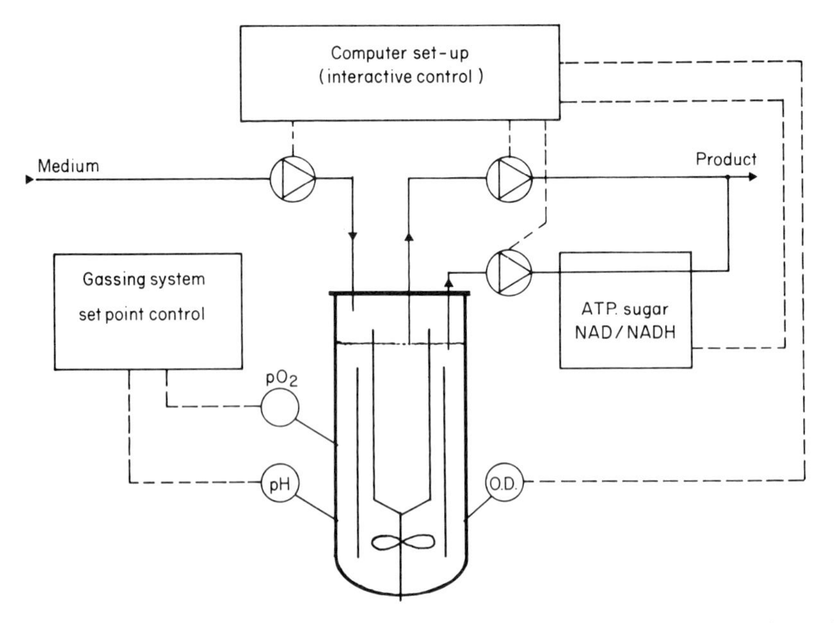

Fig. 10. Schematic principle of interactive process control using calculations based on cell density and biochemical analyses.

a microsparger, allowing cell densities of more than 3×10^7 ml^{-1}. Perfusion rates up to 3 volumes per day have been applied with no decrease in separation potency. Vessels sized from 5 to 65 litres net volume have been in long-term routine operation. With a 200-litre vessel, we could also show the proper functioning of the separation sieve.

The system is equipped with a turbidimetric probe (*13*) and, beside the medium output, a separated line for cell suspension removal. By variation of the two outlet pumps and the feeding pump according to turbidity and biochemical guide parameters (*14*) or product concentrations, any culture condition can be maintained continuously. This also allows the predetermination of growth rate between maximum and almost zero at predetermined cell densities many times higher than in non-retention fermenters (Fig. 10).

The application of computerized interactive process control would be an ideal supplement, but the on-line measurement of biochemical guide parameters is still in an immature state (*3, 14*). In Fig. 11, a typical run with a hybridoma line is presented.

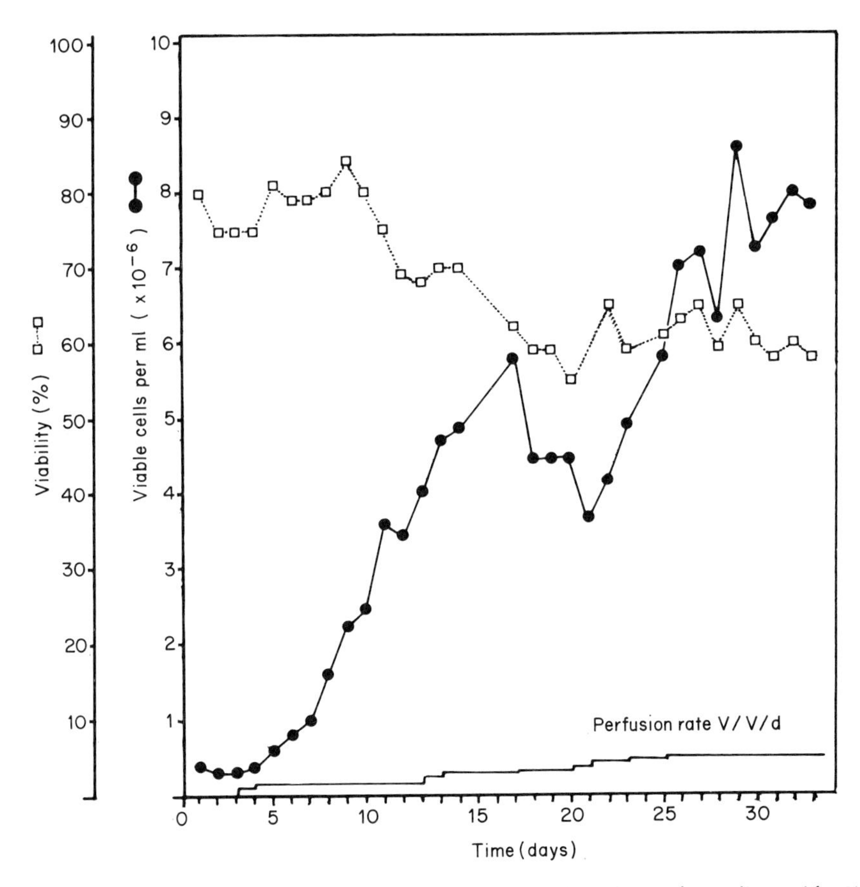

Fig. 11. Example of a fermentation diagram using a hybridoma line and a medium with 5% foetal calf serum. Culture volume was 23 400 ml.

5. CONCLUSION

As mentioned at the beginning of the chapter, new insights into the physiological and physical characteristics of animal cells during the production of biologicals call for new technological principles in fermentation. These can be established with modern separation technology and have a reasonable scale-up potential. Using these systems together with some basic characterization and optimization work, there is a very good chance of establishing a process with economic and reproducible characteristics. The scale which has been reached well meets the demand for pilot production, e.g. one may expect for a hybridoma cell line of medium quality some 20–50 g of raw monoclonal antibody per day from a 65-litre system or from a fully loaded Membroferm. Further scale-up seems to be possible. With the rotating sieve fermenter we have evidence of similar scale-up behaviour to that of the air-lift-type fermenters, which have excellent scale-up characteristics (*9, 16*). In general, cell retention technology shows many advantages and can help to solve some problems which have prompted workers in the past to overlook the technical possibilities in animal cell fermentation.

REFERENCES

1. McAleer, W. J., Markus, H. Z., Bailey, F. J., Herman, A. C., Harder, B. J., Wampler, D. E., Miller, W. J., Keller, P. M., Buynak, E. B., and Hilleman, M. R. (1987). Production of purified Hepatitis B surface antigen from Alexander Hepatoma cells grown in artificial capillary units *J. Virol. Method.* **7**, 263–271.
2. Birch, J. R. (1986). The large scale production of monoclonal antibodies in airlift fermenters. Paper delivered at "Process Possibilities for Plant and Animal Cell Cultures", University of Manchester, U.K.
3. Fleischaker, R. J., Weaver, J. G., and Sinskey, A. J. (1981). Instrumentation for process control in cell culture. *In* "Advances in Applied Microbiology", Vol. 27 (D. Perlman and A. I. Laskin, eds.), pp. 137–167. Academic Press, Orlando.
4. Glacken, M. W., Fleischaker, R. J., and Sinskey, A. J. (1983). Large-scale production of mammalian cells and their products. Engineering principles and barriers to scale-up. *Ann. N.Y. Acad. Sci.* **413**, 355–372.
5. Himmelfarb, P., Thayer, P. S., and Martin, H. E. (1969). Spin filter culture: The propagation of mammalian cells in suspension. *Science* **164**, 555–557.
6. Hu, W. S., and Flickinger, M. C. (1986). Personal communication.
7. Karkare, S. B., Phillips, P. G., Burke, D. H., and Dean, R. C. (1985). Continuous production of monoclonal antibodies by chemostatic and immobilized hybridoma culture. *In* "Large Scale Mammalian Cell Culture" (J. Feder and W. P. Tolbert, eds.), pp. 127–149. Academic Press, Orlando.
8. Katinger, H., and Scheirer, W. (1985). Mass cultivation and production of animal cells. *In* "Animal Cell Biotechnology", Vol. 1 (R. E. Spier and J. B. Griffiths, eds.), pp. 167–194. Academic Press, London.
9. Katinger, H. W. D., Scheirer, W., and Krömer, E. (1979). Bubble column reactor for mass propagation of animal cells in suspension culture. *Ger. Chem. Eng.* (Eng. Transl.) **2**, 31–38.
10. Klement, G., Scheirer, W., and Katinger, H. W. D. (1987). Construction of a large-scale membrane reactor system with different compartments for cells, medium and product. *Dev. Biol. Stand.* **66**, 221–226.
11. Merk, W. A. M. (1982). Large-scale production of human fibroblast interferon in cell-fermenters. *Dev. Biol. Stand.* **50**, 137–141.
12. Merten, O. W. (1987). The influence of culture growth rate on product generation rate with animal cells. Institut Pasteur, Paris, France. In preparation.
13. Merten, O. W., Palfi, G. E., Stäheli, J., and Steiner, J. (1987). Invasive infrared sensor for the determination of the cell number in a continuous fermentation of hybridomas. *Dev. Biol. Stand.* **66**, 357–366.

14. Merten, O. W., Palfi, G. E., and Steiner, J. (1986). On-line determination of biochemical/physiological parameters in the fermentation of animal cells in a continuous or discontinuous mode. *In* "Advances in Biotechnological Processes", Vol. 6 (A. Mizrahi, ed.), pp. 111–178. Alan R. Liss, New York.
15. Nilsson, K., Scheirer, W., Merten, O. W., Östberg, L., Liehl, E., Katinger, H. W. D., and Mosbach, K. (1983). Entrapment of animal cells for production of monoclonal antibodies and other biomolecules. *Nature (London)* **302**, 629–630.
16. Onken, K., and Weiland, P. (1983). Airlift fermenters: construction, behaviour and use. *In* "Advances in Biotechnological Processes", Vol. 1 (A. Mizrahi and A. L. Wezel, eds.), pp. 67–95. Alan R. Liss, New York.
17. Panina, G. F. (1985). Monolayer growth systems: Multiple processes. *In* "Animal Cell Biotechnology", Vol. 1 (R. E. Spier and J. B. Griffiths, eds.), pp. 211–242. Academic Press, London.
18. Peehl, D. M., and Stanbridge, E. J. (1981). Anchorage independent growth of normal human fibroblasts. *Proc. Natl. Acad. Sci. U.S.A.* **78**, 3053–3057.
19. Rupp, R. G. (1985). Use of cellular microencapsulation in large-scale production of monoclonal antibodies. *In* "Large Scale Mammalian Cell Culture" (J. Feder and W. P. Tolbert, eds.), pp. 19–38. Academic Press, Orlando.
20. Seaver, S., Rudolph, J. L., Ducibella, T., and Gabriels, J. E. (1984). Hybridoma cell metabolism/antibody secretion in culture. Paper at BIOTECH 84, U.S.A.
21. Spier, R. E. (1985). Monolayer growth systems: heterogeneous unit processes. *In* "Animal Cell Biotechnology", Vol. 1 (R. E. Spier and J. B. Griffiths, eds.), pp. 243–265. Academic Press, London.
22. Tolbert, W. R., Hitt, M. M., and Feder, S. (1980). Cell aggregate suspension culture for large-scale production of biomolecules. *In Vitro* **16**, 486–490.
23. Tolbert, W. R., Lewis, C., White, P. J., and Feder, J. (1985). Perfusion culture systems for production of mammalian cell biomolecules. *In* "Large Scale Mammalian Cell Culture" (J. Feder and W. P. Tolbert, eds.), pp. 97–123. Academic Press, Orlando.
24. Van Wezel, A. L. (1985). Monolayer growth systems: homogeneous unit processes. *In* "Animal Cell Biotechnology", Vol. 1 (R. E. Spier and J. B. Griffiths, eds.), pp. 226–283. Academic Press, London.
25. Whiteside, J. P., Whiting, B. R., and Spier, R. E. (1979). Development of a methodology for the production of FMD-virus from BHK21-C13 monolayer cells grown in a 100 l (20 m^2) glass sphere propagator. *Dev. Biol. Stand.* **42**, 113–119.
26. Whiteside, J. P., Farmer, S., and Spier, R. E. (1985). The use of caged aeration for the growth of animal cells on microcarriers. *Dev. Biol. Stand.* **60**, 283–290.

12

A Comparative Review of Microcarriers Available for the Growth of Anchorage-dependent Animal Cells

M. BUTLER
Department of Biological Sciences,
Manchester Polytechnic,
Manchester, U.K.

ANIMAL CELL BIOTECHNOLOGY VOL. 3
ISBN 0-12-657553 3

1. INTRODUCTION

The mass vaccination campaigns of the 1950s against viral diseases provided the initial impetus for the large-scale production of anchorage-dependent animal cells. The need for such large-scale cultures has increased even further with the possibilities of generating an extensive list of useful biologicals including viral vaccines, interferons, hormones, and enzymes.

Of paramount importance to the development of this technology is the provision of a suitable substratum for cell attachment and growth. Initially, this was solved by cell growth on the inner surface of large numbers of roller bottles. Although this can provide mass cell culture, there are a number of disadvantages to this approach. Each roller bottle is an independent culture unit requiring a series of sterile manipulations. In a large industrial operation, this necessitates an intensive labour demand which is expensive and exposes the cultures to excessive danger of contamination.

The principle of a unit culture system capable of scale-up is far more attractive. Such a system suitable for anchorage-dependent cell growth was originally suggested by van Wezel (*55*). This involves the suspension of microscopic particles (microcarriers) in a slowly stirred liquid medium. Such particles offer a large surface area per unit culture volume for cell attachment. Originally, chromatographic grade DEAE-Sephadex A-50 was used, but this proved unsuitable above a limited concentration. In 1977, Levine *et al.* (*29*) reported that if the surface charge density of the DEAE-Sephadex micro-carrier was optimized, then cell growth would occur at high microcarrier concentrations.

Since this discovery, a wide range of microcarriers has been successfully used for the cell culture of anchorage-dependent cells and many of these are now available commercially. The purpose of this chapter is to consider the criteria that are important for the formulation of microcarriers, to evaluate the various microcarriers that have been developed commercially, and finally to consider the potential of microcarrier technology.

2. CELL ADHESION

Most non-transformed animal cells grow as a monolayer on a surface which allows cell attachment and spreading. The interaction between the cell membrane and the solid surface is critical in allowing optimal growth and involves a combination of electrostatic attraction and van der Waal's forces (*1*).

Traditionally, cell culturists have used glass, and more recently plastic, as static surfaces provided by Petri dishes, T-flasks, Roux or roller bottles. Both glass and the plastic (polystyrene) normally used have negatively charged

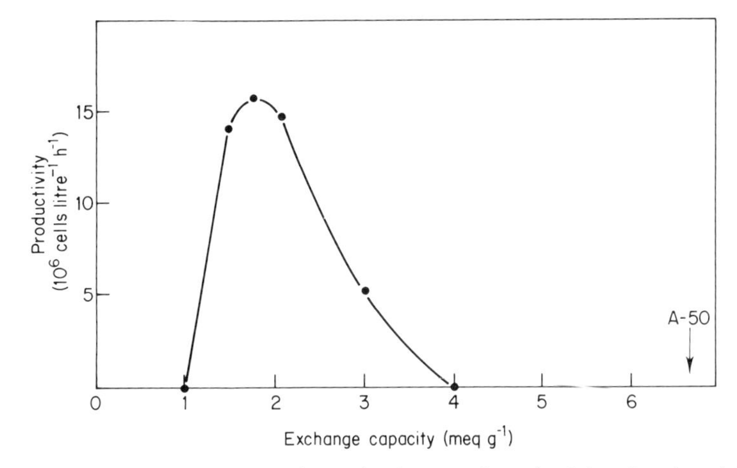

Fig. 1. The effect of microcarrier charge density on cell productivity. A series of dextran microcarriers was produced by varying the degree of substitution of DEAE. The cell productivities (10^6 litre^{-1} h^{-1}) of human diploid fibroblasts (HEL 299) were determined in microcarrier cultures at 5 g dextran litre^{-1}. The arrow indicates the exchange capacity of DEAE-Sephadex A-50 (*30*).

surfaces, as do the outer membranes of cells. In order to allow adhesion between such surfaces, positively charged molecules such as divalent cations or basic proteins are required to form a layer between the substratum and the cells. The requirement is provided by the culture medium. In particular, proteins can form a 2.5 nm thick layer on the substratum prior to cell attachment. Fibronectin is found on the surface of many cells and has been implicated as a specific cell adhesion molecule, particularly if collagen is present on the substratum (*25*). However, in most cases of cell–substratum interaction, adhesion is provided by a range of non-specific proteins (*17, 50*).

The types of surfaces suitable for cell adhesion have been studied in great detail. The density of electrostatic charges on the substratum is critical. Alkali treatment of glass and sulphonation of polystyrene used in tissue culture grade containers to a charge density of 2–5 negatively charged groups per nm^2 can optimize the surface for cell attachment and growth (*32, 33*). Evidence also suggests that preferential cell adhesion occurs on wettable substrata (*18*). The principles of cell attachment to a dynamic substratum such as microcarriers are similar. The positive charge on the surface of the DEAE-Sephadex originally used in cell cultures was too high for cell attachment. Coating the surface with serum proteins or synthesized proteins such as polylysine improves the growth-promoting properties of the microcarrier. However, acceptable cell yields are only obtained at charge densities around a value of 2 milli-equivalents g^{-1}, which has proved optimal for a range of cell lines (Fig. 1).

3. GENERAL CRITERIA FOR MICROCARRIER DESIGN

Since the development of the dextran microcarrier with an optimal charge, microcarriers have been prepared from various materials. Many parameters have been considered important in the design of these microcarriers (*46, 56*). Of these, some of the key characteristics common to all commercially produced microcarriers are as follows.

3.1. Size

A large surface area per unit volume of culture can maximize the potential for cell growth. This suggests the need for particles of the smallest possible dimensions. However, this is limited by the requirement for each microcarrier to be of sufficient size to support significant growth and multiplicity of the initial cell inoculum. A further limitation relates to the curvature of the substratum. Various cell lines have been observed to show preference for cell spreading on surfaces offering the least curvature, a phenomenon which may be explained by the rigidity of cytoplasmic microfilaments (*18*). Observations of the growth of BHK cells on glass beads of various diameters have shown that below a diameter of 50 μm, cell growth decreases (*31*).

Most microcarriers have been optimized at a diameter of 100–200 μm, so that the available surface area of 7×10^4 μm^2 can support a monolayer of over a hundred cells. Thus, by suspending 12.6×10^6 dextran microcarriers in a 1-litre culture, a surface area of 10^4 cm^2 is provided for cell growth and is equivalent to the surface available in 50 Roux bottles or 20 roller bottles (*6*). Typically, this microcarrier concentration can support the growth of cells from an initial concentration of 2×10^5 cells ml^{-1} to a final yield of 2×10^6 cells ml^{-1}.

3.2. Density

In order to provide a satisfactory environment for cell growth, the stirring rate of cultures is reduced to a minimum. This prevents unnecessary shear forces which may damage cells, minimizes foaming of the medium, and reduces the energy consumption necessary for operation. To allow low stirring rates, the microcarriers are required to be sufficiently light to allow easy suspension in the medium. Microcarrier densities of 1.03–1.05 g cm^{-3} are considered optimal. However, it should be recognized that as cells attach and grow, the density of the microcarrier gradually increases. Stirring rates of 40–50 r.p.m. are normally adequate throughout a culture.

3.3. Charge

Cells have been grown on both negatively and positively charged microcarriers of various compositions. Charge density is a critical parameter at either polarity. If the charge density is too low, cell attachment will be inadequate, whereas at too high a charge density, an apparent toxic effect will limit cell growth.

4. MICROCARRIER TYPES

These general criteria necessary for microcarrier function can be provided by a variety of formulations. The basic microcarrier design may be altered in two respects whilst still maintaining the capacity for cell growth:

1. The basic matrix may be changed.
2. The surface coating material which provides the necessary electrostatic charge may be changed.

Variations with respect to these have led to numerous microcarrier types, many of which are available commercially (Table I). The physical characteris-

TABLE I Microcarriers Which are Commercially Available

Type	Trade name	Company	Country	Composition
Dextran	Cytodex 1	Pharmacia	Sweden	DEAE-dextran
	Cytodex 2	Pharmacia	Sweden	Quaternary amine-coated dextran
	Superbeads	Flow Labs	USA	DEAE-dextran
	Microdex	Dextran Products	Canada	DEAE-dextran
	Dormacell	Pfeifer & Langen	Germany	DEAE dimers-dextran
Plastic	Biosilon	Nunc	Denmark	Polystyrene—charged
	Biocarriers	Biorad	USA	Polyacrylamide/DMAP
	Cytospheres	Lux	USA	Polystyrene—charged
	Acrobeads	Galil	Israel	Polyacrolein—various coatings
	Micarcel G	Reactifs IBF	France	Polyacrylamide/collagen/ glucoglycan
	Bioplas	Solohill Eng.	USA	Polystyrene—cross-linked
	Mica	Muller-Lierheim	Germany	Epoxy resin
Gelatin	Geli-beads	KC Biologicals/ Hazelton Labs	USA	Gelatin
	Ventregel	Ventrex	USA	Gelatin
	Cytodex 3	Pharmacia	Sweden	Gelatin-coated dextran
Glass	Bioglas	Solohill Eng.	USA	Glass-coated plastic
	Ventreglas	Ventrex	USA	Glass-coated plastic
Cellulose	DE-52/53	Whatman	UK	DEAE-cellulose

TABLE II Physical Characteristics of Microcarriers

Type	Name	Density ($g\,cm^{-3}$)	Diameter (μm)	Surface area ($cm^2\,g^{-1}$ of microcarriers)	Number of microcarriers g^{-1}
Dextran	Cytodex 1	1.03	131–220	6000 (d)	6.8×10^6 (d)
	Cytodex 2	1.04	114–198	5500 (d)	5.8×10^6 (d)
	Dormacell	1.05	140–240	7000 (d)	6.2×10^6 (d)
	Microdex	1.03	150	250	3.5×10^5
Plastic	Biosilon	1.05	160–300	225	3.4×10^4
	Acrobeads	1.04	100–200	500	7.1×10^5
	Cytospheres	1.04	160–230	250	2.1×10^5
Gelatin	Cytodex 3	1.04	133–215	4600 (d)	4.8×10^6 (d)
	Ventregel	1.03 (e)	150–250	4300 (d)	3.6×10^6 (d)
Glass	Bioglas	1.03	150–210	350	5.0×10^5
Cellulose	DE-53	1.03 (e)	(40–50) × 80–400	1000	2.7×10^6
Liquid	Fluorocarbon	ND	100–500	ND	ND

ND = not determined; e = estimated; d = dry weight of microcarrier.

tics of some of these microcarriers are listed in Table II. It should be noted that some microcarriers (e.g. dextran) require swelling before use and their characteristics are often quoted on a unit dry weight basis.

4.1. Dextran

Dextran beads of various types have been used extensively for chromatographic techniques. Dextran is a polymer of glucose consisting of 95% 1–6 linkages and some 1–3 short side-branches. Treatment with reagents such as epichlorohydrin results in cross-linking which stabilizes the structure, which is non-rigid and non-toxic to cells. Cross-linked dextran spheres of 150 μm diameter are suitable as microcarriers because of the ease of suspension in liquid media and the large surface area offered for growth.

For cell adhesion, charged groups are applied to the dextran. The positively charged ion-exchange material diethylaminoethyl-dextran in the form of DEAE-Sephadex A-50 has a surface charge density of 6 mequivalents g^{-1}, which has proved to be too high to support cell growth. However, a reduction of the proportion of DEAE groups attached to the dextran can lower the charge to a level suitable for cell attachment and growth (*29*). Reaction of 2 mmol of DEAE with 1 g of dry dextran beads gives a charge density of 2 mequivalents g^{-1} and has been found suitable for various cell lines (*54*). By carefully controlled reaction, the charged DEAE is evenly distributed as monomeric groups throughout the dextran matrix. This formulation is commercially available as Superbeads, Microdex, or Cytodex 1, which are probably the most widely used of all microcarriers (*44*).

A slight modification of this formulation is applied to Dormacell microcarriers with the use of dimeric DEAE groups which are attached to the surface of the dextran beads. The suggested advantage of the dimeric group is its pK value of 6.5 compared to 9.2 for the monomer. The pK of 6.5 is closer to the pH (~7.4) generally used under normal culture conditions and the dimer should therefore maintain a relatively constant charge even with the slight pH changes associated with cell growth (*53*). The advantage of a charged group which only attaches to the microcarrier surface is that the overall charge capacity of the bead is lower than in the case of a group which penetrates the bead. This reduces the possibility of the adsorption of nutrient or product molecules which may be of limited concentration in the culture medium. Such an advantage is also claimed for the formulation of Cytodex 2, which has a quaternary amine group—trimethyl-2-hydroxyaminopropyl—attached to the dextran bead surface (*44*). A thin layer of charge on the surface of the bead reduces the total charge to 0.6 meq g^{-1}, which is considerably lower than the value of 2.0 meq g^{-1} associated with Cytodex 1. Cytodex 3 is the third of Pharmacia's range of microcarriers. This is a dextran-based microcarrier which is coated with a gelatin layer and is discussed in the later section on gelatin microcarriers. Figure 2 shows examples of the charged groups used in combination with dextran microcarriers.

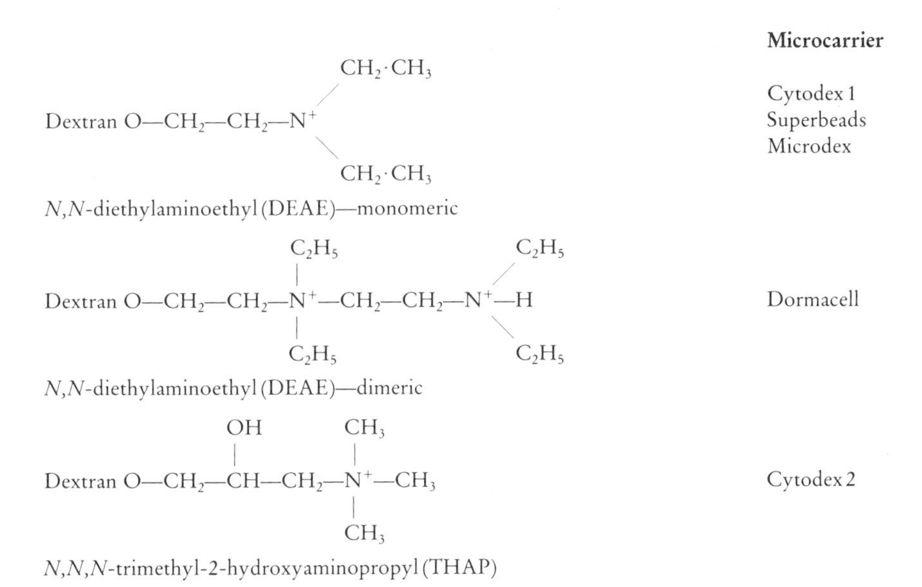

Fig. 2. Charged groups that have been used in combination with dextran microcarriers

Dextran is a strong and stable matrix which is compressible, a property which may reduce cell damage at high microcarrier concentrations. It is also non-toxic to cells. The possibility of leakage of charged groups from Cytodex 1 microcarriers was examined in concentrated polio vaccine prepared from microcarrier cultures (*64*). No contamination of any of the microcarrier components was found in the produced vaccine above 20 p.p.m., which was the limit of detection by mass spectroscopy.

Transfer of cells from bead to bead would be particularly advantageous for scale-up of microcarrier cultures. This would allow a large culture containing new microcarriers to be inoculated by the addition of a few confluent microcarriers from a previous smaller culture. This has been found possible with dextran microcarriers for a limited number of cell lines (*53*) and under selected conditions, such as low calcium concentrations (*8*). However, for most cells the attachment to dextran beads is too great to allow such transfer to unoccupied beads. In this situation, cells must be harvested by trypsinization before re-inoculation into a larger culture. This may result in relatively low yields of viable cells particularly after trypsinization at physiological pH. It may, however, be possible to increase yields by trypsinization at higher pH values, ~8.7 (*20*).

4.2. Plastic

Disposable tissue culture flasks and plates have now become commonplace in cell-culture laboratories. They may be purchased in sterile packages, thus saving labour costs associated with cleaning and sterilizing the more traditional glass containers. Treatment of the growth surface is essential to allow good cell attachment. Sulphonated polystyrene with a negative surface charge of $2\text{–}10 \times 10^{14}$ charges cm^{-2} has been found to be optimal (*33*) and is commonly used as "tissue culture grade" plasticware. Microcarriers based on this formulation have been produced as Biosilon (Nunc), Cytobeads (Lux), and Bio-beads (Biorad). The solid spheres of diameter 160–300 μm have a density of 1.05 g cm^{-3} and can easily be suspended in culture medium under gentle stirring rates (*22*). The beads can be supplied as dry packages, sterilized by irradiation.

The merits of using polystyrene microcarriers are that they are shape-stable, non-swelling, and relatively resistant to breakage. The non-porous nature of the matrix and the relatively low electrostatic charge eliminates any possibilities of adsorption of components from the medium. Good growth characteristics have been reported for a number of cell lines on these microcarriers, including primary chick, human diploid and bovine endothelial cells (*9, 40*). However, a comparison of human diploid cell growth on several microcarrier types showed that, although high cell yields could be attained with polystyrene microcarriers, the cells were found to detach from the beads too readily after confluence (*38*). Another point worthy of consideration is that, although microscopic examination of the attached cells is possible, the polystyrene is not as translucent as dextran-based beads and this may result in difficulties in observing cell growth (*53*).

The microcarriers mentioned above are based on sulphonated polystyrene. However, charged groups may be added to the polystyrene matrix by other reagents. Glycine-derivatized polystyrene microcarriers have been produced and found suitable for the growth of several cell lines (*27*).

A range of primary amine molecules of varying chain lengths have been used for the derivatization of polyacrylamide microcarriers (*47, 49*). Cell line differences were found with respect to attachment and growth rate, but a diaminohexane-derivatized polyacrylamide microcarrier proved to be as good as those derivatized with the tertiary amino group, DEAE (*47*). On a commercial basis, dimethylaminopropyl-(DMAP-)substituted polyacrylamide microcarriers have been produced as Bio-Carriers (from Biorad) but have not proved particularly successful for general use (*19*).

Other novel plastic matrices can be used to produce microcarriers. Lazar *et al.* (*28*) described the preparation of agarose-polyacrolein microsphere beads which are available commercially as Acrobeads (from Galil). This involves the encapsulation of small (0.15 μm diameter) polyacrolein beads with agarose. Coated beads of 100–200 μm diameter are then treated with divinyl sulphone to strengthen the matrix by cross-linkage. The cross-linked beads can then be derivatized with one of a number of reagents including DEAE, diaminohexane (DAH), polylysine, gelatin or collagen. Rapid derivatization of the beads can be achieved because of the high concentration of aldehyde groups on the polyacrolein.

Slightly higher cell yields were demonstrated with the DEAE-polyacrolein beads compared with DEAE-dextran beads for two established cell lines (*28*). The use of the polyacrolein beads conjugated to proteins is particularly recommended for primary cell cultures. The beads are not rigid, and this may reduce cell damage caused by collisions between microcarriers. However, like other plastic microcarriers, they are not transparent and this may cause some difficulties in microscopic examination compared to other microcarrier types.

Trisacryl beads can be formed by polymerization of monomers to a selected density. Coating these beads with gelatin or heparin can produce a substratum capable of supporting good growth rates of BHK cells (*42*). Porous beads with a surface epoxy group have been produced recently—Mica (from Müller Lierheim, see Table I). These allow a strong attachment of growth factors from the culture medium.

4.3. Gelatin/collagen

Cell attachment *in vivo* normally occurs to a matrix consisting of collagen, which is a long fibrous protein existing in many different types. Collagen surfaces can also be used for the growth of a variety of cell types *in vitro* (*3*) and are particularly useful for establishing primary cell cultures or cells that are difficult to grow (*10*). Collagen surfaces may also allow cell differentiation *in vitro* (*45*) and delay cell detachment (*26*).

Ultrastructural studies on the mode of cell attachment to plastic microcarriers coated with collagen shows a distinct morphology of binding compared to other surfaces (*13, 21*). This may be related to specific receptors on the cell membrane (*4*). Fibronectin on the cell surface may also be important in the membrane–collagen interaction, which is more specific than attachment to other charged surfaces (*11, 12*).

Gelatin is denatured collagen and has been used more often than native collagen as a cell-culture substratum. Cell specificity for a particular type of collagen is much less apparent for the denatured protein, which can therefore be used for a wider range of cell types (*43*). Denatured collagen can be covalently cross-linked to the surface of dextran beads to provide a suitable microcarrier as Cytodex 3 (from Pharmacia). The covalent cross-linkage is important to prevent the problem of protein leakage that can be encountered. A gelatin coat of 60 $\mu g\ cm^{-2}$ was found optimal for the growth of Vero cells (*14*). Bovine embryo kidney cells showed greater attachment and spreading on Cytodex 3 compared to Cytodex 1 and this eventually led to a higher final cell yield (*14*). The greater plating efficiency of such problematic cells on a collagen surface is an advantage in that lower initial cell densities may be used for inoculation. The protein surface can be digested by treatment with trypsin or collagenase which can lead to the recovery of cells from the microcarriers with a 95% viability (*43*).

Microcarriers can be made entirely of a gelatin matrix—Gelibeads (from KC Biologicals) or Ventregel (from Ventrex). The advantage of such microcarriers is that they are completely soluble by treatment with proteolytic enzymes. This solves the problem of detachment and separation of cells from microcarriers. A number of enzymatic reagents can be used and selection is dependent on the vulnerability of the cell line to membrane damage. A cocktail of trypsin, dispase, and EDTA is recommended for most general applications (*43*).

Macroporous gelatin microcarriers have been produced by allowing gelatin spheres to solidify around droplets of toluene, which is later removed by a series of washing procedures (*41*). Each microcarrier contains a large number of cavities and can offer twice the surface area for cell attachment compared to an equivalent-size solid gelatin bead. Thus, an increased number of cell doublings is achievable per bead. Vero cells have been grown up to a density of 3×10^6 cells ml^{-1} in culture with such microcarriers (*41*).

4.4. Glass

The negatively charged surface of glass has been found to be a suitable substratum for cell attachment and growth from the early days of the development of small-scale laboratory culture techniques. Recently, packed bed bioreactors have been designed for large-scale cell cultures using solid glass beads of 2–3 mm diameter (*62, 63*). However, glass is not a suitable material for microcarriers because of its high density—1.5 g cm^{-3}. At moderate stirring rates, solid glass particles cannot easily be maintained in suspension. However, a microcarrier consisting of a plastic matrix and a glass surface is commercially available as Bioglas (from Solohill Eng.) or Ventreglas (Ventrex). These microcarriers have a density of 1.03–1.04 g cm^{-3} and this allows for suspension in slowly stirred cultures.

Electron microscopic examination of attached cells (Fig. 3) shows that slender filipodia extend from the cells onto the surface of the glass (*58*). This is different from attachment on dextran beads, where a cell surface edge is involved. Such a difference in attachment may be responsible for the ease of removal of cells from the glass-covered microcarriers (*59*). Short trypsinization times (~3 min) are sufficient to detach cells in a viable state. The

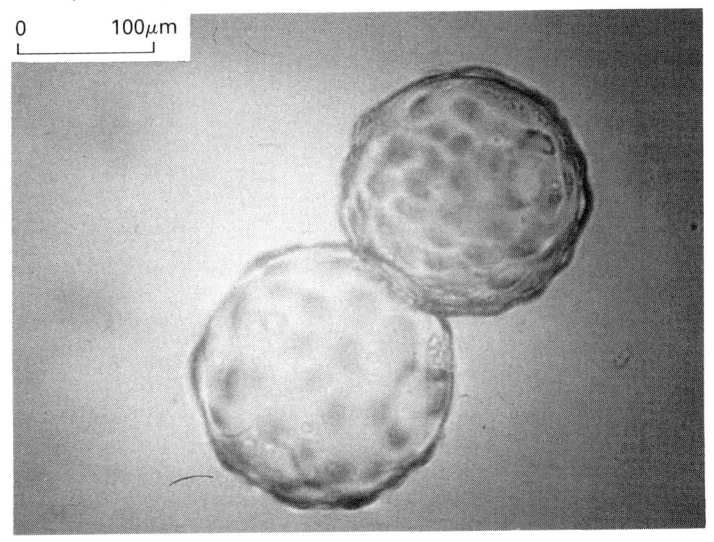

Fig. 3(*a*). Confluent monolayers of MDCK cells on the surface of DEAE-dextran microcarriers (Superbeads). (*Continued*)

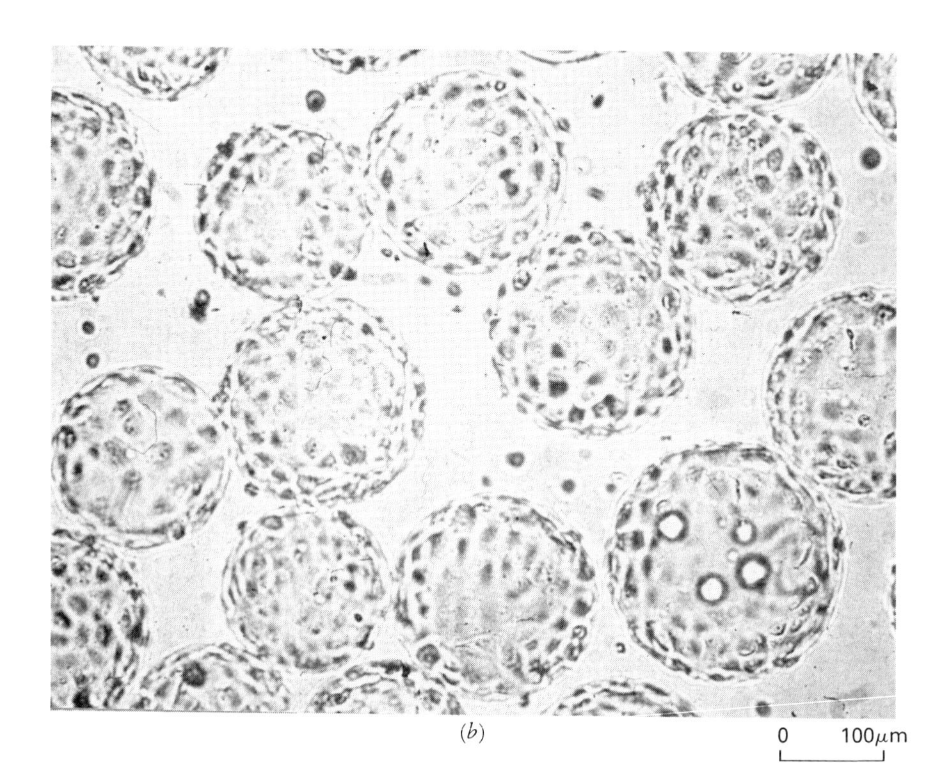

(*b*)

(*d*)

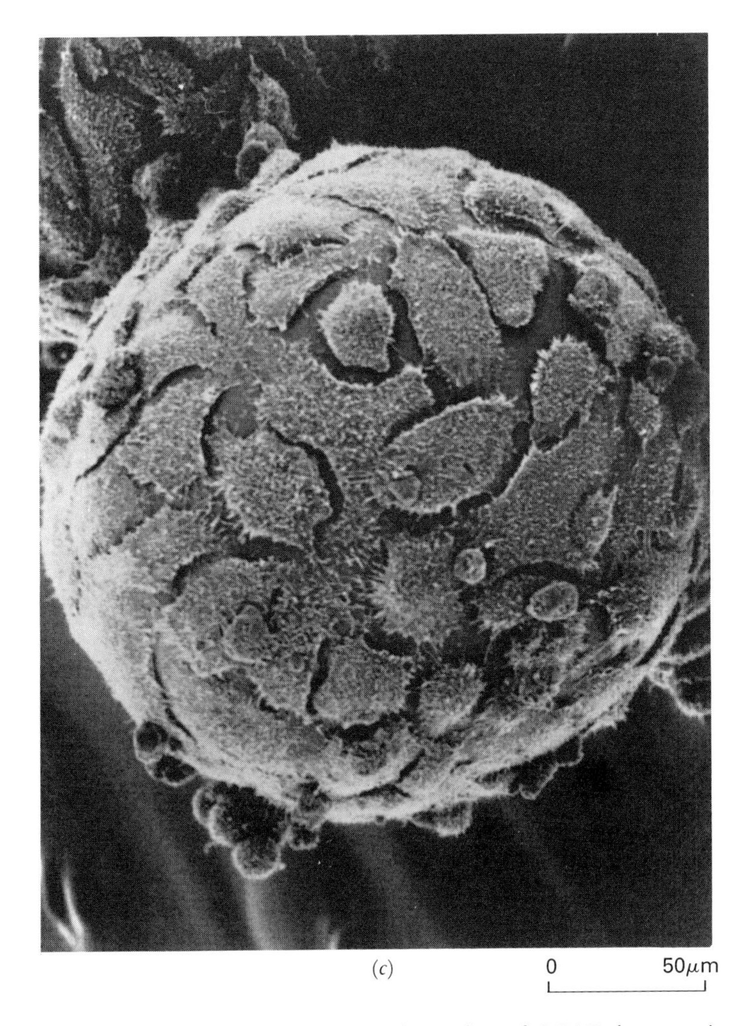

Fig. 3. *continued* (*b*) Vero cells growing on the surface of DEAE dextran microcarriers (Dormacell). (This photograph was kindly supplied by Pfeifer & Langen.) (*c*) A scanning electron micrograph showing human carcinoma KB cells attached to the surface of glass-coated microcarriers (Ventreglas). (This photograph was kindly supplied by Ventrex Laboratories Inc.) (*d*) Mouse fibroblast L-929 cells growing on plastic microcarriers (Biosilon). (This photograph was kindly supplied by Nunc.)

microcarriers can be cleaned of all cell material by treatment with trypsin and chromic acid. The beads can be re-used up to ten times without loss of cell growth capacity (*57, 60*).

Growth of human fibroblasts on glass microcarriers results in cell surface densities of up to 2×10^5 cells cm^{-2} which is a factor of 2 greater than that obtained with DEAE-dextran (*58*). These cell densities on glass-coated microcarriers are comparable with values obtained with static surfaces such as Roux bottles. However, productivity comparisons made on the basis of cell numbers obtainable per unit volume of medium are unlikely to be different between these microcarrier types.

4.5. Cellulose

Cellulose has been used successfully as a microcarrier matrix in the form of the microgranular DEAE-cellulose anion exchange material—Whatman DE-52 and DE-53 (*13, 21, 28, 48*). The cellulose fibres are of a dimension $(40–60) \times (50–300)$ μm with a hydrophilic surface with an estimated surface charge density of 0.68–1.36 meq m^{-2}. DE-52 has a charge density of 1 meq g^{-1} and has been shown to allow the efficient growth of BHK and MDCK cells (*21*). However, DE-53 has a charge density close to the optimal value determined for dextran microcarriers (*29*) and will support the growth of a wider range of cell types including primary fibroblasts (*21*). The yield of a range of cells grown on these microcarriers is roughly equivalent to DEAE-dextran microcarriers (*48*).

Cell growth occurs along the line of the cellulose fibres which form elongated cylindrical shapes. These tend to form aggregates, particularly when they are covered with a cell layer. Transfer to newly added cellulose fibres has been found possible if stirring of the culture is temporarily stopped (*28*).

The major disadvantage of these cellulose-based microcarriers is the large size distribution of the particles, which can give rise to an uneven distribution of cells in culture. Also, confluent fibres tend to aggregate and this can cause problems in stirring.

4.6. Liquid Microcarriers

A completely novel type of microcarrier has been produced by forming fluorocarbon droplets in aqueous medium (*23, 24*). A typical preparation is formed by combining FC-70, which is a 15-carbon perfluoro tertiary amine, with an alkaline solution of polylysine. Agitation of this mixture results in a stabilized emulsion consisting of droplets of 100–500 μm diameter having an outer layer of polylysine. These droplets can be added to serum-supplemented culture medium so that the surface of the droplets provides a suitable substratum for cell attachment and spreading.

Normal growth rates have been reported for a number of cell lines including mouse L cells and human diploid fibroblasts (*24*). Centrifugation of the culture will break the component phases into a heavy fluorocarbon fluid and an upper aqueous layer. The cells are of intermediate density and form a visible layer at the phase boundary. This can be used to harvest cells free of microcarriers without the use of trypsin.

These liquid microcarriers form the basis of an extremely useful experimental system for cell culture, but a number of disadvantages may be recognized in their use in large-scale cultures. The large size distribution and instability of the droplets formed could cause problems of cell-to-microcarrier distribution during culture and lead to undesirably inconsistent cell yields. Further, the advantages of cell harvesting without trypsinization are offset by the need for centrifugation and phase separation which would increase the difficulty of scale-up.

5. CHOICE OF MICROCARRIER

For any culture system, the choice of microcarrier will depend upon the cell line used and the purpose or objectives of the culture. For relatively robust established cell lines, the dextran microcarriers have proved efficient and reliable. Under a microscope, the cells can easily be seen on the microcarrier surface and staining with, for example, Crystal Violet is particularly useful in highlighting the nuclei against the transparent background of the microcarrier surface. Such observations are important in gaining an immediate assessment of the state of a culture. Crystal Violet reagents have also been useful in allowing cell counts of microcarrier cultures by enumeration of stained nuclei (*51*).

Dextran microcarriers have been used extensively for the large-scale production of biologicals such as vaccines (*34, 36, 37*) and interferon (*7, 15, 35*). In most of these applications, harvesting and separation of cells from the microcarriers is not necessary and the products can be extracted from the culture medium. Dextran microcarriers with a charge on the surface only (e.g. Cytodex 2 and Dormacell) may be useful if adsorption of the cell product is a problem.

Cells detached from microcarriers may be required for various operations including culture scale-up. The possibility of culture scale-up by the addition of new microcarriers and fresh medium has always been an attractive prospect.

However, bead-to-bead transfer has never proved particularly satisfactory for any microcarrier type except for certain cell lines and under special conditions (*8, 52*). Therefore, scale-up normally does require steps of enzymatic treatment to allow cell detachment before inoculation into the next scale of culture. Treatment of confluent microcarriers with a cocktail of trypsin and EDTA is sufficient for the detachment of many cell lines, but in some cases difficulties have been found because of the strength of cell binding. For example, human diploid fibroblasts are not detached easily by trypsin, although release can be enhanced by trypsinization at elevated pH values (*20*).

If it is necessary to isolate high cell yields of fragile or sensitive cell lines with good viability, then microcarriers other than charged dextran should be considered. Glass or gelatin microcarriers are particularly suitable for such instances, because detachment of cells can be achieved efficiently and without significant cell damage by either trypsin or collagenase treatment. Detached cells can then be separated from microcarriers by sieving through a nylon mesh of appropriate size. Microcarriers composed entirely of gelatin can be completely dissolved by proteolytic enzymes, leaving the cells in suspension.

A further value of collagen- or gelatin-coated microcarriers may be considered for the culture of primary cells. Such cells have fastidious growth conditions, and attachment to a "natural" matrix by specific cell-released proteins may be advantageous.

6. STRATEGIES AND COSTS OF MICROCARRIER CULTURE OPERATIONS

All cell-culture work is labour-intensive, and this is normally reflected in the cost of large-scale culture operations. However, with scale-up, material costs become increasingly significant as a proportion of the overall cost of the operation. Some calculations show that beyond 10^3 litres, materials exceed labour costs as the single highest unit cost of culture (*2*).

Table III shows the cost per gram of some of the microcarriers that are obtainable in the U.K. Cellulose microcarriers are particularly cheap, but they are not recommended for large-scale culture because of their large size distribution and their tendency to aggregate. It should be noted that DE-52 cellulose is produced commercially as chromatography material rather than as microcarriers for cell growth. Of the other commercially produced microcarriers, the price differential varies by a factor of 10 between the low-cost plastic microcarriers (Biosilon) and the gelatin-coated microcarriers (Ventregel). However, most of the microcarriers are priced closely to the mean of £2.5 per g. At a typical microcarrier concentration of $5\,mg\,ml^{-1}$, this can amount to 50% of the material costs of the culture. This is based on the use of 10% horse or bovine serum in standard commercial medium. Serum costs are

TABLE III Microcarrier Prices in the U.K.

Type	Trade name	Price (£ per g)
Dextran	Cytodex 1	2.39
	Cytodex 2	2.21
	Superbeads	2.48
	Dormacell	2.87
Plastic	Biosilon	0.71
	Cytospheres	3.00
Gelatin	Ventregel	7.20
	Cytodex 3	3.38
Glass	Bioglas	2.83
	Ventreglas	2.13
Cellulose	Whatman	0.33

significant in terms of these calculations, which would have to be adjusted if either the more expensive foetal calf serum were used or a lowered serum content reduced the overall cost (*16*).

In assessing the commercial costs of an operation requiring the use of microcarrier cultures, it is important to consider all possible strategies. The productivity of such cultures has been variously assessed and compared with alternatives by cells produced per surface area or per unit volume of culture. However, the important parameter to an industrial enterprise must be product generated per unit cost of operation. Any means of increasing this productivity value would be desirable.

In many cases it may be possible to increase such productivity by consideration of cell-selection or cell-culture strategy. These may indeed be alternatives to the culture volume scale-up often considered as the only option. In the first instance, careful selection of a cell line for high protein production or amplification of an appropriate gene may make productivity differences of 10 to 100 times (*61*). The choice of media perfusion rather than batch culture can increase cell yields 10 times to densities of 10^7–5×10^7 cells ml^{-1} (*5, 39*). Even though perfusion may require the expenditure of larger volumes of media, it may be possible to use diluted media after initial cell attachment to microcarriers (*5, 39*). Considerations of this sort in planning culture operations may significantly alter the bioreactor volumes necessary and reduce the overall unit cost of microcarrier cultures.

7. CONCLUSIONS

Anchorage-dependent cells are frequently chosen for culture instead of the more easily grown free-suspension cells because of their genetic stability and normal diploid nature. These factors have been considered to be particularly important in the production of human-injectable products, for which all aspects of product safety are under constant review. Microcarrier technology offers the most efficient method for culture of these cells in which yields of $\sim 2 \times 10^6$ cells ml^{-1} may be expected routinely in batch cultures. A plethora of microcarrier types have been developed commercially in the last few years and many of these have advantages for the growth of problematic cells and for use in specialized operations. However, final cell yields in batch cultures have not been significantly improved above the level of 2×10^6 cells ml^{-1}. In fact, it would seem unlikely that any further developments of novel microcarriers could result in significant increases in cell productivity.

However, the potential for higher cell yields from microcarrier cultures may be recognized from experiments with media perfusion. The relationship between cell yields and perfusion rates has shown that the supply of nutrients and the removal of toxic products from these cultures can increase cell productivity by an order of magnitude (*5, 39*). The positive identification of those factors that allow such yield increases must be the key to further major developments in microcarrier technology. By the formulation of optimal growth media or the development of a regime of selected nutrient addition and inhibitor removal, there would seem to be no logical reason why yields of $\sim 5 \times 10^7$ cells ml^{-1} could not routinely be obtained in batch microcarrier cultures. Such yields would lead to a new generation of development in microcarrier technology.

REFERENCES

1. Barngrover, D. (1986). Substrata for anchorage-dependent cells. *In* "Mammalian Cell Technology" (W. G. Thilly, ed.), pp. 131–149. Butterworth, London.
2. Birch, J. R., Thompson, P. W., Boraston, R., and Lambert, K. (1986). The large scale production of monoclonal antibodies in airlift fermenters. In "Plant and Animal Cells: process possibilities" (C. Webb and F. Mavituna, eds.), pp. 162–171. Ellis Horwood, Chichester.
3. Bornstein, P. (1980). Structurally distinct collagen types. *Annu. Rev. Biochem.* **49**, 957–1003.
4. Burke, D., Brown, M. J., and Jacobson, B. S. (1983). HeLa cell adhesion to microcarriers in high shear conditions: evidence for membrane receptors for collagen but not laminin or fibronectin. *Tissue Cell* **15**, 181–192.
5. Butler, M., Imamura, T., Thomas, J., and Thilly, W. G. (1983). High yields from microcarrier cultures by medium perfusion. *J. Cell Sci.* **61**, 351–364.
6. Butler, M. (1987). Growth limitations in microcarrier cultures. *Adv. Biochem. Eng.*, **34**, 57–84.
7. Clark, J. M., and Hirtenstein, M. D. (1981). High yield culture of human fibroblasts on microcarriers: a first step in the production of fibroblast derived interferon, human beta interferon. *J. Interferon Res.* **1**, 391–400.
8. Crespi, C. L., and Thilly, W. G. (1981). Continuous propagation using low charge microcarriers. *Biotechnol. Bioeng.* **23**, 983–994.
9. Davies, P. F. (1981). Microcarrier culture of vascular endothelial cells on solid plastic beads. *Exp. Cell Res.* **134**, 367–376.

10. Elsdale, T., and Bard, J. (1972). Collagen substrata for studies on cell behaviour. *J. Cell Biol.* **54**, 626–636.
11. Engvall, E., and Rouslahti, E. (1977). Binding of soluble form of fibroblast surface protein, fibronectin to collagen. *Int. J. Cancer* **20**, 1–5.
12. Engvall, E., Rouslahti, E., and Miller, E. J. (1978). Affinity of fibronectin to collagens of different genetic types and to fibrinogen. *J. Exp. Med.* **147**, 1584–1595.
13. Fairman, K., and Jacobson, B. S. (1983). Unique morphology of HeLa cell attachment, spreading and detachment from microcarrier beads covalently coated with a specific and nonspecific substratum. *Tissue Cell* **15**, 167–180.
14. Gebb, C., Clark, J. M., Hirtenstein, M. D., Lindgren, G., Lindskog, U., Lundgren, B., and Vretblad, P. (1982). Alternative surfaces for microcarrier culture of animal cells. *Dev. Biol. Stand.* **50**, 93–102.
15. Giard, D. J., Loeb, D. H., and Thilly, W. G. (1979). Human interferon production with diploid fibroblasts grown on microcarriers. *Biotechnol. Bioeng.* **21**, 433–442.
16. Griffiths, B. (1986). Can culture medium costs be reduced? Strategies and possibilities. *Trends Biotechnol.* **4**, 268–272.
17. Grinnell, F., and Minter, D. (1978). Attachment and spreading of baby hamster kidney cells to collagen substrata. *Proc. Natl. Acad. Sci. U.S.A.* **75**, 4408–4412.
18. Grinnell, F. (1978). Cellular adhesiveness and extracellular substrata. *Int. Rev. Cytol.* **53**, 65–144.
19. Hirtenstein, M., Clark, J., Lindgren, G., and Vretblad, P. (1979). Microcarriers for animal cell culture: a brief review of theory and practice. *Dev. Biol. Stand.* **46**, 109–116.
20. Hu, W-S., Giard, D. J., and Wang, D. I. C. (1985). Serial propagation of mammalian cells on microcarriers. *Biotechnol. Bioeng.* **27**, 1466–1476.
21. Jacobson, B. S., and Ryan, U. S. (1982). Growth of endothelial and HeLa cells on a new multipurpose microcarrier that is positive, negative or collagen coated. *Tissue Cell* **14**, 69–84.
22. Johansson, A., and Nielsen, V. (1980). Biosilon: a new microcarrier. *Dev. Biol. Stand.* **46**, 125–129.
23. Keese, C. R., and Giaever, I. (1983). Cell growth on liquid interfaces: role of surface active compounds. *Proc. Natl. Acad. Sci. U.S.A.* **80**, 5622–5626.
24. Keese, C. R., and Giaever, I. (1983). Cell growth on liquid microcarriers. *Science* **219**, 1448–1449.
25. Kleinman, H. K., Klebe, R. J., and Martin, G. R. (1981). Role of collagenous matrices in the adhesion and growth of cells. *J. Cell Biol.* **88**, 473–485.
26. Koninsberg, I. R. (1979). Skeletal myoblasts in culture. *Methods Enzymol.* **58**, 511–527.
27. Kuo, M. J., Lewis, Jr. C., Martin, R. A., Miller, R. E., Schoenfeld, R. A., Schuck, J. M., and Wildi, B. S. (1981). Growth of anchorage-dependent mammalian cells on glycine-derivatized polystyrene in suspension culture. *In Vitro* **17**, 901–906.
28. Lazar, A., Silverstein, L., Margel, S., and Mizrahi, A. (1985). Agarose-polyacrolein microsphere beads: a new microcarrier culturing system. *Dev. Biol. Stand.* **60**, 457–466.
29. Levine, D. W., Wong, J. S., Wang, D. I. C., and Thilly, W. G. (1977). Microcarrier cell culture: new methods for research scale application. *Somatic Cell Genetics* **3**, 149–155.
30. Levine, D. W., Wang, D. I. C., and Thilly, W. G. (1979). Optimization of growth surface parameters in microcarrier cell cultures. *Biotechnol. Bioeng.* **21**, 821–846.
31. Maroudas, N. G. (1972). Anchorage dependence: correlation between amount of growth and diameter of bead, for single cells grown on individual glass beads. *Exp. Cell Res.* **74**, 337–342.
32. Maroudas, N. G. (1975). Adhesion and spreading of cells on charged surfaces. *J. Theor. Biol.* **49**, 417–424.
33. Maroudas, N. G. (1977). Sulphonated polystyrene as an optimal substratum for the adhesion and spreading of mesenchymal cells in monovalent and divalent saline solutions. *J. Cell Physiol.* **90**, 511–519.

34. Meignier, B., Maugeot, H., and Favre, H. (1980). Foot and mouth disease virus production on microcarrier grown cells. *Dev. Biol. Stand.* **46**, 249–256.
35. Merck, W. A. M. (1982). Large scale production of human fibroblast interferon in cell fermenters. *Dev. Biol. Stand.* **50**, 137–140.
36. Mered, B., Albrecht, P., Hopps, H. E., Petricciani, J. C., and Salk, J. (1981). Propagation of poliovirus in microcarrier cultures of 3 monkey kidney cell lines. *J. Biol. Stand.* **9**, 137–146.
37. Montagnon, B. J., Fanget, B., and Nicolas, A. J. (1981). The large scale cultivation of African green monkey kidney Vero cells in microcarrier culture for virus vaccine production: preliminary results for killed poliovirus vaccine. *Dev. Biol. Stand.* **47**, 55–64.
38. Morandi, M., Bandinelli, L., and Valeri, A. (1982). Growth of MRC-5 diploid cell on 3 types of microcarriers. *Experientia* **36**, 668–670.
39. Nahapetian, A. T., Thomas, J. N., and Thilly, W. G. (1986). Optimization of environment for high density Vero cell culture: effect of dissolved oxygen and nutrient supply on cell growth and changes in metabolites. *J. Cell Sci.* **81**, 65–104.
40. Nielsen, V., and Johansson, A. (1980). Biosilon: optimal culture conditions and various research scale culture techniques. *Dev. Biol. Stand.* **46**, 131–136.
41. Nilsson, K., Buzsaky, F., and Mosbach, K. (1986). Growth of anchorage-dependent cells on macroporous microcarriers. *Biotechnology* **4**, 989–990.
42. Obrenovitch, A., Maintier, C., Sene, C., Boschetti, E., and Mosigny, M. (1983). Microcarrier culture of fibroblastic cells on modified tris acryl beads. *Biol. Cell* **46**, 249–256.
43. Paris, M. S., Eaton, D. L., Sempolinski, D. E., and Sharma, B. P. (1983). A gelatin microcarrier for cell culture. 34th Ann. Meet. Tissue Culture Assoc., Orlando, Florida.
44. Pharmacia (1981). Microcarrier cell culture: Principles and methods. Trade publication.
45. Reid, L. M., and Rojkind, M. (1979). New techniques for culturing differentiated cells: reconstituted basement membrane rafts. *Methods Enzymol.* **58**, 511–527.
46. Reuveny, S., Silberstein, L., Shahar, A., Freeman, E., and Mizrahi, A. (1982). Cell and virus propagation on cylindrical cellulose based microcarriers. *Dev. Biol. Stand.* **50**, 115–124.
47. Reuveny, S., Mizrahi, A., Kotler, M., and Freeman, A. (1983). Factors affecting cell attachment, spreading and growth on derivatized microcarriers. 2. Introduction of hydrophobic elements. *Biotechnol. Bioeng.* **25**, 2969–2980.
48. Reuveny, S. (1983). Microcarriers for culturing mammalian cells and their applications. *Adv. Biotechnol. Process.* **2**, 1–32.
49. Reuveny, S., Mizrahi, A., Kotler, M., and Freeman, A. (1983). Factors affecting cell attachment, spreading and growth on derivatized microcarriers. 1. Establishment of working system and effect of the type of amino charged groups. *Biotechnol. Bioeng.* **25**, 469–480.
50. Rubin, K., Hoeoek, M., Oebrink, B., and Timpl, R. (1981). Substrate adhesion of rat hepatocytes: mechanism of attachment to collagen substrates. *Cell* **24**, 463–470.
51. Sanford, K. K., Earle, W. R., Evans, V. S., Waltz, J. K., and Shannon, J. E. (1951). The measurement of proliferation in tissue cultures by enumeration of cell nuclei. *J. Natl. Cancer Inst.* **11**, 773–795.
52. Schulz, R., Krafft, H., and Lehmann, J. (1986). Experiences with a new type of microcarrier. *Biotechnol. Lett.* **8**, 557–560.
53. Spier, R. E. (1980). Recent developments in the large scale cultivation of animal cells in monolayer. *Adv. Biochem. Eng.* **14**, 119–162.
54. Thilly, W. G., and Levine, D. W. (1979). Microcarrier culture: a homogeneous environment for studies of cellular biochemistry. *Methods Enzymol.* **58**, 184–195.
55. van Wezel, A. L. (1967). Growth of cell-strains and primary cells on microcarriers in homogeneous culture. *Nature (London)* **216**. 64–65.
56. van Wezel, A. L. (1985). Monolayer growth systems: homogeneous unit processes. *In* "Animal Cell Biotechnology", Vol. 1 (R. E. Spier and B. Griffiths, eds.), pp. 265–282. Academic Press, London.

57. Varani, J., Dam, M., Beals, T. F., and Wass, J. A. (1983). Growth of 3 established cell lines on glass microcarriers. *Biotechnol. Bioeng.* **25**, 1359–1372.
58. Varani, J., Dam, H., Rediske, J., Beals, T. F., and Hillegas, W. A. (1985). Substrate-dependent differences in growth and biological properties of fibroblasts and epithelial cells grown in microcarrier culture. *J. Biol. Stand.* **13**, 67–76.
59. Varani, J., Bendelow, M. J., Chun, J. H., and Hillegas, W. A. (1986). Cell growth on microcarriers: comparison of proliferation and recovery from various substrates. *J. Biol. Stand.*, 331–336.
60. Ventrex (1986). Trade publication.
61. Weymouth, L. A., and Barsoum, J. (1986). Genetic engineering in mammalian cells. *In* "Mammalian Cell Technology" (W. G. Thilly, ed.), pp. 9–62. Butterworth, London.
62. Whiteside, J. P., Whiting, B. R., and Spier, R. E. (1979). Development of methodology for the production of foot and mouth disease virus from BHK21 C13 monolayer cells grown in a 100 l (20 m^2) glass sphere propagator. *Dev. Biol. Stand.* **42**, 113–120.
63. Whiteside, J. P., and Spier, R. E. (1981). The scale up from 0.1 to 100 liter of a unit process system based on 3 mm diameter glass spheres for the production of four strains of FMDV from BHK monolayer cells. *Biotechnol. Bioeng.* **23**, 551–565.
64. Windig, W., Haverkamp, J., and van Wezel, A. L. (1981). Control on the absence of DEAE-polysaccharides in DEAE-Sephadex purified poliovirus suspensions by pyrolysis mass spectrometry. *Dev. Biol. Stand.* **47**, 169–177.

13

Large-scale Fluidized-bed, Immobilized Cultivation of Animal Cells at High Densities

PETER W. RUNSTADLER, Jr.
STEPHEN R. CERNEK
Verax Corporation,
Lebanon, New Hampshire 03766, U.S.A.

ANIMAL CELL BIOTECHNOLOGY VOL. 3
ISBN 0-12-657553 3

1. INTRODUCTION

Advances in biotechnology, particularly in genetic engineering, in the past two decades have made possible the production of diagnostic and therapeutic biochemicals that promise the potential to control or eradicate some of mankind's most devastating diseases. Recombinant DNA technology has been employed with great success to manipulate the genetic codes of microorganisms to manufacture valuable cellular products. These products include a wide variety of proteins, particularly those of value to animal and human health care, such as vaccines, blood chemistry compounds, hormones and endocrines, monoclonal antibodies, anti-cancer agents, and other therapeutic biomolecules. A new and highly competitive biotechnology industry has emerged as a consequence of these technology developments in order to exploit the commercialization of these products. Much of the activity in the biotechnology industry during the past decade has focused on the early stages of product development—recombinant DNA genetic engineering, the production of small amounts of diagnostic or therapeutic products, and preclinical or early clinical trials to test the efficacy and safety of these biochemicals. Now the biotechnology industry is beginning to focus upon the large-scale bioprocessing—the manufacture of finished biochemical products made by living cells—that is necessary for the successful commercial introduction of natural and recombinant proteins at reasonable cost.

Initially, it was thought that the majority of recombinant proteins would be cultured in prokaryotic organisms using traditional batch-processing systems. However, the many complex therapeutic proteins now under development require post-translational processing, e.g. glycosylation and folding, that cannot be achieved in prokaryotic organisms. Consequently, hybridoma cells and genetically altered animal cells are increasingly being used to serve as host organisms for the production of these health proteins. As a consequence of the generally low specific productivity per cell by mammalian cells for these proteins, and the large volume of therapeutic-grade material that will often be required, the biotechnology industry is now beginning to address the need for cost-efficient, large-scale, mammalian cell culture technologies. As of early 1987, in human therapeutic products alone, there were at least 200 biotechnology companies developing more than 200 different genetically-engineered products. Yet, only five such products had been approved and licensed for commercial sale in the U.S.A. A reasonable estimate of the world-wide production of health protein products for clinical use in 1986 is only in the tens of kilograms. This includes both research and commercial production of therapeutic and diagnostic monoclonal antibodies, vaccines, insulin, growth hormones, and anti-cancer agents. Within a few years, however, the demand for clinical-grade material will be in the many hundreds

TABLE I Estimated Annual U.S. tPA Demand

Heart attack victims	1.67 million
Treatable cases	0.84 million
Average dose	50 milligrams
Total tPA needed (purified)	42 kilograms
Total tPA needed (unpurified)	84 kilograms[a]

[a] Assumes a purification process efficiency of 50%.

of kilograms per year. A review of the anticipated demand for but one of these recombinant products, tissue plasminogen activator, illustrates this point.

Tissue plasminogen activator (tPA) is a recombinant protein intended to be used in the treatment of thrombolytic blood clots in heart-attack patients. Currently being developed by an estimated 14 companies world wide, the first company should receive marketing approval from the FDA (Food and Drug Administration) in the U.S.A. sometime in 1987. In the United States in the year 1990 there will be an estimated 1.67 million heart attack victims (*5*). If it is assumed that 50% of these victims will be treated with tPA, using the regimen and dosage reported in the literature (*2*), the yearly tPA demands in the United States alone would be as indicated in Table I.

The world-wide demand for tPA is likely to be at least double the United States demand, bringing to over 165 kilograms the amount of unpurified tPA required to meet projected world needs. While demand for annual quantities of this magnitude may not appear until the early 1990s, 165 kilograms is an enormous amount of product to be made from mammalian cells whose cell specific productivity is generally measured in nanograms per 10^6 cells per hour. Using traditional batch-fermentation technology (stirred or airlift vessels) and available productivity data, well over 200 000 litres of reactor vessel volume would be required to produce the 165 kilograms of unpurified tPA.

This chapter describes a unique but logical cell-culture system that has been developed to provide the high space–time productivity, high product yield and scalability necessary for commercial production of these natural and recombinant proteins. The system is an efficient, economic system for continuous cell culturing of immobilized animal cells.

2. CULTURING SYSTEM OVERVIEW

The continuous-culture, fluidized-bed culturing technology and bioreactor hardware described in this chapter have been developed to produce large quantities of medical proteins economically from hybridomas and genetically engineered mammalian cells. Computer monitoring and control of the bioreactor maintains an optimized cell environment permitting high cell densities and high-volume productivity. The continuous-culture, fluidized bed reactor operates for months at a time and the hardware must therefore have highly reliable aseptic operation. The fundamental basis of the technology—the fluidized bed of microspheres—is what allows the predictable scalability that is required for large-scale production systems.

Attachment-independent, attachment-dependent, or attachment-preferred mammalian cells can be cultured in the fluidized-bed bioreactor by first immobilizing them inside weighted, porous, essentially sponge-like microspheres. A vertically oriented recycle flow through the bioreactor vessel suspends the microspheres as a fluidized bed, while a separate flow stream continuously adds medium and removes harvest liquor containing the cell products. The excellent mass transfer characteristics of the bioreactor fluidized bed, and of the membrane gas exchanger in a recycle loop, provide oxygen supply and removal of carbon dioxide at the very high rates demanded by high cell densities. The fluid velocities in the bioreactor are sufficient to suspend the cell-containing microspheres in the culture liquor, but do not damage the often fragile mammalian cells.

This fluidized-bed, continuous-culturing process has been specifically designed to meet three critical requirements for cost-effective manufacturing of biomolecules:

1. To grow cells to the high densities necessary to optimize productivity.
2. To retain high-producing cells for long periods of time under continuous culture.
3. To optimize conditions for maximum product yield.

This system has also been designed to address a vexing problem within the biotechnology industry, namely how to scale-up protein production predictably.

3. THE FUNDAMENTAL ELEMENT—THE MICROSPHERE

The basic element of the fluidized-bed, continuous-culture system is the porous microsphere. It is made of collagen and weighted with a non-cytotoxic material to achieve a high specific gravity (typically 1.6 and higher), so that it will remain suspended in the high-velocity (order of 70 cm min^{-1}), upward-flowing, fluidizing culture liquor. The typical sponge-like microsphere has a diameter of 500 μm and contains interconnected pores and channels of the order of 20–40 μm wide, allowing cells to enter easily and populate the largely empty interior of the sphere. Several thousand cells will typically populate a single microsphere and the size of the microsphere relative to the cell size is what enables the cells to have the high oxygen and nutrient transfer rates required by the cells—the cubic relationship between the diameter and volume of the microsphere means that 7/8 of the cell population inhabits the outer half of each sphere.

SEM (scanning electron microscopy) micrographs of the microspheres, as shown in Fig. 1, reveal a leaf-like morphology which provides large surface areas for cell attachment, or capture, and proliferation. The microsphere's internal volume, which is nearly 85% open, enables cell population throughout. This characteristic, coupled with the biocompatibility and biological attractiveness to the cells of the collagen, encourages cells to grow to high densities. Typical cell densities per millilitre of microsphere volume for hybridomas and animal cells are shown in Table II.

In the SEM micrograph shown in Fig. 2, attachment-dependent cells proliferate on the microsphere, and also attach to the weighting particles, which are the large solid spheres visible in this figure. In the transmission electron micrograph shown in Fig. 3, the dark areas are cross-sections of a number of attachment-preferred cells closely packed in the collagen microsphere. The cell nuclei as well as the cell membrane boundary are visible and the collagen is the fibrous, white material. The remaining open areas are the channels through which nutrients reach the cells and products excreted by the cells are removed.

The sponge-like microspheres are derived from native bovine collagen using a proprietary process. The microspheres are classified to establish an acceptable fluidization velocity range and are strongly cross-linked by a proprietary process to ensure durability and long life in the turbulent, fluidized-bed reactor environment. The collagen morphology is controlled during manufacture so that the microsphere can be engineered to meet the needs of various cell lines' production characteristics. To date, the production process has made microspheres ranging from 100 to 1000 μm diameter and having pore openings ranging in width from 5 to 100 μm. The specific gravity of the microspheres is established by the number, size, and composition of the weighting particles and can be varied to provide microspheres ranging in specific gravity from 1.05 to 3.0. Once prepared to designed specifications, the microspheres can be stored aseptically for many months prior to use in the reactor system.

TABLE II Typical Cell Densities per Millilitre of Microsphere

Cell type	Cell density (cells ml^{-1})
Hybridoma	$1–2 \times 10^8$
Chinese hamster ovary (CHO)	$3–4 \times 10^8$
Fibroblast	$3–4 \times 10^8$

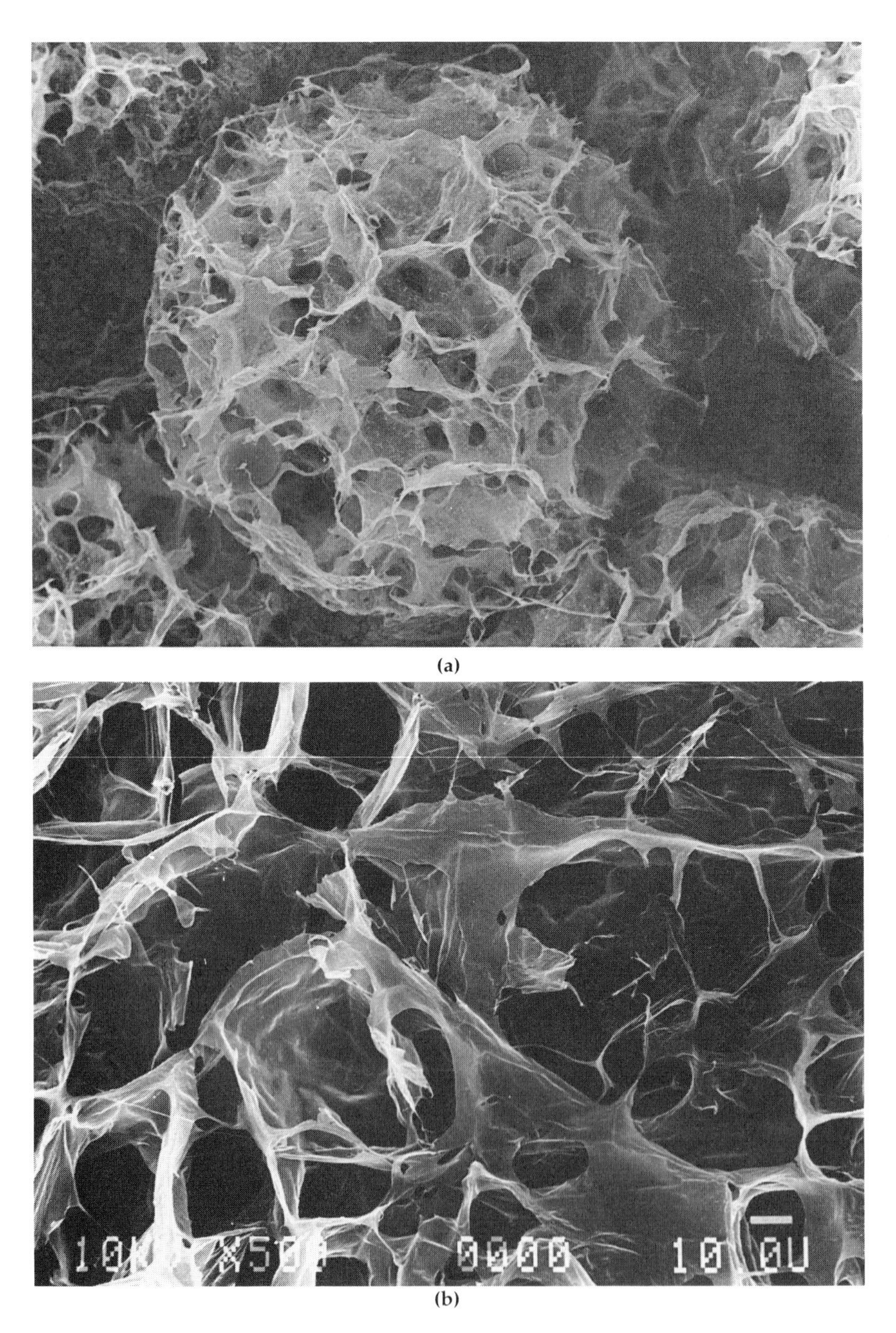

(a)

(b)

Fig. 1. Scanning electron micrographs of collagen microspheres: (*a*) 500 μm-diameter collagen microsphere (*b*) leaf-like collagen morphology. (bar = 10μm)

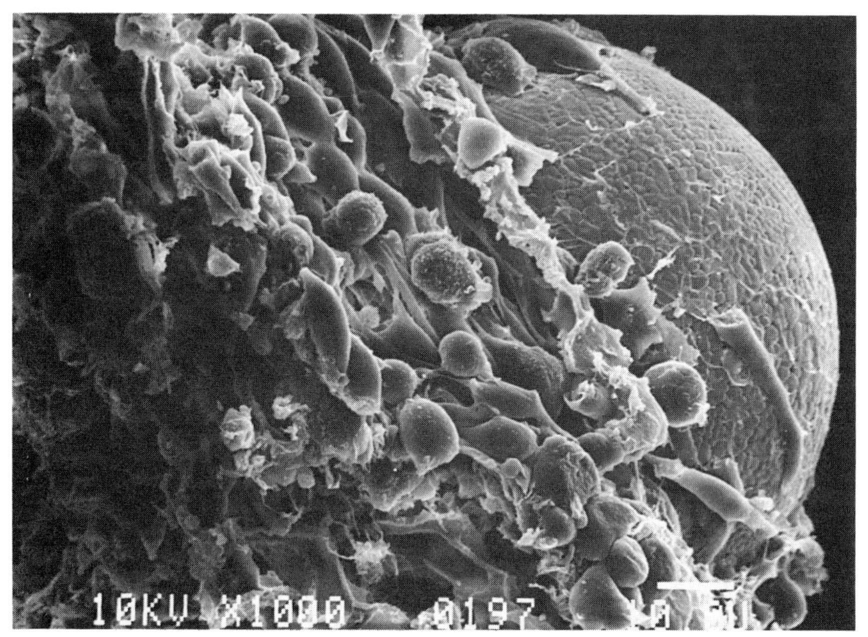

Fig. 2. Scanning electron micrograph of attachment-dependent cells populating microspheres. (Bar = 10μm)

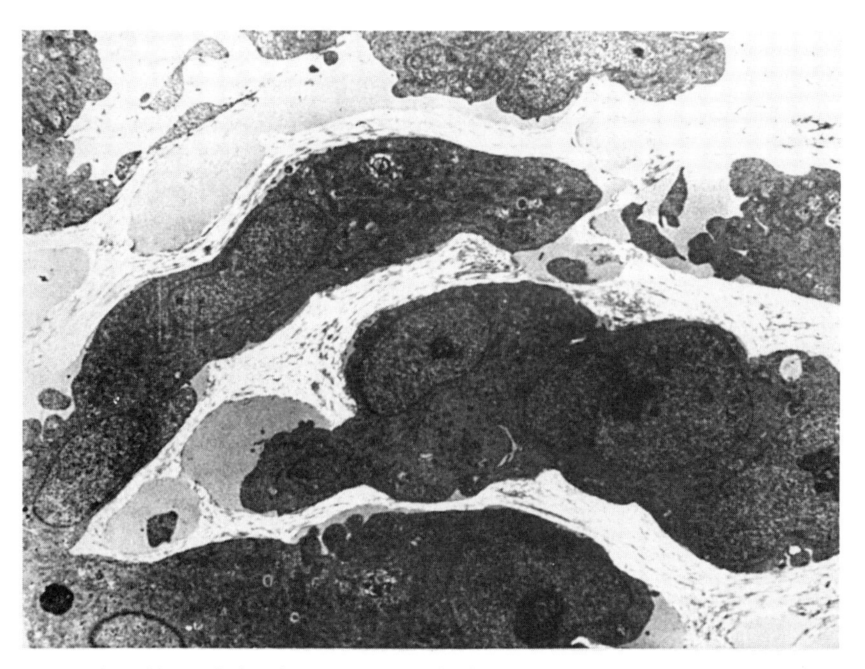

Fig. 3. Transmission electron micrograph of cells in microsphere. (× 700)

TABLE III Design Specifications for Porous Collagen Microspheres

Average diameter	500 μm
Pore size	20–40 μm
Wet specific gravity	1.6–1.7
Typical fluidization velocity	70 cm min^{-1}
Effective microsphere number density in reactor (25% solids)	3.8×10^6 microspheres $litre^{-1}$
Microsphere outer surface area based upon expanded fluidized bed (25% solids)	3 m^2 $litre^{-1}$
Estimated microsphere surface area (internal)	30 m^2 $litre^{-1}$
Biocompatible	Yes
Post-inoculate with cells	Yes

The design parameters for the microspheres are given in Table III. A very important specification is that the microspheres are post-inoculated, i.e. the open character of the sponge matrix permits inoculation with cells after the microspheres are placed in the bioreactor. This means that the microspheres do not have to be manufactured with cells in them, and can be stored for long periods of time prior to use in the bioreactor.

4. THE FLUIDIZED-BED BIOREACTOR

The post-inoculated microspheres are fluidized as a thick slurry in the bioreactor vessel. Thus, the microspheres form the fundamental element of the fluidized-bed, continuous-culture system. Each microsphere is populated with thousands of cells (depending on cell size and type) and the microspheres circulate freely throughout the bioreactor vessel, thereby providing a uniform, homogeneous, viable cell concentration and productivity of the cells throughout the reactor volume.

A schematic of the bioreactor system is shown in Fig. 4. A recycle flow of

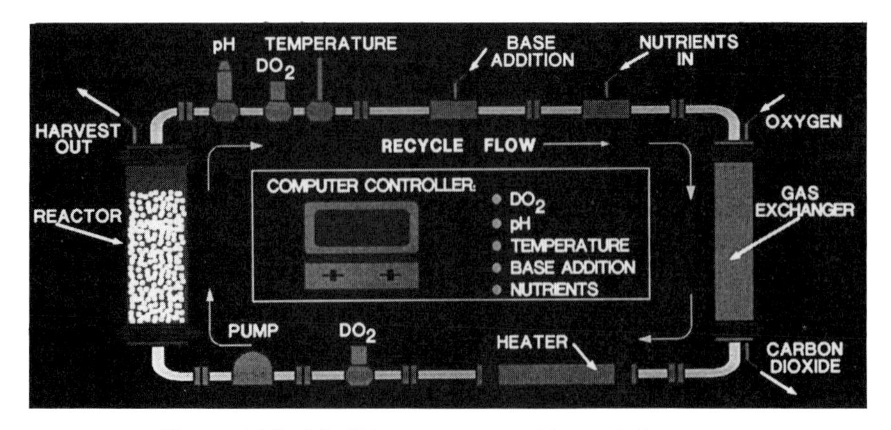

Fig. 4. Fluidized-bed bioreactor system with recycle flow.

culture liquor surrounds the microspheres in the bioreactor vessel, providing nutrients by diffusion through the microsphere pores to the cells inside. Cell products secreted by the cells are diffused out of the microsphere and into the culture liquor, along with dead cells and excess live cells. Fluidized-bed technology is used principally to achieve uniformity throughout the bioreactor vessel and to provide the high mass transfer of oxygen and carbon dioxide at the rates necessary to support the high cell densities that are achieved in the microspheres.

4.1. The Recycle Loop

The fluidization flow, which suspends the slurry of microspheres in the bioreactor vessel, is obtained by using a recycle loop. The specific gravity of the microspheres establishes the fluidization velocities required in the bioreactor which, together with the bioreactor cross-sectional area, determine the fluidization flow rates. The liquid/solid fluidized bed of culture liquor and microspheres establishes a separation horizon near the top of the bioreactor vessel, providing a clear detachment zone between the culture liquor and the slurry. The recycle loop removes culture liquor from the zone above the separation horizon at the top of the reactor, conditions it, and then returns it to the bottom of the vessel. The microspheres, which contain over 95% of the cells in the bioreactor vessel, never enter the recycle loop, but remain in the suspended slurry in the bioreactor.

The bioreactor culture liquor is conditioned in the recycle loop before it returns to the vessel. The loop consists of a recycle pump, a membrane gas exchanger (to add and extract oxygen and carbon dioxide, respectively) sensors for control of pH, dissolved oxygen and temperature, and a heater/cooler to control temperature of the system.

The recycle loop's membrane gas exchanger is a tube-and-shell design using permeable silicone rubber membranes to separate the culture liquor and the gas side of the exchanger. Oxygen is transferred by permeation through the silicone membrane from the gas (shell) side of the exchanger to the culture liquor which flows in tubes. Carbon dioxide permeates in the opposite direction from the culture liquor to the shell side. The gas exchanger must be sized to supply oxygen in amounts sufficient to meet the requirements of highly oxygen-consuming cells at the high densities maintained in the bioreactor. Typical oxygen supply rates per litre of reactor volume approach 10 mmol hr^{-1}.

Dissolved oxygen levels are monitored at the inlet and outlet of the gas exchanger in order to measure the oxygen transfer rate of the reactor and to control the oxygen flow. Carbon dioxide can also be supplied to the shell side of the gas exchanger to permit addition of carbon dioxide to the culture liquor for purposes of pH control, should the culture liquor become too basic. A computer-controlled diaphragm metering pump also provides for the addition of a base solution (such as NaOH) for the usual operating condition in which base must be added to counteract the acidic condition produced by the living cells.

The all-liquid systems (oxygen and carbon dioxide are only present in the culture liquor in the dissolved state) eliminate all gas–liquid interfaces, and the resultant foaming found in many reactor systems which has been shown to damage some proteins. In addition, the fluid dynamics of the bioreactor and recycle loop flow must minimize shear forces so that cells are not damaged and sensitive proteins manufactured by the cells are not damaged.

4.2. Medium Supply/Harvest Systems

Steam-sterilizable diaphragm metering pumps are used to supply medium continuously to the bioreactor from a chilled medium reservoir. Because the bioreactor is a wholly liquid system, medium addition and product harvest occur at the same rate. The medium supply and harvest tanks are sized to facilitate daily addition of medium and removal of product for processing.

4.3. System Monitoring and Control

Automated monitoring and control of the bioreactor system are provided by a microcomputer and an analogue–digital/digital–analogue interface module. This system provides measurement, recording and control of reactor pH, temperature, dissolved oxygen levels, medium and harvest feed rates, recycle flow rate, and base addition and oxygen transfer rates. Control is maintained against operator set-points to provide steady-state operation for long-term cultures.

5. SYSTEM PERFORMANCE

The high cell densities obtained with the use of cells immobilized in the microspheres result in a very high space–time productivity. Table IV displays results for both suspension cells and attachment-preferred cells cultured in this system. Space–time productivity data are compared to the same cell lines cultured in either chemostat or batch culture modes. The data displayed in Table IV show that for the same cell line with the same cell-specific productivity, the fluidized-bed/microsphere culture produces cell products at a rate

TABLE IV Fluidized-bed/Microsphere Culture Performance

Culture type	Cell type	Cell density (viable cells per ml of reactor)	Volume productivity (mg per litre of reactor per hr)
Fluidized-bed[a]/microsphere culture	Hybridoma	0.3×10^{8} [a]	30[b]
Fluidized-bed[a]/microsphere culture	Attachment-dependent	1.3×10^{8}	1.2[a]
Chemostat suspension	Hybridoma	2×10^{6}	2[b]
Microcarrier/stirred-tank culture	Attachment-dependent	2×10^{6}	0.02[c]

[a] Fluidized bed at 30% solids.
[b] Cell specific productivity = 1 μg/10^6 viable cells/hr.
[c] Cell specific productivity = 10 ng/10^6 viable cells/hr.

15–60 times greater than that produced by conventional chemostat or microcarrier/stirred-batch modes. This means, simply, that a 1-litre fluidized-bed/microsphere continuous culture bioreactor has the same production capability for attachment-dependent cells as a 60-litre microcarrier/stirred batch system using the same cell line. Unlike other continuous culture techniques, however, the fluidized-bed/microsphere system can be scaled-up indefinitely. To produce from one bioreactor system the same amount of product harvested from a 1000-litre microcarrier/stirred batch system requires a scale-up to only 16 litres.

The scalability inherent in the microsphere/fluidized-bed system is illustrated in Table V. Volume productivity for a monoclonal antibody for a 500-ml and a 10-litre fluidized-bed culture using the same cell line are almost identical. This represents a scaling ratio of 20 : 1.

TABLE V Comparable Productivity Data of a Hybridoma Cell in Different-size Fluidized-bed Reactor Systems

Reactor vessel size	500 ml	10 000 ml
Volume of packed microspheres	200 ml	4 000 ml
Packed bed height	10 cm	60 cm
Expanded bed volume	400 ml	6 000 ml
Solids fraction	0.25	0.33
Observed volume productivity	36 mg/LPB/hr[a]	38 mg/LPB/hr[a]
Daily productivity	173 mg day^{-1}	3.6 gm/day
Relative scale	1	20

[a] (LPB = litres of packed bed).

Maximum cell density in the fluidized bed is typically achieved within 10–15 days from inoculation of the bed with cells at a density of 2–5 × 10^5 viable cells per ml of reactor. The length of continuous culture runs made with 500 ml and 10-litre bioreactor vessels have been from 3000 to over 4000 hours. There is no inherent reason why significantly longer runs cannot be obtained.

6. CULTURE STABILITY

Genetic mutations in a living cell population always pose the possibility that a low-producing, fast-growing cell (low doubling time compared to the higher-producing cell population) can evolve and eventually take over the entire cell population, leading to decline of culture production (culture instability). In hybridoma continuous-suspension cultures (e.g. a chemostat), for example, this will often occur after 30–40 generations (cell doublings) have taken place. The microsphere/fluidized-bed continuous-culture system has proved to be immune to this behaviour. Microsphere/fluidized-bed and chemostat hybridoma cultures have been run side-by-side after having been started with inoculum from the same frozen cell seed bank. After 30–40 generations, the chemostat culture will often lose most or all of its monoclonal antibody productivity, but the fluidized-bed system productivity has run undiminished for over several hundred cell generations.

The inherent reasons for this observed culture stability are not known, but two hypotheses are proposed. One hypothesis is that immobilization biologically stabilizes the cell against mutations, as cited by Bailey (*1*) and Dykhuizen and Hartl (*4*). The second hypothesis is that a single mutation tends to be isolated to the single microsphere in which it occurs because the fluidized-bed operation and high dilution rates at which these systems operate greatly mitigates against such a mutation spreading to other microspheres. Work is now in progress to confirm these hypotheses.

7. INCREASE IN CELL SPECIFIC PRODUCTIVITY

The immobilization of microorganisms can lead to an increase in the specific productivity of the organism. For example, Doran and Bailey (*3*) have reported such effects for immobilized bacteria and yeast. Some evidence exists that just such an increase is found in optimized, fluidized-bed, immobilized cultures of hybridomas compared to the cell specific productivity for an optimized chemostat culture. Increases in cell specific productivity of 2–4 times are indicated. Work is continuing to confirm these increases by repeated measurements of cultures under long-term, steady-state conditions, since such increases will be significant in reducing the commercial production costs of products from immobilized-type cultures.

8. HARDWARE SYSTEMS

Fluidized-bed/microsphere continuous-culture systems have been built in sizes ranging from 100-ml research bioreactors to large-scale, 24-litre, commercial production systems.

8.1. Research Reactors

Figure 5 shows a small 100–800-ml fluidized-bed research and development system. These systems have been used for culture optimization to maximize cell specific productivity and product yield. Such systems have also been used to obtain performance data and to demonstrate system scalability using a number of hybridoma and mammalian cell lines.

8.2. 10-Litre Research/Production Systems

Figure 6 is a photograph of a research/production reactor system with a 10-litre fluidized-bed vessel. These systems have been used for process development, contract manufacturing of mammalian cell products, and as systems for obtaining scale-up performance data.

8.3. Verax Large-scale Production Bioreactor

Figure 7 shows a large-scale continuous-culture bioreactor designed around fluidized-bed/microsphere technology for economically producing kilogram quantities per year of medical proteins on a production scale. This system was engineered for reliable aseptic operation using industrial-grade components. Designed in accordance with GMP/FDA guidelines, the production bioreactor is operated as a clean-in-place (CIP) and sterilize-in-place (SIP) system. Computer instrumentation controls the environment within the reactor and logs and displays pertinent process information during start-up and operation.

The system has a 24-litre fluidized-bed volume and for attachment-dependent cells is equivalent to well over 1200 litres of microcarrier/stirred-tank culture capacity.

Fig. 5. Small research 100–800 ml fluidized-bed bioreactor system.

Fig. 6. 10-litre research/production fluidized-bed bioreactor system.

Fig. 7. Verax System-2000 large-scale 24-litre production fluidized-bed bioreactor system.

9. SUMMARY

The features of the microsphere/fluidized-bed continuous-culturing system are unique among large-scale mammalian cell technologies. The immobilization of attachment-dependent and attachment-preferred cells, as well as attachment-independent cells, inside the microsphere provides very high cell densities and consequent high volume (space–time) productivity from these mammalian cells. Continuous steady-state operation of these systems permits the optimization of culture conditions to maximize productivity per cell and product yield per litre of medium used. Fluidized-bed technology using porous microspheres, the core of the bioreactor design, achieves high mass transfer rates and provides a uniform growth environment, cell density, and cell viability throughout the culture vessel. Most importantly, this system provides a basic design with well established scale-up capabilities. The recycle-loop membrane gas exchanger supplies the high oxygen transfer rates required with the reactor high cell densities and accomplishes this with minimal shear damage effects and with the elimination of all liquid–gas interfaces and potential foaming conditions.

For developers of therapeutic recombinant protein products, the ability to maximize productivity per cell and rapidly and confidently to develop large-scale animal cell-culture production systems will be crucial to commercial success in a developing and highly competitive marketplace.

REFERENCES

1. Bailey, Kevin (1986). PhD Thesis. Rutgers University, New Brunswick, New Jersey 08903.
2. Builder, S. E., and Grossbard, E. (1986). "Transfusion Medicine: Recent Technological Advances", pp. 303–313. Alan R. Liss, New York.
3. Doran, P. M., and Bailey, James E. (1986). Effects of immobilization in growth, fermentation properties, and macromolecular composition of *Saccharomyces cerevisiae* attached to gelatin. *Biotechnol. Bioeng.* **28**, 73–87.
4. Dykhuizen, D. E., and Hartl, D. L. (1983). Selection in chemostats. *Microbiol. Rev.* **47**(2), 150–168.
5. Eisenberg, M. S. *et al.* (1986). Sudden cardiac death. *Sci. Am.* **254**(5), 37–43.

14

Employing a Ceramic Matrix for the Immobilization of Mammalian Cells in Culture

GUY J. BERG
Chemap AG,
Volketswil, Switzerland

BERTHOLD G. D. BÖDEKER
Bayer, AG,
Wuppertal, West Germany

Advances in the use of a variety of biomolecules, including monoclonal antibodies, in human diagnostics and therapeutics have rapidly increased the capacity requirement for mass production of mammalian cells and their

ANIMAL CELL BIOTECHNOLOGY VOL. 3
ISBN 0-12-657553 3

products. There have been many approaches to the problem of providing a mammalian cell culture system that can be scaled not only to meet the needs of the research and pilot plant scales, but can also be employed at the production plant scale (*18*). One of the newer technologies that has been developed employs ceramic matrices to immobilize cells, suspension or attachment-dependent, within a continuously recycled perfusion culture loop (*2, 4, 12, 13*). This technology has been commercially developed under the trade name "Opticell".

The use of the ceramic matrix as a method in mammalian cell culture, along with the control it employs, can simplify many aspects of maximizing the culture in terms of productivity and allow for direct scaling-up to production-sized cultures. This describes the basis of the Opticell technology and provides details of results gathered using the system.

1. THE OPTICELL CULTURE SYSTEM

1.1. The Ceramic Matrix

The ceramic matrices are of two types, one possessing a smooth surface and one possessing a porous surface (Fig. 1). The smooth-surfaced ceramic is generally employed in those instances when it is imperative that the cells are harvested, be it by trypsinization, cold-shock, or changes in pH. The other ceramic contains pores having diameters of ca. 50 μm throughout 40% of its surface. These pores entrap either suspension or adherent cells. This type of core is generally used for the production of secreted biomolecules in low-serum or serum-free medium. Because of the porous nature of this ceramic, it offers a larger surface area for the same volume than the smooth ceramic, but does not allow for complete cell harvest.

The smooth-surfaced ceramic matrix is commercially available with three available surface areas, 0.425 m^2, 4.25 m^2, and 12 m^2. The ceramics are cylindrical in shape and have square channels running their length. These channels are approximately 1 mm square with a wall thickness of 0.15 mm, or about 25–40 $cm^2\ cm^{-3}$. To increase surface area from 4.25 m^2 to 12 m^2, the diameter, not the length, of the ceramic cylinder is increased, although final testing has not been done to determine the ultimate length one could use before unacceptable gradients would occur in the medium. The porous ceramic matrix is available with two different approximate available surface areas, 1 m^2 and 11 m^2. The ceramic matrices are housed in a plastic cylinder which contains flow diverters at each end, to ensure that the medium flow is equally distributed throughout all channels.

(*a*)

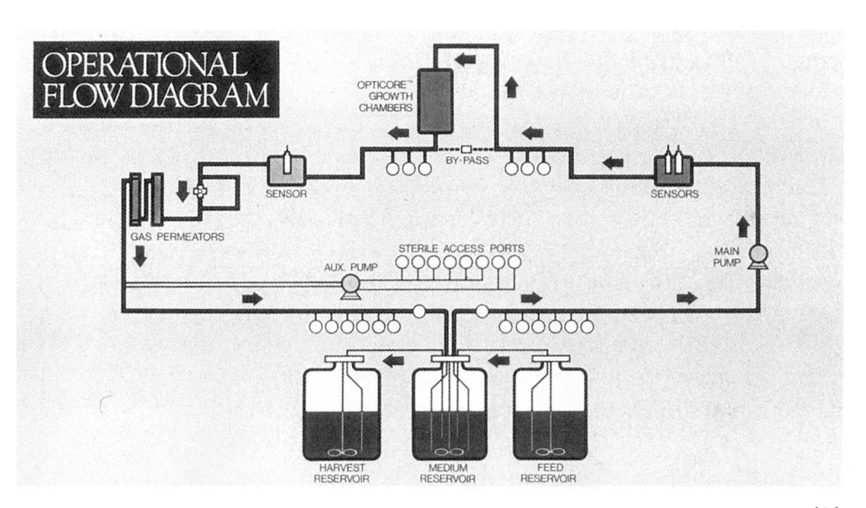

(*b*)

Fig. 1. (*a*) Showing the two different ceramic matrices employed and the three different sizes. Left, the smooth-surfaced 12 m^2 ceramic, middle, the 11 m^2 porous ceramic, and right, the 1 m^2 porous ceramic. (*b*) Schematic of bioreactor. Closed-loop process for medium circulation. pH and pO_2 are monitored before and after the medium has perfused through the culture. Gas permeators maintain controlled levels of O_2 and CO_2. Auxiliary pump is for the addition of basic solution for pH control. The two secondary flow paths are for the continuous feed and harvest (CFH) of medium.

1.2. The Perfusion Loop

Figure 1 illustrates diagrammatically the loop into which the matrix is built. The main loop provides for the constant recirculation of the medium through the culture, while the two secondary loops provide for the continuous replenishment with fresh medium and continuous removal of the exhausted medium containing products and/or cellular toxins (continuous feed and harvest, CFH). The pO_2 and the pH of the medium are measured upstream of the immobilized culture and these values are utilized in the constant feedback control for these parameters. This is accomplished by the continuous mixing of the gases (N_2, O_2, CO_2) in the correct percentages within the gas permeators in order that the desired values or set-points for the pH and pO_2 are maintained. In addition, this pH electrode also activates the auxiliary pump for the dose-addition of a base solution to raise the pH value, when required. The changes in these parameters after the medium has passed through the culture are determined downstream from the culture. This downstream pO_2 value is used in automatically altering the rate of medium flow through the matrix in order that any gradient that forms throughout the length of the culture is within an acceptable range, as determined by the user.

The sizes of the different components, i.e. pump, tubing diameter, gas permeators, etc., must obviously be varied depending upon which size of matrix is being used or when matrices are being used in parallel. There are presently three different models of Opticell in which these components are varied to compensate for the differences in available surface area.

1.3. Uniform Cell Attachment and Growth

Cell inoculation is a four-step procedure in which one-quarter of the cells are pumped into the matrix at each step. The cells are allowed to settle onto the ceramic surface, attaching within 10 min, regardless of serum concentration or temperature. After this seeding–settling process is accomplished for each of the four channel sides, the system is allowed to “rest” for 1–2 hr before the main circulation through the ceramic begins (*8*). After this period, the cells are firmly attached, and flow rates of more than 1 l min^{-1} will not detach them (*12*).

The fact that from this point on the cells will grow uniformly throughout the ceramic core by using total protein determinations has been published previously (*13, 14*). To obtain this uniform growth throughout the length of the core and maintain it throughout the culture period, the flow rate of the medium must increase proportionally in response to the increase of number of cells immobilized within the ceramic during the growth phase, and plateau just as the cell growth plateaus. It has been found that the pO_2 value of the medium is the best parameter to determine medium flow rates in a perfusion system, as its utilization is much more rapid than that of any other nutrient, while the liberation of toxic material is slower than its utilization (*12*). Because of this, the downstream pO_2 value is used to determine the flow rate.

1.4. Determination of Cellular Growth

Because the cells are immobilized in the closed ceramic core, it is impossible to visualize the cells or to determine cell numbers by counting. Therefore, an alternative method is employed to determine indirectly the cell number in the ceramic core as well as the metabolic state of these cells once the plateau has been reached. This is done by calculating the rate at which the biomass consumes oxygen. This oxygen consumption rate (OCR) proves to be a reliable, continuous, on-line method for determination of cell numbers of immobilized cells.

The pO_2 of the medium is measured both before and after it flows through the chamber, and the flow rate of the medium through the core is known. Using a simple mathematical equation based upon Henry's law (*9*), the OCR is then continuously determined on-line by the controller and is constantly graphed, allowing for an immediate overview of the phase of cellular growth and the condition of the culture. It is also possible to evaluate cell growth in the ceramic matrix by using glucose consumption. However, this measurement must be done off-line in order to determine indirectly the biomass, making this secondary or complementary determination of growth more complex and, therefore, slower.

It has previously been published that there is a direct correlation of the OCR value with the number of cells throughout the growth phase of the culture (*9, 14*). It was shown that the doubling times for BHK-21 cells and for MRC-5 cells was 16 hr and 36 hr, respectively. These are also the exact doubling times for these cells in respect to the doubling of their OCR values. However, once the plateau stage of growth has occurred, the direct linearity between cell number and the OCR value is not as reliable in indicating total cell numbers. This is not an unexpected finding, because once cells become confluent, the rate of oxygen utilization per cell is decreased (*1*).

This plateau OCR value is also used in media management and in determination of the ideal pO_2 of the medium for each particular cell grown. Because the OCR is constantly being monitored, it is easy to determine when the exchange rate of the continuous feed and harvest (CFH) of medium must be decreased or increased (*4, 15*). Any drop in the OCR value at this stage, at a pre-set pO_2 concentration, indicates that the exchange rate must be altered.

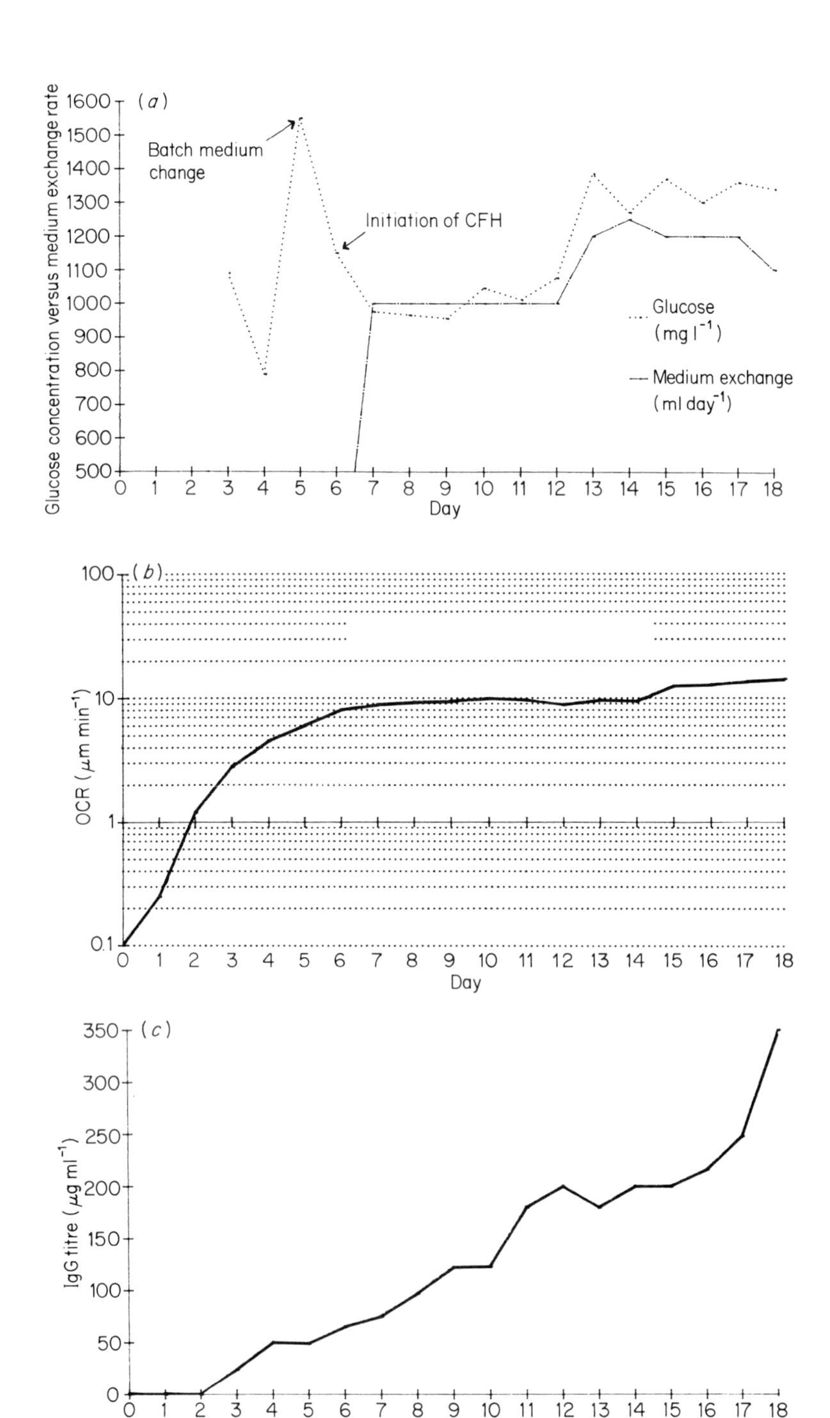

(a)
Glucose concentration versus medium exchange rate
Batch medium change
Initiation of CFH
Glucose (mg l⁻¹)
Medium exchange (ml day⁻¹)
Day
(b)
OCR (μm min⁻¹)
Day
(c)
IgG titre (μg ml⁻¹)
Day

In addition, this plateau OCR value itself can in many cases be raised by increasing the pO_2 of the medium. This increase in the plateau OCR value caused by an increase in pO_2 in the medium is not an artificial increase, however, but most probably indicates a higher metabolic activity of the cells. This can be deduced by the fact that the OCR increase is most often mirrored by an increase in cellular productivity as seen, for example, in higher titres of MAbs obtained in the case of hybridomas grown in higher pO_2 concentrations (*9*, *16*), as seen in Fig. 2(*c*).

2. EXPERIMENT AND RESULTS WITH OPTICELL

2.1. Production of Monoclonal Antibodies

A variety of hybridomas as well as transformed lymphocytes have been grown in the porous ceramic matrix (*3*, *6*, *16*, *19*) for the production of monoclonal antibodies (MAbs). As an example of MAb production, results obtained from the L243 cell line will be reported. This is an NS-1 derived mouse hybridoma producing IgG2a at reported average titres (ATCC) of 40 $\mu g\ ml^{-1}$ and is available from the ATCC.

For this cell line, 10^8 cells were seeded in an S-51 Opticore which had a total available surface area of approximately 1 m^2 and the Opticell Culture System 5200R was used as the control unit. DMEM containing 10% FCS (foetal calf serum) was used as the growth medium and this was completely replaced after five days with medium containing 1% MPS, a serum-free medium supplement (Hazelton, Denver, PA). The CFH of this serum-free medium commenced at day 6 with an exchange rate of approximately 1 litre day^{-1} or about 30% of the total medium in the system.

Figure 2(*b*) shows the OCR curve for this cell line. The OCR apparently reaches a plateau at day seven into the run. However, at day 14 the OCR plateau again rises by approximately 10%. This is achieved by increasing the available pO_2 in the medium at the inlet to the immobilized cells as well as by increasing the exchange rate of the CFH, resulting in the concomitant increase in glucose concentration of the medium (Fig. 2(*a*)). It has been reported previously that either of these manipulations can independently raise the plateau level (*9*, *16*) for a variety of cell lines. Figure 2(*c*) shows an increase in

---

Fig. 2. L243 mouse hybridoma cells were grown over an 18-day period in the 1 m^2 ceramic matrix using the Opticell Model 5200R. At day 5, the medium containing 10% FCS was completely exchanged with medium containing 1% MPS. On day 6, CFH was commenced with this serum-substituted medium at the rates shown. (*a*) Glucose concentration remains relatively constant at a set CFH rate. (*b*) The OCR (in $\mu M\ O_2/min^{-1}$) shows an approximate 10% increase at day 15 when the CFH rate increased and the pO_2 set-points were elevated. (*c*) Titres of IgG rose continuously throughout this 18-day run using this cell line. Upon increasing the CFH rate and pO_2 set points, the IgG titre increased significantly.

MAb titre accompanying this increase in OCR, as is the case with all cell lines so far tested.

Parallel spinner cultures of the L243 cell line were maintained as controls. The cells in the spinner flasks were seeded at 2×10^4 cells ml^{-1} DMEM. As in the case of the ceramic, 1% MPS was employed as a production medium. Cultures were re-fed every 2–3 days by replacing 20% of the spent medium containing cells with fresh medium. Throughout this 18-day experiment, $4–6 \times 10^5$ cells ml^{-1} were maintained in the 250-ml spinner flask.

Using the ceramic matrix, the IgG titres ranged from an initial 25 $\mu g\ ml^{-1}$ up to more than 300 $\mu g\ ml^{-1}$. In the spinner cultures, the titres fluctuated between 30 and 60 $\mu g\ ml^{-1}$ throughout the culture period. The average titres for the 18 days were 45 $\mu g\ ml^{-1}$ in the spinners and 185 $\mu g\ ml^{-1}$ in the Opticell.

In addition to the titres, there is an obvious difference in the total amount of MAb that is harvested from the two culture methods. Correcting for the differences in total volumes of medium harvested per day, i.e. 20 ml per 2 days in the spinners versus 1 litre per day from the Opticell (a factor of 100), the daily production rate in the spinner controls would range from 35 mg day^{-1} to 70 mg day^{-1} IgG, while the production in the Opticell ranged from 75 mg day^{-1} to 380 mg day^{-1} IgG. Again, correcting for the total volumes harvested, 680 total mg IgG would have been harvested in 18 days from a spinner in which 1 litre of medium per day was exchanged. In the case of the ceramic matrix, 3494 total mg of IgG was actually harvested.

These results have been repeated both with the L243 cell line as well as with a variety of other hybridomas (*3*). Table I summarizes results obtained with two other mouse cell lines using the S-51 ceramic. In each case, the controls produced similar results to those reported above for the L243.

Results using a larger ceramic, Opticore S-451, a porous ceramic with an available surface area of approximately 10 m^2, in the Opticell Model 5300 have been previously published (*6, 16*). Here it was reported that using the 20-8-4S mouse hybridoma immobilized on this large ceramic, an average of 1.7 g IgG

TABLE I MAb Production Data from Three Different Mouse Hybridoma Cell Lines

Cell type	Product	Average titres ($\mu g\ ml^{-1}$)	Average per day production ($mg\ day^{-1}$)	Length of production[a] (days)	Total IgG Produced (mg)
20-8-4S	IgG 2a	83.6	129	7	903
T1B 109	IgG 2b	54.0	59	22	1300
L243	IgG 2a	185.0	291	12	3494

[a] Cells were initially grown in medium containing 5–10% FCS for 5 days. Production is considered to be only those days in which the FCS content of the medium was 1% or less.

per day was harvested over a 30-day period with an average titre of 114 $\mu g\ ml^{-1}$. By comparing these values with what is shown in Table I for that same cell line but with the different-sized ceramic, one can see that the titres are similar in both cases and the total amounts of IgG harvested daily are proportional.

2.2. Production of Tissue-type Plasminogen Activator

The production of tissue-type plasminogen activator (tPA) is an example of a biomolecule that is secreted from an adherent cell line, human Bowes melanoma (*17*), which is immobilized on the ceramic matrix. This cell line was cultured on the smooth-surfaced 4.25 m^2 AD-451 Opticore ceramic matrix. A two-step process was used, starting with an initial growth phase in medium containing 7% FBS followed by a production phase in serum-free medium. This latter was run semi-continuously, harvesting the medium every second day and replacing it with fresh medium. The serum-free medium was DMEM/Ham's F12 (50/50, v/v) supplemented with BSA, insulin, transferrin, oleic acid, phosphoethanolamine, and vitamin E acetate (*5*).

A typical culture of the melanoma cells in the Opticell is summarized in Fig. 3 (OCR) and Fig. 4 (tPA activity and glucose concentration of the medium, a supplemental indicator of the metabolic activity of the culture). As controls,

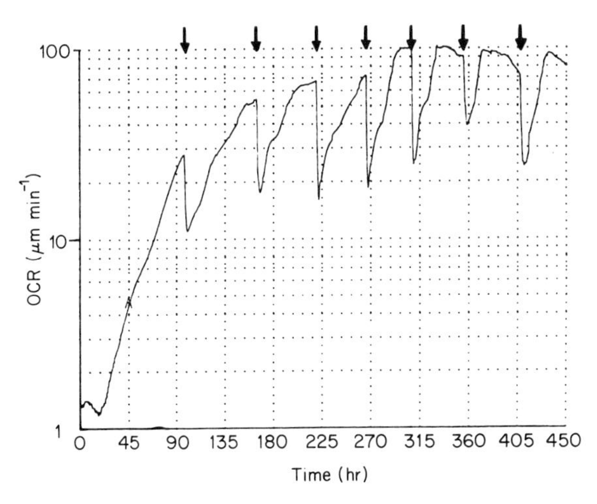

Fig. 3. OCR (in $\mu M\ O_2\ min^{-1}$) of Bowes melanoma cells in a semicontinuous culture in the Opticell Model 5300 using the 4.25 m^2 ceramic. The culture was started with 7×10^8 cells and 10 litres of medium containing 7% FCS. At the times indicated, the medium was harvested and replaced by 20 litres of fresh protein-rich, serum-free production medium.

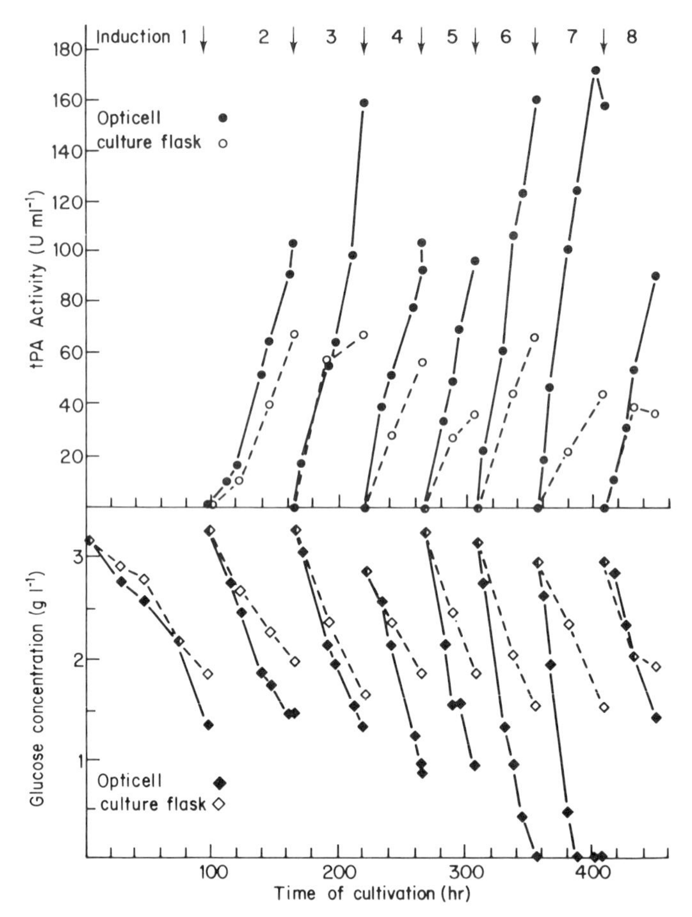

Fig. 4. Glucose consumption and tPA release from Bowes melanoma cells grown in the ceramic and in culture flasks as controls. The Opticell culture (see Fig. 3 for details) and cultures in 25 cm^2 tissue culture flasks were run under the same basic conditions, i.e. initial cell concentration, media composition, medium exchange interval.

tissue culture flasks were run in parallel under the same basic conditions, i.e. initial cell concentrations per volume and per surface area, medium composition and medium exchanges. The corresponding control kinetics of glucose utilization and product release are integrated into Fig. 4.

After an initial lag phase of 20 hr in the Opticell, the cells started to multiply at a population doubling time of 20–24 hr (as estimated from the OCR doubling time). In the beginning of the serum-free production phase, the cells continued to grow until they attained an OCR of 100 μM min^{-1} at the end of the fifth harvest period, or 300 hr into the culture. Until this point, cell detachment was negligible, as seen by the small number of cells circulating in the medium. During the sixth harvest period (350 hr), the maximal OCR neither increased nor decreased, although the medium was renewed. However, the number of circulating, detached cells increased, indicating that a steady state between cell growth and cell detachment had been reached in the growth chamber. These detached cells remained viable while circulating in the loop. This steady-state OCR value of 100 μM min^{-1} corresponded to the maximal OCR which could be attained with the Bowes melanoma cells in this serum-free formulation at a pO_2 of 100% saturation in air of the medium entering the growth chamber; tPA was continuously released during all medium changes.

The effect of a substrate limitation was tested in the seventh harvest phase. Under limiting conditions, i.e. when the substrates of the medium had been consumed, the OCR decreased, tPA activity reached a maximal level, and the number of circulating cells greatly increased. Obviously, these cells detached from the ceramic and were no longer viable. However, even when the culture continued under these limiting conditions for more than 1 day, once fresh medium was added (harvest 8), the cells again started to grow and produce tPA. Therefore, the culture could be reconstituted simply by a complete exchange of medium.

After 450 hr the cells were harvested from the ceramic surface by trypsinization. The total cell number of attached cells was 5×10^{10}, corresponding to a cell density of 1.2×10^6 cells cm^{-2} or 2.5×10^6 cells ml^{-1} medium. In total, 1.85×10^7 units of tPA activity were harvested in 150 litres of serum-free culture supernatant, yielding an average concentration of 125 U ml^{-1}. The average production rate was 60–80 U ml^{-1} day^{-1} or 1.2–1.6 $\times 10^6$ U day^{-1} in a single ceramic growth chamber with a surface area of 4.25 m^2.

The comparison between the ceramic system and tissue culture flasks showed that the glucose consumption and tPA release were favoured in the Opticell. The tPA production rate was 2.5-fold higher in the technical culture system than in flasks. This indicates that under controlled conditions, the cells grew better, had a higher metabolic activity, and produced more product than under uncontrolled conditions.

Compared to tPA production from Bowes melanoma cells on collagen-coated microcarriers, where maximal activities of 15–20 U ml^{-1} were obtained from cultures in a 40-litre fermenter using medium with 0.5% FBS (*11*), tPA productivity was enhanced in the Opticell. However, it cannot be directly concluded from these data whether this beneficial effect was solely due to the different culture system or also due to the improved production medium.

Since serum-free medium could also be used for the initial growth phase of the Bowes melanoma cells (*4*), a completely serum-free cultivation process

TABLE II Influence of the Addition of Protein to the Serum-free Medium on the Maximal OCR and tPA Production of Bowes Melanoma Cells

Serum-free medium	Maximal OCR[a] (μmol min^{-1})	tPA production U ml^{-1} day^{-1}
Protein-rich[b]	100	52
Protein-free[c]	25	4

[a] Corresponding to the number of attached cells in the growth chamber; cultures were run at a volume of 20 litres and a pO_2 of 100%-saturation in air.
[b] Containing 1 g l^{-1} BSA, 10 mg l^{-1} transferrin, 10 mg l^{-1} insulin.
[c] Basal medium: DMEM/HAMs F12 (1 + 1).

was possible when employing the ceramic matrix to immobilize the cells. The composition of this serum-free medium, however, is very important for cell growth and tPA product yield, especially in terms of the protein, in this case BSA, content. When Bowes melanoma cells were cultured either in the BSA-supplemented, protein-rich medium as described above or in a serum-free basal medium with no BSA, both the maximal OCR as an indicator for the maximal cell density in the growth chamber and the tPA production were strongly enhanced in the protein-containing medium (Table II). The beneficial effect was solely due to BSA and not to the other medium supplements such as insulin or transferrin. The minimal BSA concentration for unaffected growth and tPA production in the Opticell was found to be 100 mg l^{-1}. This addition of the BSA was much more important using the ceramic as a substrate than for tissue culture flask cultures (*7*).

2.3. Production of Other Biomolecules

A variety of biomolecules have been produced from mammalian cells grown on the ceramic matrix (see Table III). These were produced either from genetically engineered cell lines, from cells that inherently secrete the product, or from primary cells that required an induction step for production.

With most of these biomolecules, the results remain proprietary; however, in the case of pro-urokinase, a direct comparison using three different culture methods has been made and the results have been published (*10*). In this study pro-urokinase was continuously produced over a period of 6 weeks in roller bottles, gelatin-coated microcarriers, and using the ceramic matrix. No

TABLE III Biomolecules Produced from Mammalian Cells Immobilized on a Ceramic Matrix

Vaccines
Urokinase
Pro-urokinase
Tissue-type Plasminogen Activator
Monoclonal antibodies, murine and human
Factor VIII
Interferons
Interleukin-2
Colony-stimulating factor
Superoxide dismutase
Human growth hormone

significant production differences were observed between the three systems when the volumetric productivity was compared, based upon the total amount of medium utilized.

3. CONCLUSIONS

The employment of a ceramic matrix has proved to be a successful alternative for the cultivation of mammalian cells. The results with both MAbs and tPA indicate that the ceramic matrix employed in the Opticell Culture System is suited for their production. The ceramic matrix provides a large surface area in a small volume; it can be used for either suspension or attachment-dependent cells; it minimizes shear forces on the cells; and serum-free medium can easily be employed. The Opticell is a unit process in which the medium can be harvested either by batch or by a pre-programmed continuous feed and harvest mode, and the monitoring of the culture does not require sampling. However, the true scalability of the ceramic matrix has yet to be convincingly shown in terms of very large production units, i.e. those comparable with ⩾1000-litre stirred-tank fermenters. Research must be carried out on not only multiple ceramic matrices, but also on ceramics with larger surface areas. In addition, for a true production facility employing the ceramic for cell immobilization, one must be able to include, on-line, all aspects of production, from medium preparation/storage, sterilization, adaptable scaling, to final concentrated-product recovery. Once these points are addressed, the ceramic matrix should prove to be an efficient option for the production of biomolecules.

REFERENCES

1. Balin, Arthur K., Goodman, David B., Rasmussen, Howard, and Cristofalo, Vincent (1976). The effect of oxygen tension on the growth and metabolism of WI-38 cells. *J. Cell Physiol.* **89**, 235–249.
2. Berg, G. J. (1985). An integrated system for large scale cell culture. *Dev. Biol. Stand.* **60**, 297–303.
3. Berg, G. J., and Pugh, G. G. (1987). A comparison of monoclonal antibody production in spinner flasks versus a 0.425 m^2 ceramic matrix. Poster presentation at ESACT meeting, Tiberias, Israel, April.
4. Bödeker, Berthold, G. D. (1985). Einsatz des Opticell-Kultursystems. *LaborPraxis* **9**, 970–980.
5. Bödeker, B. G. D., Berg, G. J., Hewlett, G., and Schlumberger, H. D. (1985). A screening method to develop serum-free culture media for adherent cell lines. *Dev. Biol. Stand.* **60**, 93–100.
6. Bödeker, B. G. D., Hübner, G. E., Hewlett, G., and Schlumberger, H. D. (1985). Production of human monoclonal antibodies from immobilized cells in the Opticell Culture System. Poster presentation. ESACT, Baden bei Wien, October.
7. Bödeker, B. G. D., Klimetzek, V., Klein, U., Hewlett, G., and Schlumberger, H. D. (1987). Cultivation of Bowes Melanoma cells in the Opticell System: influence of the addition of protein supplements to the serum-free medium on the production of plasminogen activator. *Dev. Biol. Stand.* **66**, 291–197.
8. Bognar, E. A., Pugh, G. G., and Lydersen, B. K. (1983). Large scale propagation of BHK-21 cells using the Opticell Culture System. *J. Tissue Culture Methods* **8**, 147–154.
9. Bognar, E. A., Putnam, J. E., Pugh, G. G., Noll, L. A., and Lydersen, B. K. (1985). Monitoring of large scale animal cell cultures by analysis of oxygen consumption rate. Presented at 36th Meeting of the Tissue Culture Association, New Orleans, LA.
10. Hu, W. S., Tao, T. Y., Bohn, M. A., Ji, G. Y., and Einsele, A. (1987). Kinetics of pro-urokinase production by mammalian cells in culture. Presented at 8th ESACT Meeting, Modern Approaches to Animal Cell Technology, Tiberias, Israel.
11. Kluft, C., van Wezel, A. L., van der Welden, C. A. M., Emeis, J. J., Verheijen, J. H., and Wijngaards, G. (1983). Large scale production of extrinsic (tissue-type) plasminogen activator from human melanoma cells. *In* "Advances in Biotechnological Processes," Vol. 2 (A. Mizrahe and A. L. van Wezel, eds.), pp. 97–110. Alan R. Liss, New York.
12. Lydersen, B. K. Perfusion cell culture system based on ceramic matrices. (Submitted for publication.)
13. Lydersen, B. K., Pugh, G. G., Duncan, E. C., Overman, K. T., Johnson, D. M., and Sharma, B. P. (1983). A novel ceramic material for large scale animal cell cultures. Presented at 34th Meeting of the Tissue Culture Association, Orlando, Florida.
14. Lydersen, B. K., Pugh, G. G., Paris, M. S., Sharma, B. P., and Noll, L. A. (1985). Ceramic matrix for large scale animal cell culture. *Bio/Technology* **3**, 63–67.
15. Lydersen, Bjorn K., Putnam, James, Bognar, Ernest, Patterson, Michael, Pugh, Gordon G., and Noll, Lee A. (1985). The use of a ceramic matrix in a large scale cell culture system. *In* "Large-Scale Mammalian Cell Culture" (Joseph Feder and William R. Tolbert, eds.), pp. 39–58. Academic Press, Orlando.
16. Putnam, J. E., Wyatt, D. E., Pugh, G. G., Noll, L. A., and Lydersen, B. K. (1985). Maximizing antibody production using the Opticell Culture System. Presented at 4th Annual Congress for Hybridoma Research, San Francisco, CA.

17. Rijken, D. C., and Collen, D. (1981). Purification and characterization of the plasminogen activator secreted by human melanoma cells in culture. *J. Biol. Chem.* **256**, 7035–7041.
18. Spier, R. E., and Griffiths, J. B. (eds.) (1985). "Animal Cell Biotechnology," Vol. 1. Academic Press, London.
19. von Wedel, R. J., Peterson, J. A., and Pugh, G. G. (1987). A scale-up technique for production of monoclonal antibodies to tumor antigens from bench top perfusion cultures of immobilized cells. Presented at 6th Annual Congress for Hybridoma Research, San Francisco, CA.

15

Culture of Animal Cells in Hollow-fibre Dialysis Systems

O. T. SCHÖNHERR
P. T. J. A. VAN GELDER*
Cell Culture Department,
R&D Diosynth,
Oss, The Netherlands

1. INTRODUCTION

Production of biologicals by mammalian cell cultures is still an expensive process with many pitfalls. To develop more efficient production

*Present address: Intervet, Box Meer, The Netherlands

ANIMAL CELL BIOTECHNOLOGY VOL. 3
ISBN 0-12-657553 3

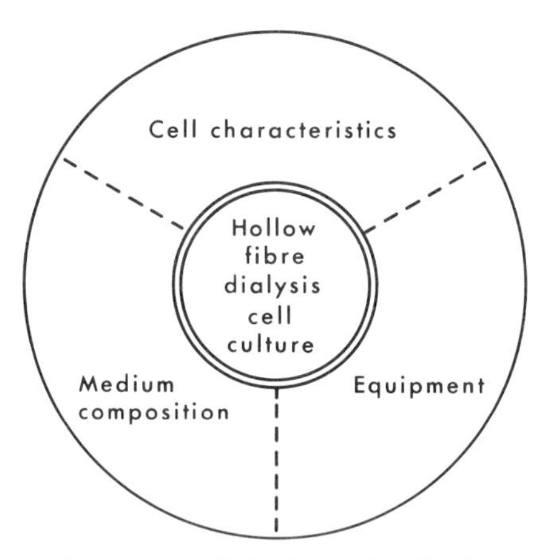

Fig. 1. Inter-relating factors in cell cultures.

technologies, each of the following inter-related aspects of cell culture (as shown in Fig. 1) should be considered:

1. Cell characteristics: cell lines should have a short generation time, a high product synthesis per cell for a prolonged period of time, suspension-like cell characteristics and a firm cell membrane.
2. Medium composition: a serum-free, and preferably a protein-free, chemically defined medium should be used initiating high rates of cell growth, as well as increased cell numbers and product expression per litre of culture medium.
3. Equipment: the bioreactor, the core of the cell culture equipment, should provide a suitable physical environment for the cells, combining low shear forces to avoid damage to the fragile cells, with high transfer rates of oxygen, carbon dioxide, nutrients and wastes to create a well-controlled microenvironment.

A hollow-fibre dialysis bioreactor aims to improve the conditions for cell culture both on the level of required medium composition, because a protein-free medium can be used, and of equipment design, because shear-free conditions and high mass transfer rates are reached. As a result, high cell and product concentrations are obtained.

Since the early 1970s, hollow-fibre membranes have been used by Knazek (*13, 14*) and others (*6, 24*) for the culture of mammalian cells in high densities. Their aim was to provide an environment for the cells that simulated the *in vivo* situation. In a network of capillaries, cells grew as solid tissue masses and could be maintained for weeks. These cultures were used to study responses of hormone-producing cells under more physiological conditions than in conventional *in vitro* cell cultures. Since that time, hollow-fibre bioreactors have been used for the production of cancer (*7, 9*) and viral antigens (*11, 17, 19*). This cell-culture system became especially important after the development of hybridomas for the synthesis of monoclonal antibodies (MCA) (*15*) and the need to produce purified MCAs on a kilogram level (*21*). A hollow-fibre dialysis cell-culture technique was described by Schönherr (*22*) as an alternative to producing MCAs via ascitic fluids in mice. MCAs were produced for commercial purposes (e.g. pregnancy tests) in high concentrations using serum-free and protein-free culture medium. The MCA-concentration was a hundred- to thousand-fold higher than in conventional *in vitro* cultures and comparable to MCA concentrations found in ascitic fluids. This chapter describes various hollow-fibre systems and special emphasis is placed on the use of dialysis membrane bioreactors for the production of highly concentrated MCAs which are used for diagnostics and therapeutics.

2. PHYSICAL ASPECTS OF HOLLOW-FIBRE CULTURE SYSTEMS

Hollow-fibre membranes are applied in the medical field, for instance for haemodialysis of patients with kidney failures and in laboratory and industrial processes like separation and purification of gases, liquids, and solids (*5*). More recently, hollow-fibre reactors have been used for compartmentalization of enzymes and for the culture of microbial, plant, and animal cells (*26*). Membranes and reactors with specific properties have been developed for these purposes. Some physical aspects of hollow-fibre equipment used in cell culture will be discussed.

2.1. Membranes

Hollow-fibre membranes in bioreactors are made of cellulose, modified cellulose, cellulose acetate, polypropylene, polysulphone and other polymers. Both isotropic membranes, which have a homogeneous structure, and anisotropic membranes, which consist of a multilayer, are used. Isotropic membranes, like the recently developed Cuprophan dialysis fibres, have a wall thickness of only 5 μm. Nevertheless, their mechanical strength can withstand an internal pressure of more than 2 bar. Cuprophan, based on modified cellulose, has high biocompatibility. Tests with endothelial cells show low cytotoxicity and complement activation in blood compared to other membranes. These features make them the material of choice for application in haemodialysis (*3*) as well as for mammalian cell cultures (*22*). Anisotropic membranes, such as the polysulphone fibres in Amicon's Vitafiber cartridges,

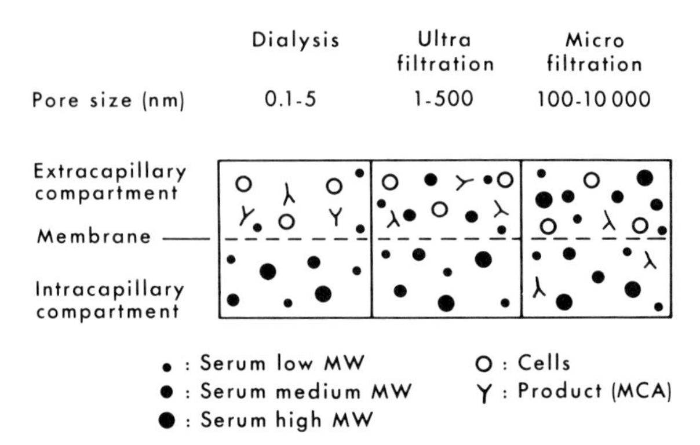

Fig. 2. Separation characteristics of membranes used in cell culture.

consist of a 1–2 μm ultrafiltration layer and a spongy support layer of 50–75 μm (*10*).

Hollow-fibre membranes in bioreactors separate the cells from the main medium compartment. A choice can be made of membranes with different pore sizes resulting in dialysis, ultrafiltration, or microfiltration characteristics. Figure 2 illustrates that the medium composition in the cell compartment is determined by the pore size of the membranes. If the membranes are only permeable for low-molecular-weight molecules, the cells will be deprived of serum and other components of the supply medium of high molecular weight. Ultrafiltration membranes with higher cut-off values than dialysis membranes are permeable to a wider range of serum components, whereas all serum components can freely pass microfiltration membranes. Table I shows the permeability, determined under cell-culture conditions, of three commercial types of membranes with cut-off values of 5000, 10 000, and 70 000 Daltons. It was found that under these conditions, glucose passed freely across the membranes, but immunoglobulins were retained nearly completely. Depending on the type of membrane used, insulin and bovine serum albumin were retained to different degrees. These experiments show that the different pore sizes of the commercial cartridges strongly influence the composition of the medium and the retention of products in the cell compartment of the bioreactor. Thus, a careful choice has to be made in the type of membrane in hollow-fibre cartridges for cell culture.

For efficient cell culture, high mass transfers over the membrane are necessary. A reduction of wall thickness is an important development, as the diffusive permeability is inversely proportional to the wall thickness. The membrane thickness of Cuprophan has been reduced from 16 μm to 5 μm during the last decade, resulting in a 3-fold increase in transfer rates for small molecules (*3*).

TABLE I Retention Values (%) of Dialysis and Ultrafiltration Membranes in Hollow-Fibre Cartridges[a]

	Organon Teknika Andante	Amicon VF_2P10	CD Medical II
Glucose (180 Da)	0	0	0
Insulin (6000 Da)	52	15	7
BSA[b] (65 000 Da)	103	90	15
IgG[c] (150 000 Da)	96	95	92

[a] Dialysis membranes (5000 Da), in an Organon Teknika, Andante cartridge, have been compared with ultrafiltration membranes, in the Amicon VF_2P10 cartridge (10 000 Da) and the CD Medical cartridge model II (70 000 Da) under cell-culture conditions (see Section 2.1). Retention values of four substances with different molecular weights were determined under cell culture conditions. The extracapillary compartment of the cartridges were filled with solutions at a concentration of 1 g l^{-1}. Twenty-five litres of buffered saline were continuously circulated through the capillaries for three days. The substances were determined colorimetrically and the percentage of the remaining concentrations in the extracapillary compartment were calculated.
[b] Bovine serum albumin.
[c] Immunoglobulin class G.

Mass transfer can be reduced during cell culture as precipitates and aggregates may clog the hollow fibres. Firstly, precipitates in serum-containing supply medium could obstruct the lumen of the fibres with a diameter of about 200 μm. Secondly, the pores of the membranes can be clogged by the supply medium or by cell lysates. Mass transfer in diffusion processes is less sensitive to clogging effects than in filtration processes. It was found that in dialysis processes, which are based on diffusion, hollow-fibre membranes retain high transfer rates even during prolonged cell cultures.

Next to the hydrophilic membranes, which are used for mass transfer of molecules in liquid–liquid exchange, special membranes like silicone-rubber (*13*) or hydrophobic propylene polymers are available for gas–liquid exchange.

2.2. Hollow-fibre Cartridges

Several bioreactors are commercially available as shown in Table II. Most are based on cartridges already in use either as haemodialysis (CD Medical, Endotronics, Organon Teknika), ultrafiltration (Amicon, Queue) or microfiltration (Endotronics) cartridges (see technical leaflets from the manufacturers). The physical characteristics vary a hundred-fold in terms of volume and membrane surface area; a number of fibre materials with different

TABLE II Physical Characteristics of Commercially Available Hollow-fibre Cartridges

System	Amicon Vitafiber I	II	III	CD Medical Inc. Cell-pharm I	II	Endotronics Acusyst S/M	Jr/P	Organon Teknika Andante	Allegro	Queue/Monsanto Hybrinet
Dimensions										
Volume (ml) extracapillary	2.5	25	250	75	90	20	100	100	200	45
Surface area (m^2)	0.01	0.1	1.0	1.4	1.8	0.2	1.3	0.8	1.7	0.2
Fibres										
Material	Polysulphone or acrylic polymer			Cellulose	Cellulose acetate	Mixture of cellulose acetate and cellulose nitrate		Cellulose		Polysulphone
Pore size (nm)						200				30
(kDa)	10 or 50 or 100			10	70		7	5		
Sterilization	EO[a]			EO[a]		EO[a]		EO[a]		Autoclave

[a] Ethylene oxide.

pore sizes are available. Only one of the cartridges is autoclavable. This first generation of bioreactors has largely the same configuration: they contain only one type of hydrophilic membrane. These types of bioreactors are easily produced in large numbers as they fit in with the existing production lines for other purposes.

In 1974 Knazek (*13*) had already recognized gradient formation in hollow-fibre bioreactors as one of the main problems. This is especially critical for dissolved oxygen, which is only present in the medium in concentrations of 0.2 mmol l^{-1} at 37°C and 0.21 bar oxygen partial pressure. Therefore, the driving forces for mass transfer are much lower than for nutrients like glucose and glutamine which are present in concentrations up to 20 and 5 mmol l^{-1} culture medium respectively. Knazek (*13*) tried to overcome oxygen limitation by using a mixed-bed hollow-fibre bioreactor. In addition to hydrophilic fibres transporting nutrients, silicone rubber capillaries were used for oxygen transfer. In hollow-fibre reactors presently commercially available, dissolved oxygen is transported across hydrophilic membranes. High mass transfer rates over the membrane to the cells in the extracapillary compartment are achieved by a dense fibre matrix, resulting in short diffusion distances. *In vivo* the characteristic diffusion radius between blood capillaries and cells is less than 100 μm (*25*), which is similar to the distance in some of the hollow-fibre cartridges. The ultrafiltration flow, caused by pressure differences over the length of the cartridges, also contributes to oxygen transport, although this process plays only a minor role compared to diffusion in dialysis membrane cartridges (*3*).

Prototypes of bioreactors for further research have been developed to improve mass transfer efficiency. The Monsanto group (*16*) developed a flat-bed reactor in which the medium flows through a screen with 2 μm pores. Oxygen is provided via the lumen of hollow fibres to the cell culture compartment. Tharakan (*25*) used a radial flow hollow-fibre cartridge. Medium is supplied into the cartridge by a central feed distribution tube and then flows radially across the cartridge to an annular bed of hollow fibres. An air/CO_2 mixture flows through these fibres, supplying oxygen to the cells. This bioreactor configuration has some advantages over conventional cartridges, since the feed is uniform along the length of the fibres. As there are no axial gradients, the cartridge is not limited in length.

In conclusion, a number of different hollow-fibre bioreactors have been developed. The current commercial cartridges contain only one type of fibre. More complicated prototypes have been described using two types of fibres to improve efficiency of mass transfer. The need for more advanced mixed-bed bioreactors should be supported by more experimental evidence before it becomes commercially attractive to produce these difficult-to-construct bioreactors in large numbers.

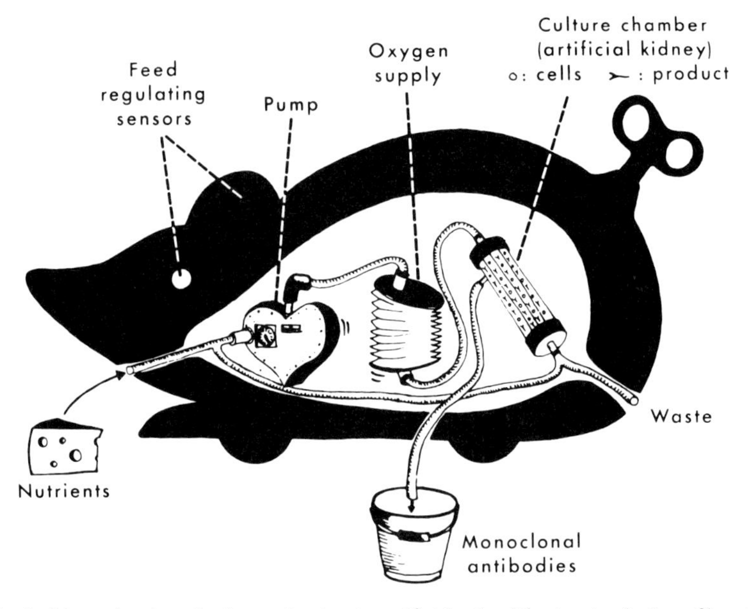

Fig. 3. Monoclonal antibody production in artificial mice. The *in vitro* hollow-fibre dialysis culture method imitates the *in vivo* production in mice. Sense-organs, heart, blood, and lungs in mice are used for efficient growth of hybridomas in the abdominal cavity from which MCAs are harvested.

2.3. Systems

Classical MCA production was done *in vivo*. About 20 000 mice had to be sacrificed for 1 kg of MCA. These animals have been used as "culture chambers" for hybridomas in the production of highly concentrated MCA solutions. About 2×10^6 hybridoma cells were injected intraperitoneally per mouse and grew as tumours within a period of a few weeks. A prerequisite was that the hybridomas had to be histocompatible with the animal strain used. The abdominal cavity in such an *in vivo* system was supplied with nutrients and oxygen from the recirculating blood to the cells via diffusion.

This *in vivo* culture system has been used as a basis for the *in vitro* hollow-fibre dialysis cell culture system, as illustrated in Fig. 3. In the extracapillary compartment of hollow-fibre cartridges, the cells are supplied with nutrients and oxygen by diffusion from the well-controlled intracapillary medium over dialysis membranes. The culture medium is oxygenated and pH-adjusted before it enters the cartridge.

Three different configurations of medium-supply systems are described for hollow-fibre bioreactors in Fig. 4. The linear flow system, in which medium

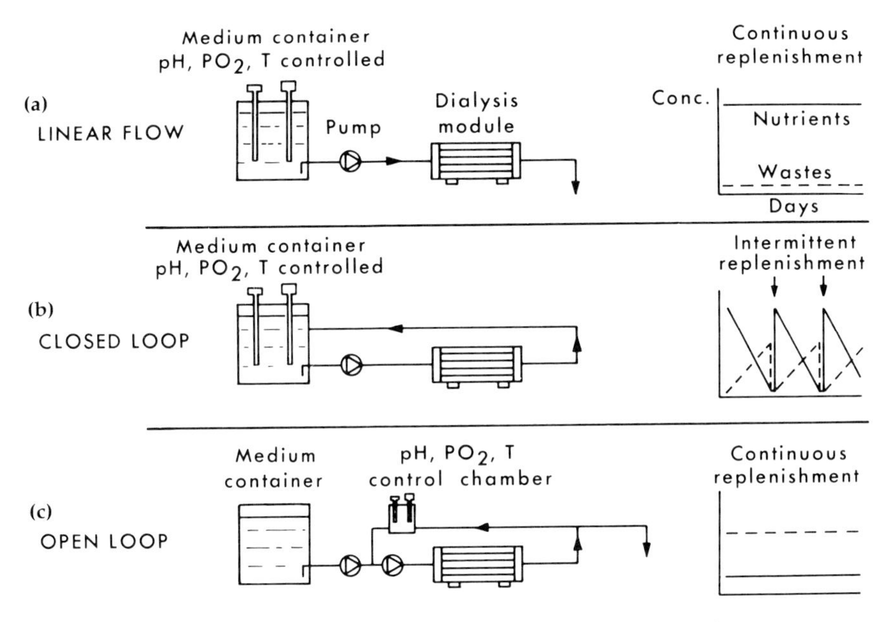

Fig. 4. Three medium-supply systems for hollow-fibre cultures. (*a*) *Linear flow:* temperature-, pH-, and pO_2-controlled fresh medium is pumped through the fibres of the cartridge. The concentrations of nutrients and wastes in the medium are at a constant level during culture. (*b*) *Closed loop:* the controlled medium is pumped through the cartridge and recirculated to the medium container. Nutrients are consumed and waste products accumulate. The medium is replenished batch-wise at pre-defined levels of nutrient or waste concentrations. (*c*) *Open loop:* a combination of the systems described above in which a small volume of medium is controlled and recirculated. A constant level of nutrients and waste levels is controlled by sensors.

passes the fibres only once, is the simplest. Unless large quantities of medium are used, only a low flow rate can be established. This leads to gradients in the capillaries, both in axial and radial direction, and consequently to low transfer rates of nutrients and oxygen over the membranes. These aspects are improved in a loop system, in which medium is recirculated through the cartridges. The medium in the loop is continuously adjusted for pH and enriched for oxygen. The gradients in the lumen of the capillaries are reduced to a minimum by applying high recirculation flow rates.

Two options are available for replenishment of supply-medium in a recirculation system, a closed-loop and an open-loop, for batch and continuous medium supply respectively:

1. The closed-loop is applied by all companies who sell complete systems. In these systems, the culture medium is conditioned for pH, pO_2 and temperature in the medium container. Gas-exchange cartridges are

usually applied in the loop for oxygen and pH correction (see technical leaflets from CD Medical and Endotronics). The medium has to be monitored for nutrient depletion and waste accumulation to indicate the point of batch-wise replenishment. As a consequence, culture conditions change drastically during a medium-replenishment cycle.

2. In the open-loop system medium is conditioned in a small recirculation circuit. The medium is replenished continuously at a rate controlled by sensors for nutrients and/or waste detection. Pre-defined culture conditions can be maintained continuously.

For efficient operation of cell-culture production systems, knowledge of metabolism of cells, medium component limitations, and inhibiting effects of wastes are a prerequisite. Monitoring of cell growth by non-selective sampling from the extracapillary compartment is practically impossible. As an indication of cell growth, the consumption rate of nutrients can be used or the oxygen consumption rate can be determined using sensors directly in front of and behind the bioreactor. A disadvantage of most commercial hollow-fibre bioreactors is that they are not autoclavable. This is due to the use of non-autoclavable housing or embedding materials for the fibres, since the membranes themselves are autoclavable. It is expected that the sterilization problem will be solved in the next generation of bioreactors.

3. CELL CULTURE IN HOLLOW-FIBRE DIALYSIS CARTRIDGES

Hollow-fibre dialysis culture systems have been successfully used for the growth of mouse hybridomas. This system was based on haemodialysis cartridges in a recirculation system (*22*).

In this section, the growth of monoclonal-antibody-producing hybridomas in dialysis cartridges is described. Mass transfer rates, consumption of nutrients and the absence of serum in the hollow-fibre dialysis cultures will be discussed in relation to other membrane bioreactors.

3.1. Medium and Equipment

In the dialysis culture, a chemically defined, protein-free medium was used; it consisted of a 1 : 1 mixture of Dulbecco's Modified Eagle Medium and Ham F12 with some supplements as described previously (*22*). In other types of culture, this medium was supplemented with 10 mg l^{-1} human transferrin or 10% foetal calf serum (FCS) where indicated. The dialysis culture recirculation system as described in Fig. 5 with loops as indicated in Fig. 4 was used.

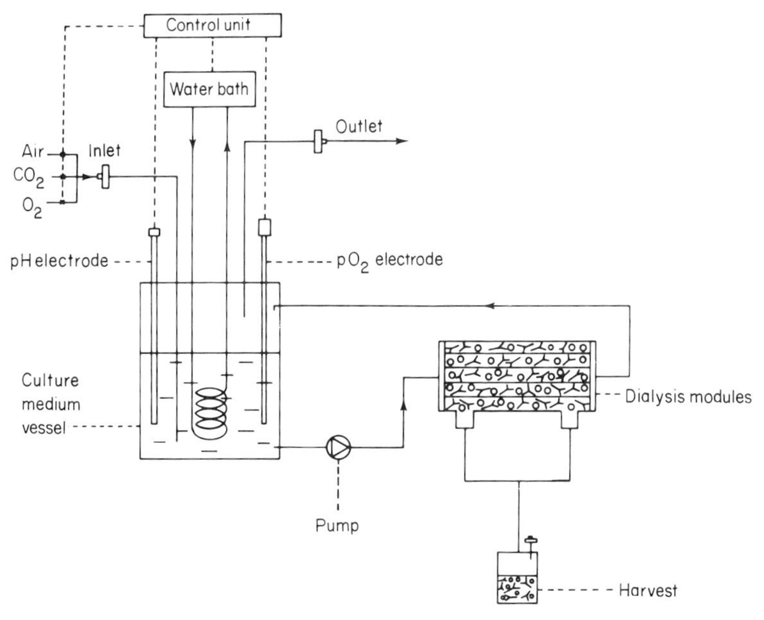

Fig. 5. Scheme of a hollow-fibre dialysis culture system. The system consists of a storage vessel for culture medium and a haemodialysis module. The conditions for the culture medium are adjusted in the 30-litre vessel by a heating coil and supply of carbon dioxide, oxygen, and air. The culture medium is recirculated through the fibres of the dialysis module. Non-adherent cells and products are harvested from the outer compartment through the side-ports. From Schönherr *et al.* (*22*).

TABLE III Specifications of Haemodialysis Units Organon Teknika

	Andante 3G	Allegro 2G
Housing		
Material	Polystyrene	Polystyrene
Diameter (cm)	3.2	4.7
Length (cm)	26	26
Extracapillary volume (ml)	100	200
Fibres		
Material	Cuprophan (cellulose)	Cuprophan (cellulose)
Number	5300	10 000
Diameter (μm)	200	200
Length (cm)	22	23
Wall thickness (μm)	8	11
Surface area (m^2)	0.8	1.7
Sterilization	Ethylene oxide	Ethylene oxide

Two or more ethylene-oxide-sterilized hollow-fibre haemodialysis cartridges, Organon Teknika, type Allegro or Andante (see Table III) were connected in parallel to this system, containing protein-free medium supply, sensors and a control unit. The pH was kept at 7.2 ± 0.1 at a temperature of 37 ± 0.5°C with a dissolved oxygen concentration between 30 and 100% air saturation. The culture medium was pumped through the hollow fibres at a flow rate of 200 ml min^{-1} per cartridge.

For the determination of monoclonal antibodies, a sol particle immunoassay (SPIA) was used; glucose concentrations were measured using the GOD-Perid method (*22*).

3.2. Culture

Mouse hybridomas, after recovery from liquid nitrogen, were grown in the medium (see Section 3.1) containing 10% FCS as suspension cultures using conventional systems (spinner flasks or roller bottles). The cells were inoculated in the extracapillary compartment of the haemodialysis cartridges at a density of 5×10^6 cells ml^{-1}. The cell compartment, initially inoculated with 10% serum, was gradually washed with protein-free medium during harvesting two or three times a week. The culture could be maintained for at least two months as a continuous cell culture system. Figure 6 shows a typical example of the monoclonal antibody production and glucose consumption by a hybridoma cell line (clone D) over a production period of 47 days. The monoclonal antibody production rate for this hybridoma was in the range of 230–600 mg per Organon Teknika Allegro haemodialysis cartridge per day.

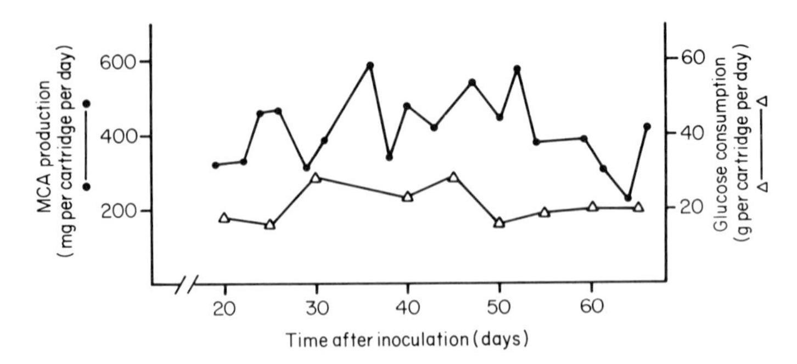

Fig. 6. MCA production and glucose consumption in hollow-fibre dialysis culture. Hybridoma, clone D, was grown in Organon Teknika haemodialysis cartridges (type, Allegro): 200 ml was harvested two or three times a week from the extracapillary compartment. The mean MCA production (●, mg per cartridge per day) and glucose consumption (△, g per cartridge per day) are indicated.

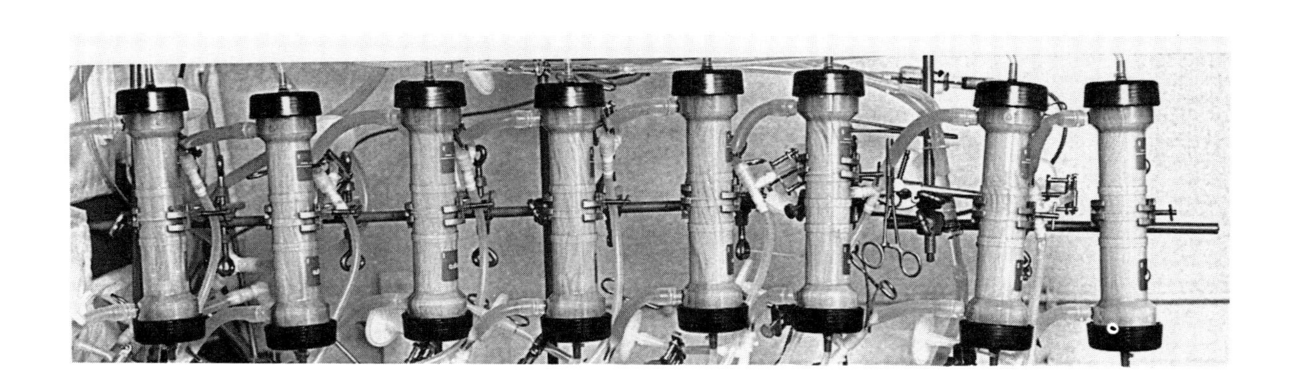

Fig. 7. Eight Organon Teknika Allegro haemodialysis cartridges connected in parallel for cell culture.

TABLE IV The Mean MCA Production of Five Hybridomas in Hollow-fibre Dialysis Cultures[a]

	Monoclonal antibody production		
Hybridoma	Concentration ($g\,l^{-1}$)	Productivity ($g\,l^{-1}\,day^{-1}$)	Total production per culture (g)
A	2.9	0.9	47
B	4.9	1.5	66
C	3.1	0.8	53
D	7.8	2.2	136
E	13.1	2.8	202

[a] Five mouse hybridomas were grown independently for 1–2 months in hollow-fibre dialysis cell cultures using up to eight Organon Teknika haemodialysis cartridges (100 or 200 ml extracapillary volume) connected in parallel. The extracapillary volume was harvested two or three times a week. The mean MCA concentration ($g\,l^{-1}$), the productivities ($g\,l^{-1}$ extracapillary volume per day) and the total MCA harvests (g) of the systems during the culture periods are indicated.

The glucose consumption rate was in the range of 15–30 g per cartridge per day.

Five mouse hybridomas were grown independently in dialysis systems (Table IV and Fig. 7). The mean concentration of harvested monoclonal antibodies varied between 3 and 13 $g\,l^{-1}$. These concentrations are at least a hundred-fold higher than those found in conventional cultures for the hybridomas and correspond to the higher cell densities of hybridomas in dialysis cultures. The average monoclonal antibody productivity of the various hybridomas ranged between 0.8 and 2.8 $g\,l^{-1}$ of extracapillary volume per day. Total monoclonal antibody amounts of 50–200 g were harvested per production run.

3.3. Discussion

Most cell-culture media are complex. They contain more than fifty chemically defined components and a supplement of undefined serum, up to 20%. The development of serum-free media for hybridoma cultures has received much attention (*8*, *18*).

Hybridomas were grown on dialysable serum components by culturing the cells in a serum-free medium in dialysis tubing, placed in a medium supplemented with serum (*12*). A more than ten-fold increase in cell density and monoclonal antibody concentration was found compared to conventional batch culture. Serum could not be excluded, indicating that low-molecular-weight serum components were necessary for growth. Similar results were obtained by others (*1, 23*).

The effect of the molecular weight of serum components on hybridoma cell growth was also studied in hollow-fibre reactors. Cultures in Amicon cartridges with ultrafiltration membranes of different cut-off values (10 000, 50 000, and 100 000 Daltons) showed higher cell densities and monoclonal antibody yields with increasing membrane pore sizes (*2*). Thus, serum components of low and medium-high molecular weight had to be supplied for cell growth support in these experiments. A reduction in the absolute amount of serum was attained by compartmentalization of cells via encapsulation (*20*) or hollow-fibre techniques (*10*), although a concentration of 1–2% still remained essential.

In contrast to the results described above, it was shown (*22*) with hybridoma cultures in haemodialysis cartridges that serum can be omitted completely from both the supply medium and the extracapillary compartment, if the cell concentration is sufficiently high. The cells appear to be self-supporting owing to the synthesis of growth factors, which cannot pass the hollow-fibre dialysis membrane. This production method with a serum- and protein-free medium is apparently possible because of the well-defined low pore size of the haemodialysis membranes. The molecules with a molecular weight higher than 5000 Daltons are well retained by the dialysis membranes. If membranes are used with higher cut-off values, these factors are diluted and a serum-containing medium becomes essential. The use of well-defined membranes with a pore size tested under stringent quality control seems to be essential for the hollow-fibre dialysis bioreactor. In immobilization techniques like encapsulation (*20*), pre-tested pore sizes cannot be attained, since the process starts with encapsulation of cells, and a check on permeability has to be done during production.

Transfer of oxygen to the cells in the extracapillary compartment has been indicated as the limiting factor of the hollow-fibre system. The oxygen transfer rates of Cuprophan dialysis membranes from the liquid medium supply to the extracapillary cell compartment is of the order of 4 mmol m^{-2} h^{-1} at 37°C and 0.21 bar oxygen partial pressure. The transfer by diffusion from the capillaries to the cells is regarded as the most important barrier. However, the maximal diffusion distance in the haemodialysis cartridges is only 80 μm, which reduces the diffusion gradients to a minimum.

Exponentially growing mouse hybridomas in suspension cultures consume 0.2 mmol oxygen per 10^9 cells per hour. It was found that in a dialysis bioreactor with an extracapillary compartment of 200 ml (Organon Teknika,

Allegro) 6 mmol hr^{-1} oxygen is consumed by 30×10^9 cells (unpublished results). Thus, the oxygen consumption per cell is of the same order as was found in free-suspension cells. Moreover, the dissolved oxygen concentration has little influence on the cell growth rate between 8 and 100% saturation (*4*). In view of this, oxygen concentration has only a marginal effect on cell cultures in the dialysis cultures.

In summary, the following advantages of the hollow-fibre dialysis culture system are observed:

1. MCAs are produced under a constant supply of well-defined, simple, protein-free culture medium. The serum has no stimulatory effect on generation time and MCA production (*22*). The use of serum-free medium is an important advantage for further downstream processing of therapeutic products.
2. The MCA-production is maintained for several weeks at a constant daily mean production level of up to 0.5 g of MCA per 0.2 litre bioreactor. The MCAs are harvested at concentrations of 3–13 g l^{-1}. In comparison, the *in vivo* cultivation of hybridomas in mice results in an average batch-wise production of 50 mg after 2–3 weeks. Thus, the production capacity of one dialysis cartridge in this period is equivalent to 150–200 mice. The MCAs are harvested at similar concentrations, devoid of serum proteins.
3. The haemodialysis modules are suitable for MCA production at a level up to 100 g MCA per module per year. Further scaling-up to the kilogram level is achieved by multiplying the number of dialysis cartridges or by increasing the size of the hollow-fibre reactors.
4. The system is useful for several other animal and human cell lines, both of the monolayer and suspension type. It was shown that the production of human monoclonal antibodies in mice, even natural killer cell depleted athymic nude mice, is extremely difficult and variable. Through the use of haemodialysis cartridges, it is possible to produce gram quantities of pharmaceutical-grade human monoclonal antibodies (Haspel, personal communication).

A disadvantage of the hollow-fibre culture is the heterogeneity of the system, which makes non-selective sampling of the cell culture compartment impossible.

Many products are produced in low concentrations and are hardly detectable in a rich serum-containing cell culture. Since, in the dialysis system, membranes are used with a low cut-off and products are harvested at high concentrations in the protein-free medium, new products (in a wide range of molecular weight), may be discovered.

4. CONCLUSIONS

1. Hollow-fibre dialysis cell culture systems have been developed using cartridges as a spin-off from the medical haemodialysis equipment.
2. In the cartridges, cells have been propagated at densities more than two orders of magnitude higher than in conventional systems, yielding concentrated and less impure products.
3. Continuous dialysis with a serum- and protein-free and well-controlled (with respect to temperature, pH, pO_2) medium was used for exchange of nutrients and wastes.
4. Nutrients and oxygen gradients could be kept to a minimum owing to high mass transfer by a dense capillary network and a large surface area of very thin membranes.
5. The cartridges were suitable for animal and human cells, both for anchorage-dependent and suspension-like cell types.
6. The system was shown to be valuable for the synthesis of monoclonal antibodies, both for diagnostics and therapeutic products, devoid of contaminants from serum.

Acknowledgements

The authors wish to thank Dr. H. D. Berkeley, Dr. M. Haspel, Mr. P. J. v. Hees, Mr. E. H. Houwink, Mr. J. C. H. Oomens, Mrs. A. M. J. M. v. Os, Mr. H. W. M. Roelofs, and Mr. A. L. M. Sanders for their contributions with regard to discussion or correction of the manuscript, and Miss M. T. Krancher for her careful typing.

REFERENCES

1. Adamson, S. R., Fitzpatrick, S. L., Behie, L. A., Gaucher, G. M., and Lesser, B. H. (1983). *In vitro* production of high titre monoclonal antibody by hybridoma cells in dialysis culture. *Biotechnol. Lett.* **5**, 573–578.
2. Altshuler, G. L., Dziewulski, D. M., Sowek, J. A., and Belfort, G. (1986). Continuous hybridoma growth and monoclonal antibody production in hollow fiber reactors-separators. *Biotechnol. Bioeng.* **28**, 646–658.
3. Baurmeister, U., Pelger, M., Tretzel, J., and Henne, W. (1985). Presentation at the Symposium "Biocompatibilité des Membranes en Hémodialyse", Grenoble, France, Nov. 1984. *In* "Development and properties of Cuprophan® membranes". April 1985 Product Group Membrana, ENKA, Obernburg, Germany.
4. Boraston, R., Thompson, P. W., Garland, S., and Birch, J. R. (1984). Growth and oxygen requirements of antibody producing mouse hybridoma cells in suspension cultures. *Dev. Biol. Stand.* **55**, 103–111.

5. Brock, T. D. (1983). "Membrane Filtration." Springer-Verlag, Berlin.
6. Chick, W. L., Like, A. A., and Lauris, V. (1975). Beta cell culture on synthetic capillaries: an artificial endocrine pancreas. *Science* **187**, 847–849.
7. David, G. S., Reisfeld, R. A., and Chino, T. H. (1978). Continuous production of carcinoembryonic antigen in hollow fiber cell culture units: Brief communication. *J. Natl. Cancer Inst.* **60**, 303–306.
8. Fasekas de St.Groth, S., and Scheiddegger, D. (1980). Production of monoclonal antibodies: strategy and tactics. *J. Immunol. Method.* **35**, 1–21.
9. Hager, J. C., Spiegelman, S., Ramanarayanan, M., Bausch, J., Galletti, P. M., and Calabresi, P. (1982). Tumor-associated antigens produced by mouse mammary tumor cells in artificial capillary culture. *J. Natl. Cancer Inst.* **69**, 1359–1365.
10. Hopkinson, J. (1985). Hollow fiber cell culture systems for economical cell-product manufacturing. *Bio/Technology* **3**, 225–230.
11. Johnson, A. D., Eddy, G. A., Gangemi, J. D., Ramsburg, H. H., and Metzger, J. F. (1978). Production of Venezuelan equine encephalitis virus in cells grown on artificial capillaries. *Appl. Environ. Microbiol.* **35**, 431–434.
12. Klinman, D. M., and McKearn, T. J. (1981). Dialyzable serum components can support the growth of hybridoma cell lines in vitro. *J. Immunol. Method.* **42**, 1–9.
13. Knazek, R. A., Gullino, P. M., Kohler, P. O., and Dedrick, R. L. (1972). Cell culture on artificial capillaries: An approach to tissue growth in vitro. *Science* **178**, 65–67.
14. Knazek, R. A., Kohler, P. O., and Gullino, P. M. (1974). Hormone production by cells grown in vitro on artificial capillaries. *Exp. Cell Res.* **84**, 251–254.
15. Köhler, G., and Milstein, C. (1975). Continuous culture of fused cells secreting antibody of predefined specificity. *Nature (London)* **299**, 592–596.
16. Ku, K., Kuo, M. J., Delente, J., Wildi, B. S., and Feder, J. (1981). Development of a hollow-fiber system for large-scale culture of mammalian cells. *Biotechnol. Eng.* **23**, 79–95.
17. McAleer, W. J., Markus, H. Z., Bailey, F. J., Herman, A. C., Harder, B. J., Wampler, D. E., Miller, W. J., Keller, P. M., Buynak, E. B., and Hilleman, M. R. (1983). Production of purified hepatitis B surface antigen from Alexander hepatoma cells grown in artificial capillary units. *J. Virol. Method* **7**, 263–271.
18. Murakami, H., Masui, H., Sato, G. H., Sueoka, N., Chow, T. P., and Kano-Sueoka, T. (1982). Growth of hybridoma cells in serum-free medium: ethanolamine is an essential component. *Proc. Natl. Acad. Sci. U.S.A.* **79**, 1158–1162.
19. Ratner, P. L., Cleary, M. L., and James, E. (1978). Production of "Rapid-Harvest" moloney murine leukemia virus by continuous cell culture on synthetic capillaries. *J. Virol.* **26**, 536–539.
20. Rupp, R. G. (1985). Use of cellular microencapsulation in large scale production of monoclonal antibodies. *In* "Large-scale Mammalian Cell Culture" (J. Feder and W. R. Tolbert, eds.), pp. 19–38. Academic Press, Orlando.
21. Schönherr, O. T., and Houwink, E. H. (1984). Antibody engineering, a strategy for the development of monoclonal antibodies. *Antonie van Leeuwenhoek* **50**, 597–623.
22. Schönherr, O. T., van Gelder, P. T. J. A., van Hees, P. J., van Os, A. M. J. M., and Roelofs, H. W. M. (1986). A hollow fibre dialysis system for the in vitro production of monoclonal antibodies replacing in vivo production in mice. *Dev. Biol. Stand.* **86**, 211-220.
23. Sjögren-Jansson, E., and Jeansson, S. (1985). Large-scale production of monoclonal antibodies in dialysis tubing. *J. Immunol. Method.* **84**, 359–364.
24. Sun, A. M., Parisius, W., Healy, G. M., Vacek, I., and Macmorine, H. G. (1977). The use, in diabetic rats and monkeys, of artificial capillary units containing cultured islets of langerhans (artificial endocrine pancreas). *Diabetes* **26**, 1136–1139.

25. Tharakan, J. P., and Chau, P. C. (1986). A radial flow hollow fiber bioreactor for the large-scale culture of mammalian cells. *Biotechnol. Bioeng.* **28**, 329–342.
26. Vick Roy, T. B., Blanch, H. W., and Wilke, C. R. (1983). Microbial hollow fiber bioreactors. *Trends Biotechnol.* **1**, 135–138.

16

Large-scale Mammalian Cell Culture Utilizing ACUSYST Technology

MICHAEL A. TYO, BARBARA J. BULBULIAN,
BEVERLY ZASPEL MENKEN, and THOMAS J. MURPHY
Endotronics, Inc.,
Coon Rapids, MN, U.S.A.

1. INTRODUCTION

The need for economical, large-scale mammalian cell culture has increased as a result of recent scientific advances and pressure to provide a consistent, reliable, and inexpensive means of producing large quantities of mammalian cell-secreted products. Utilization of hollow-fibre cell-culture systems to maintain viable cells at high densities and allow production of concentrated harvest was pioneered by Knazek *et al.* in 1972 (*7*). These systems have been

ANIMAL CELL BIOTECHNOLOGY VOL. 3
ISBN 0-12-657553 3

used to produce a variety of products, including monoclonal antibodies (*1*), virus (*9*), gonadotropin (*8*), insulin (*15*), and carcinoembryonic antigen (*4*). They have not been utilized for large-scale production of cell-secreted products until recently, owing to problems associated with the formation of gradients (*6, 11, 12*), microenvironments (*10*), and anoxic pockets (*6, 12*). ACUSYST hollow-fibre technology utilizes current hollow-fibre technology combined with modifications in fluid dynamics to overcome the problems associated with conventional hollow-fibre technology and to meet the necessary criteria for the successful and economic production of cell-secreted products (*2, 5, 14*).

Re-engineering of hollow-fibre technology in the ACUSYST systems has resulted in significant advantages for large-scale production of mammalian cell products. Unique system design allows production of highly concentrated and relatively pure (60–95% of total protein) harvest with reduced serum protein contaminants, thus minimizing the time and expense of downstream processing. Continuous, automated process control allows the system to respond and adjust to changes in the internal environment of the ACUSYST instrument, reducing labour requirements. In a continuous production process, labour is invested primarily to get cells to reach maximum density. After that point, the cells remain viable and productive and minimal labour is needed to harvest the product on a continuous basis. Standardization of the culture process allows reproducible runs to be performed throughout the product development life-cycle to meet validation requirements. A printed log of all activity is available at the discretion of the user so that a hard-copy record may be maintained for future use.

2. ACUSYST TECHNOLOGY

The ACUSYST-P and the ACUSYST-Jr systems are designed to provide the versatility to support a variety of different cell types and cell-secreted products. The ACUSYST-P provides automated production of large quantities of mammalian cell-secreted products. It consists of two independent flowpaths capable of supporting up to six hollow-fibre bioreactors in each flowpath. The system components, directed by the ACUSOFT software, provide controlled conditions for the cell environment. Two cell lines can be cultured simultaneously or the same cell line can be cultured using different process control parameters for each flowpath. As many as 4.8×10^{11} cells have been cultured per instrument. Data graphing of vital parameters such as glucose consumption, lactate production, oxygen uptake, and pH are available through colour graphics in the system console; and a permanent copy may be obtained through the system printer. The ACUSYST-Jr, designed for production of smaller quantities needed for diagnostics and pre-clinical trials, contains a single flowpath with one hollow-fibre bioreactor. An average of 1–2 $\times 10^{10}$ cells can be maintained for long periods (months). The ACUSYST-Jr design utilizes the same hollow-fibre technology and process control strategy as found on the larger-scale ACUSYST-P system and includes a flexible user interface to a microprocessor-based control system. Data gathered from and used to run the ACUSYST-Jr system can be used directly to set up and run the ACUSYST-P large-scale cell culture system.

In the ACUSYST systems, mammalian cells are maintained in an internal environment that simulates normal physiological conditions *in vivo*. Cells are grown in a bioreactor which consists of thousands of individual hollow fibres (i.d. 200 μm, o.d. 256 μm, length 20 cm, 6000 Dalton MW cut-off) bundled together in a disposable cartridge. The total fibre surface area within each bioreactor is 1.4 m^2. These fibres are porous hair-like strands with a hollow centre referred to as the lumen. Cells are cultured in the space between the outside of the fibres and the inside wall of the bioreactor, called the extracapillary space (ECS). The semi-permeable nature of the hollow fibres allows nutrients and dissolved gases to cross into the ECS and waste products and depleted medium to return from the ECS into the lumen through which the medium flows continuously. The cells grow and pack themselves around the fibres, receiving nutrients from the lumen through the porous walls of the fibres. Important parameters such as pH, oxygen, temperature, medium usage, supplemental factor feed rates, harvest rate, fresh medium flowrates, glucose and oxygen uptake, and on-line hardware diagnostics are monitored and carefully controlled. The supplemental factor pumps allow the addition of base, glucose, or any other desired component to the medium. Cell populations thrive, grow to high cell densities (exceeding 10^8 cells ml^{-1}), and remain viable for extended periods of time. Owing to automated process control, cells continuously receive adequate nutrients and waste products are efficiently removed.

The ACUSYST-P and ACUSYST-Jr hollow-fibre systems incorporate the use of hollow-fibre bioreactor(s), an external expansion chamber, and an external integration chamber to provide an efficient system for the equal distribution of nutrients and the effective removal of metabolic wastes. Collectively the hollow fibres act as capillaries, simulating the mammalian circulatory system as nutrient media, oxygen, carbon dioxide, and various signal chemicals are pumped through the lumen of each fibre. The cells grow and pack themselves around the fibres, receiving nutrients from the lumen through the porous walls of the fibres. Cells and secreted product with MW $>$ 6000 Daltons are retained in the small volume of the ECS. Product can be pumped from the ECS, assuring the economical harvest of highly concentrated proteins. The expansion chamber is connected to the ECS of the

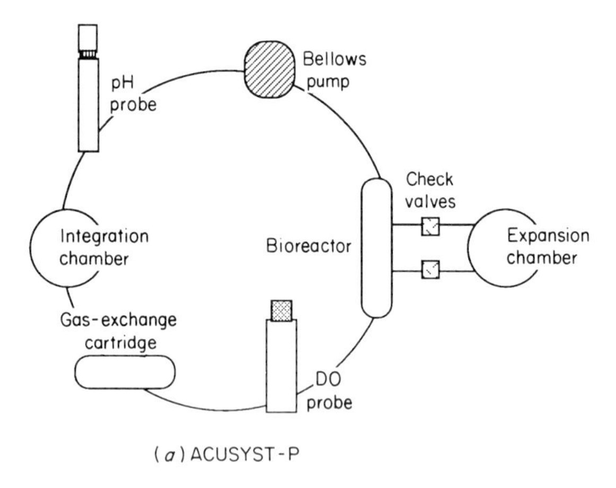

(a) ACUSYST-P

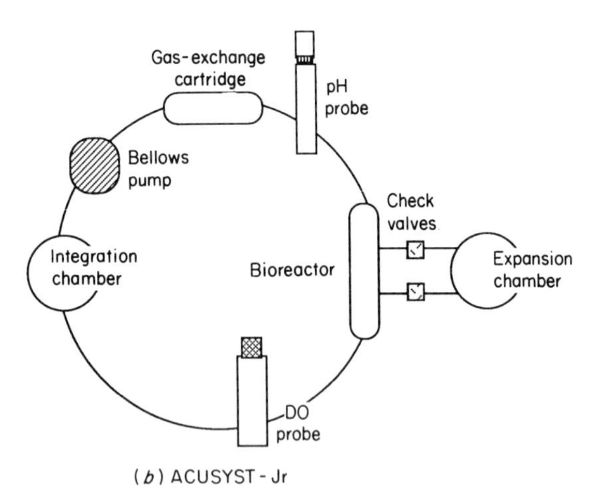

(b) ACUSYST-Jr

Fig. 1. Simplified schematic diagram of the ACUSYST-P (*a*) and ACUSYST-Jr (*b*) system flowpaths depicting components of the expansion and integration circuits.

hollow-fibre bioreactor by two fluid lines in a flowpath referred to as the expansion circuit (EC) (Fig. 1). These fluid lines have monodirectional valves that allow fluid to move from the ECS to the expansion chamber and back to the ECS. In the case of the ACUSYST-P, this ensures that all cells are exposed to the same environment in each of the bioreactors. This design improves the circulation of fluid around the packed cells, minimizing nutrient and waste product gradients and microenvironments due to areas of inadequate nutrients or concentrated waste products. Movement of fluid through a dissolved oxygen probe, the lumina of the bioreactor, a pH probe, and the integration

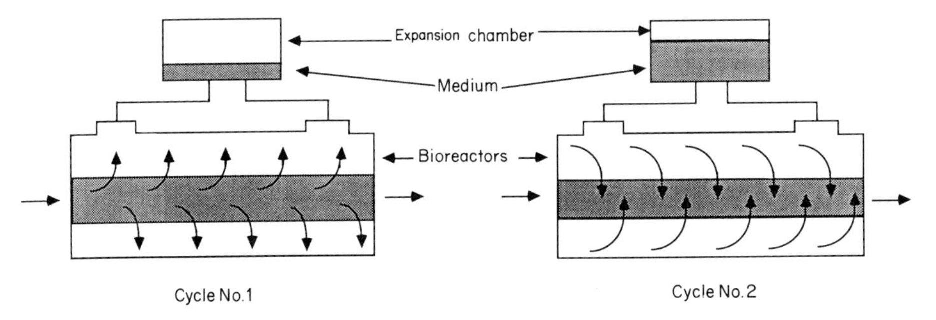

Fig. 2. Schematic diagram of the unique fluid cycling process used in ACUSYST hollow-fibre technology. This cycling ensures a consistent environment for up to 10^8 cells ml^{-1} and greater, depending upon the cell line.

chamber is referred to as the integration circuit (IC). Continual medium flow through the IC is provided by circulation pumps creating a driving pressure for ultrafiltration. Luminal fluid is mixed in an integration chamber before it is introduced to the bioreactor via a high-speed bellows pump(s). Luminal fluid is replenished with oxygen and the pH is adjusted with CO_2 in a hollow-fibre gas-exchange cartridge. Gas flow in the ECS is countercurrent to medium flow in the fibres, improving transfer efficiency. A bank of six peristaltic pumps is available for each flowpath for automated harvesting of secreted product, supplemental factor addition, addition of fresh medium, and removal of spent medium.

The fluid dynamics of the system are controlled through the use of pressure cycling (Fig. 2) which overcomes the problems inherent in conventional hollow-fibre technology (*13*). The cycling mechanisms are slightly different in the ACUSYST-P and the ACUSYST-Jr, but the effect is similar. In the ACUSYST-P, the expansion chamber is maintained at a constant pressure, approximately 130 mmHg above atmospheric pressure. The pressure in the lumen of the fibres is varied by computer control. Cycling begins by increasing luminal pressure to 200 mmHg above atmospheric pressure. This establishes a pressure differential between the lumen and the ECS, forcing ultrafiltration of nutrients and medium equally along the length of the hollow-fibre membrane into the ECS and expansion chamber. When the fluid level in the EC reaches an upper set-point, an ultrasonic level sensor signals the computer to lower the luminal pressure to atmospheric pressure, causing a pressure differential between the ECS and the lumen. Medium returns to the lumen equally along the length of the bioreactor. When the medium in the expansion chamber reaches a lower set-point, the process repeats. Cycling occurs continuously during the culture period and represents an important distinction between ACUSYST systems and other hollow-fibre systems.

3. PROCESS CONTROL STRATEGY

ACUSOFT software provides monitoring and feedback control of vital parameters required to maintain cell growth, viability and product secretion in the ACUSYST-P. The software uses computer algorithms to maintain pH, glucose, and lactic acid at predetermined levels. The control strategy relies on continual medium feed, on-line pH and dissolved oxygen monitoring, and off-line glucose and lactic acid monitoring. This, combined with data entry, provides specific control over culture conditions. The pH is controlled primarily by gassing, with base addition and medium feed rate providing secondary and tertiary control systems. The pH is maintained at a predetermined set-point by automatically increasing or decreasing the amount of CO_2 delivered to the gas exchange cartridge. Secondary control is by base (NaOH) addition to the culture medium. The addition of base rejuvenates the buffering capacity of the medium and counters lactic acid production in a growing culture. Tertiary control is provided by medium dilution via increased medium delivery to the culture. This effectively increases pH by removing waste products, i.e. lactic acid, more rapidly. Glucose is controlled utilizing a factor pump to provide fresh glucose to the system. Medium addition is available as a secondary control. The effect of lactic acid is neutralized by base addition and medium dilution. The actual glucose and lactic acid concentrations are determined off-line and entered into the system computer.

The process control strategy used in the ACUSYST-Jr is based on the same strategy used for the ACUSYST-P. Process control will maintain the pH set-point through gassing, base addition and media dilution. Variable levels of CO_2 and gas flow rates control pH levels automatically. Lactate and glucose levels are maintained through manual control of pump rates for medium dilution and glucose feed, respectively.

The process control strategy in an ACUSYST-P system is divided into two phases, growth and production. In a continuous culture system, maximum production is obtained only after cells have reached maximum density. Therefore, cell growth is promoted early in the production cycle by providing the proper environment, and production is maximized after cells reach maximum density. Once cells reach maximal density, the process control is changed to provide a new environment that induces the cells to secrete the desired end-product. Specific set-points for pH, glucose and lactate are mathematically determined to optimize these two phases. Calculations are based on cell line information, static culture data, medium composition, energy requirements and utilization, and product information (stability, feedback inhibition, etc.). This information may be as basic as cell counts, pH, glucose uptake rate, and lactic acid production rates, but the more that is known about a cell line, the more accurate is the process control. Once determined, these set-points are entered into the ACUSYST-P process control system and automatically maintained. In the ACUSYST-Jr, the pH is entered into the process control system and automatically monitored and maintained. Dissolved oxygen is also monitored on-line. Off-line analyses of glucose and lactic acid are used to monitor the bioreactor environment. Cells can be grown to high densities quickly and shifted to new control parameters to keep the cells in a stable environment and production at its peak.

4. SPECIFIC APPLICATIONS

The ACUSYST hollow-fibre cell-culture system has been used to produce a wide variety of biologicals. Two of these have been chosen, monoclonal antibody production by hybridoma cells and hepatitis B surface antigen by L cells, to demonstrate typical growth and production kinetics. These cultures were carried out for 62 days and 140 days, respectively. The length of time that mammalian cells can be maintained in a producing capacity in the hollow-fibre environment is one of the most significant advantages of the ACUSYST system.

4.1. IgA Monoclonal Antibody Production

The ACUSYST-P hollow-fibre cell-culture system is frequently used for the production of monoclonal antibodies. To examine the kinetics of growth, metabolism, and product formation, a six-bioreactor flowpath was inoculated with 1.7×10^8 viable cells per bioreactor. The inoculum was prepared in two 500-ml spinner cultures in modified DMEM containing 10% NCTC 109 medium, 0.2 U ml^{-1} bovine insulin, 0.45 mM pyruvate, 1 mM oxaloacetate, 3.7 g l^{-1} $NaHCO_3$ and 10% foetal bovine serum. Following inoculation, the medium pump was set to a continuous flow rate of 50 ml hr^{-1}. The medium was identical to the inoculum formulation with the exception of the absence of any foetal bovine serum.

Metabolic parameters were followed closely to permit evaluation of the growth of the cells. The oxygen uptake rate was monitored continuously by observing the dissolved oxygen concentration in the recirculation stream at the outlet of the bioreactors. It is known from many measurements under a variety of conditions that the bioreactor inlet dissolved oxygen concentration is very close to equilibrium with the gas passing through the gas-exchange cartridge. Consequently, the oxygen uptake rate is determined by the calculated drop in oxygen concentration across the bioreactors (expressed as mmole oxygen per litre times the recirculation flow rate through the system).

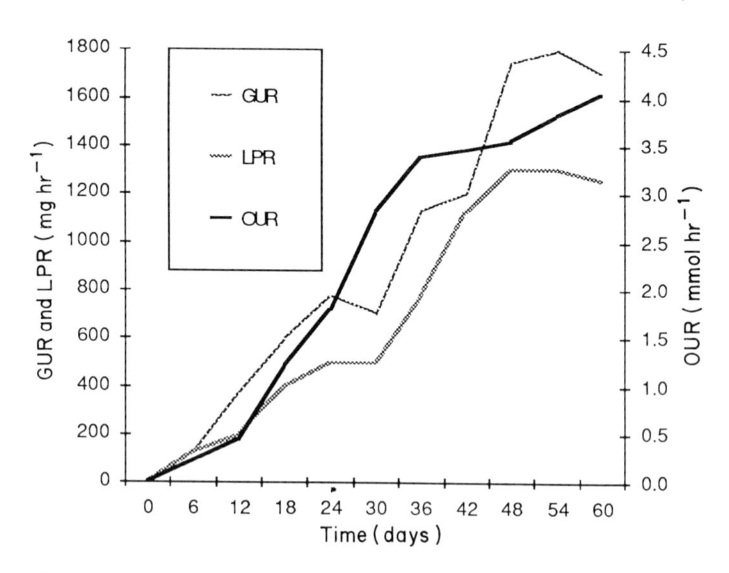

Fig. 3. Metabolic rates observed for oxygen uptake (OUR), glucose uptake (GUR) and lactic acid production (LPR) during a 60-day culture of hybridoma cells in the ACUSYST-P.

Since this parameter is measured on-line, it is very sensitive to any rapid changes in the cells' metabolic activity.

Since the ACUSYST system does not have the capability of determining glucose, lactate, or product concentrations on-line, the rates of their formation or consumption must be determined by taking successive daily samples and then applying the appropriate differential equations. With data from two successive samples and a knowledge of the flow rates and chemical concentrations in the inlet streams, one can determine an average rate for the period in question. The numerical techniques have been presented elsewhere (*5*). The ACUSOFT software which controls the ACUSYST-P will perform these calculations automatically. The glucose uptake and lactate production rates can be used to determine medium and glucose pump feed rates. Once a set-point for glucose or lactate concentration has been determined to optimize either the growth or productivity of the cells (*2*), the latest kinetic information is used to calculate the correct pump rates to maintain that concentration near the set-point in the face of increasing metabolic rates.

Figure 3 shows the pattern of metabolic rates observed in this experiment. During the first 35 days, the rates of oxygen, glucose and lactate metabolism increased steadily. The feed rate of fresh medium was increased to 900 ml hr^{-1} on day 47 and maintained at that value thereafter. Following this, the oxygen uptake rate continued to increase, reaching a maximum value of 4 mmole hr^{-1}.

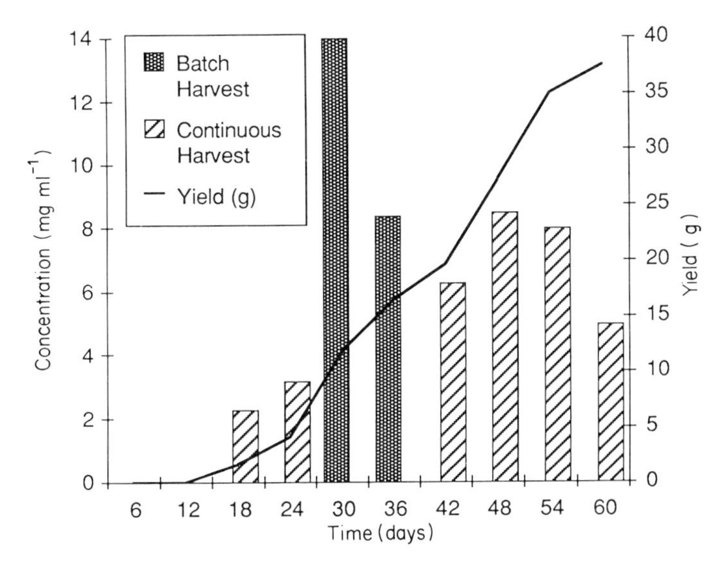

Fig. 4. Total IgA harvest from hybridoma cells cultured in the ACUSYST-P.

Note the overall parallel lines of glucose and lactate metabolism. In typical batch culture kinetics, these metabolic parameters will level off after a certain number of days (*3*), in contrast to results observed here for a perfusion culture. These hybridoma cells consistently convert about 70% of the glucose they consume to lactate (on a weight basis).

Figure 4 shows the product-formation kinetics. Harvesting began on day 14 at a rate of 6 ml hr^{-1} per flowpath. At the same time, a factor pump began adding foetal calf serum to the ECS to maintain the cells at a constant serum concentration of 10%. At day 20 the serum addition was discontinued. For two successive weeks, batch harvests were collected by drawing off the ECS volume (550 ml) over 1 hr. On day 40, continuous harvest was resumed at 6 ml hr^{-1} as concentration levels had increased sufficiently. No further increases in harvest rate were made during the run. At the end of 62 days, a total of 37.5 g of IgA were recovered in 5.5 litres. The average concentration was 6.87 mg ml^{-1}, and the average productivity was 0.60 g day^{-1}. Total medium usage during that time was 405 litres, resulting in a product yield from medium of 93 μg ml^{-1}. This figure compares very favourably with typical yields from large fermenter cultures of hybridoma cells. These fermenter cultures, however, would require over 40 litres of foetal calf serum to maintain the cells at 10% serum concentration. Through the ability of the ACUSYST system to sequester both cells and proteins in the extracapillary spaces of the bioreactors, the total serum usage during this phase was under 3 litres, less than 8% of the serum requirement of a fermenter system. The harvest from the ACUSYST-P was 73.6 times more concentrated than it would have been had the product been secreted directly into the feed medium.

4.2. Hepatitis-B Surface Antigen Production

Anchorage-dependent cells require attachment to a compatible surface. This creates a requirement for large amounts of surface area in order to grow large numbers of cells. The hollow-fibre bioreactors used in the ACUSYST systems have over 100 cm^2 of surface area per cm^3 extracapillary space, but the nature of the fibres' surface is not ideally suited for attachment and growth of anchorage-dependent cells. Pre-coating the fibres with polycationic materials seemed a reasonable approach to increasing cellular adhesion.

To study the effects of pre-treatments of the fibres with polycations, a series of fibre bundles was removed from a bioreactor, treated with various materials, washed, rinsed with medium, and placed in a Petri dish with DMEM. Mouse L cells were added to the medium over the fibres and were allowed to stand undisturbed for 24 hr. The fibre bundles were then transferred to fresh microbiological Petri dishes (upon which the cells could not grow) and observed daily through phase-contrast microscopy. Figure 5 shows the visual appearance of the fibres after 72 hr in the new dish. Although either serum (Fig. 5(*a*)) or poly-D-lysine (Fig. 5(*b*)) alone showed poor growth, the combination of the two was synergistic (Fig. 5(*c*)). The addition of fetuin to the inoculation mixture (Fig. 5(*d*)) allowed even more cells to attach to the fibres. Photographs of the edge of the fibres showed that the addition of fetuin promoted multilayered growth of the cells.

To test whether this pre-treatment scheme would allow these cells to grow and produce in the hollow-fibre environment, a bioreactor was prepared by filling the extracapillary volume (100 ml) with a solution of 0.1 mg ml^{-1} poly-D-lysine in de-ionized water and allowing the solution to stand for 1 hr. The system was then flushed with 4 litres of DMEM. An identical bioreactor was prepared without poly-D-lysine pre-treatment. Cells (2×10^8) at 92% viability were inoculated into each bioreactor. The poly-D-lysine-pre-treated bioreactor received 1 mg ml^{-1} fetuin in the inoculum. Both bioreactors received 10% NuSerum (Collaborative Research, Lexington, MA) in the extracapillary space and were maintained in an ACUSYST development system under identical conditions.

The cells selected for this experiment are mouse L cells into which a gene for the 20 nm hepatitis-B surface antigen assembly has been introduced. The plasmid is a proprietary non-oncogenic gene-expression vector system with a

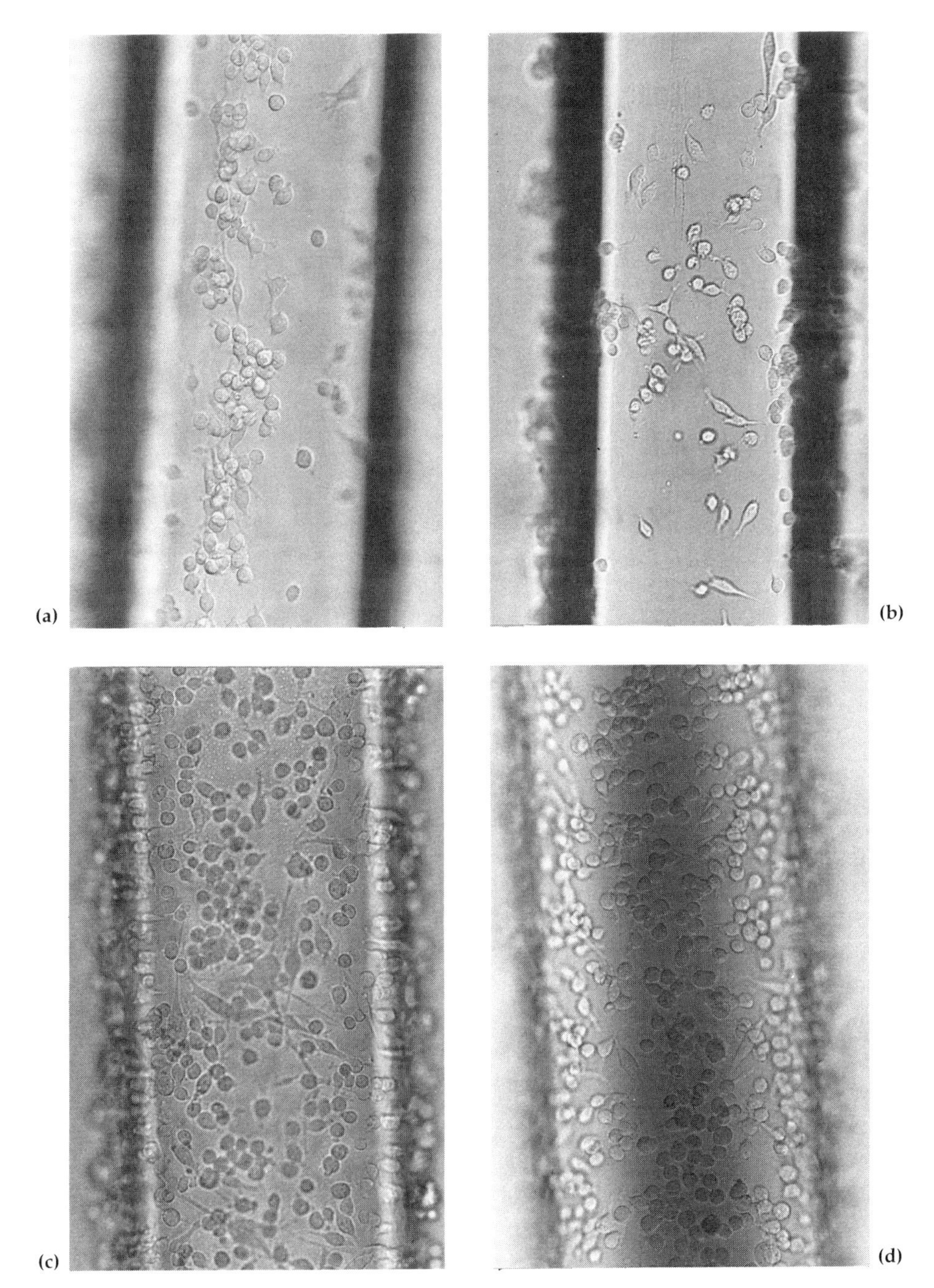

Fig. 5. Growth of attached mouse L cells on the surface of hollow fibres: (*a*) pre-treated with serum; (*b*) coated with poly-D-lysine; (*c*) pre-treated with serum and coated with poly-D-lysine; and (*d*) following serum pre-treatment, poly-D-lysine coating and exposure to fetuin.

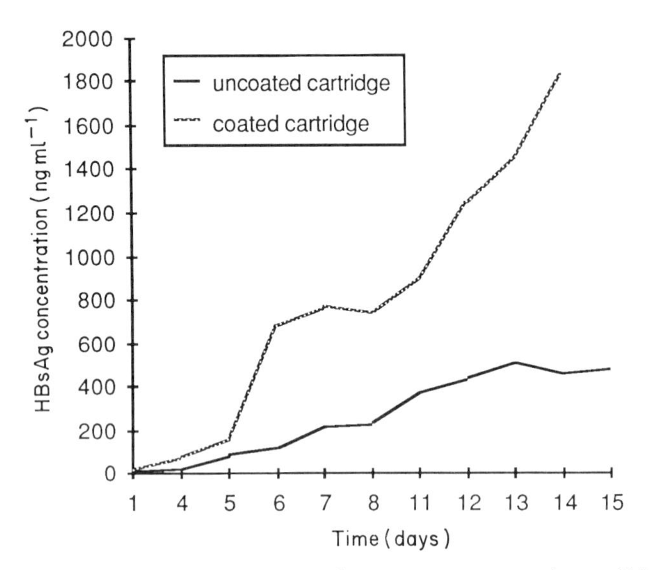

Fig. 6. Comparison of product concentration from a serum-pre-treated control bioreactor and a bioreactor coated with poly-D-lysine and exposed to fetuin.

high-efficiency metallothionine gene promoter. The cells secrete the hepatitis-B surface antigen particles through membrane budding. The concentration of the hepatitis-B surface antigen in the extracapillary space is determined by subjecting a sample to a solid-bead radioimmunoassay from Nuclear Medical Laboratories (Irving, TX) calibrated against reference antigen at 1000 ng ml^{-1} from Georg-Speyer-Haus, Frankfurt, Germany.

Figure 6 shows the results of hepatitis-B surface antigen assays performed on samples of the extracapillary fluid of the two bioreactors. Clearly, the pre-treated bioreactor demonstrated superior performance. By day 15, the pre-treated bioreactor had three-fold more product than did the control system.

Encouraged by the above results, we used an ACUSYST-Jr single bioreactor system to evaluate the potential for long-term attachment-dependent cell growth in a hollow-fibre system. The bioreactor and inoculum were prepared as described above. The system exhibited a considerable lag before beginning exponential growth. Figure 7 shows the cumulative number of moles of glucose and oxygen consumed or lactate produced as well as the cumulative yield of hepatitis-B surface antigen harvested. The most significant data are those from days 90 to 120. During this 30-day period, the cells demonstrated stable metabolic rates, indicative of a stable cell population. However, they continued to secrete product during this entire time. This

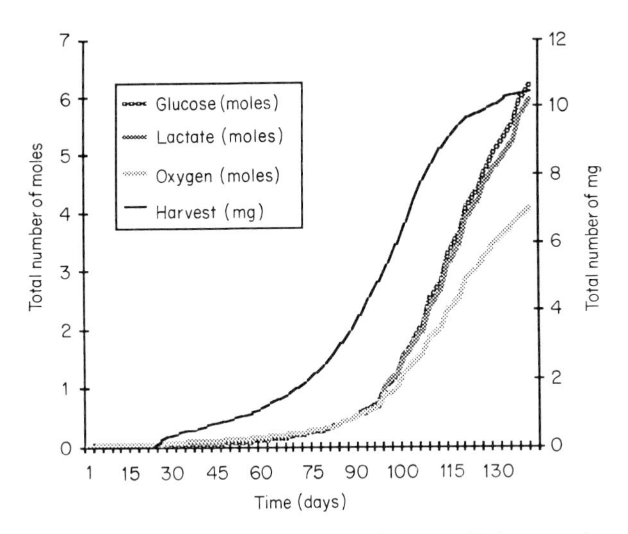

Fig. 7. Metabolic activity and product formation from L cells in a 140-day culture in the ACUSYST-Jr.

linear production rate in the absence of increasing cellular metabolic rates signifies that product formation in this system does not appear to be growth-associated. This implies that long-term production of recombinant products by anchorage-dependent cells is indeed feasible in the ACUSYST hollow-fibre cell-culture systems.

The excessive lag phase is possibly due to an insufficient inoculum. Many anchorage-dependent cells have minimum requirements for cells per unit area as well as cells per unit volume, so a larger inoculum might show a much more rapid initial growth phase. In addition, this bioreactor became congested with cells. It is conceivable that the final tapering of productivity observed after day 120 reflected a lack of mobility of this very large product (MW 2.2×10^6 Daltons). Possibly some means of limiting the growth of the cells might result in an even more extended production phase.

5. CONCLUSIONS

Successful large-scale mammalian cell culture must be capable of providing a controlled environment to support cell growth and production. The ideal technology must supply basic biological materials including medium, serum, gases such as oxygen and carbon dioxide, and energy sources such as glucose to allow cells to grow and produce secreted products. The ACUSYST-P and ACUSYST-Jr utilize similar system-component design and ACUSYST hollow-fibre technology to provide standardized process control throughout the product-development life-cycle. These instruments are able to generate sufficient quantities of product from research through production phases. Examples of long-term continuous culture of a hybridoma cell line secreting monoclonal antibody in the ACUSYST-P and an attached mouse L cell line producing hepatitis-B surface antigen in the ACUSYST-Jr demonstrate the versatility and capability of these systems to supply and maintain cellular requirements for growth and production and provide continuous harvest of product over an extended period of time.

Acknowledgements

The authors would like to express their gratitude to Dr. Wei-Shou Hu, Department of Chemical Engineering and Materials Science, University of Minnesota, for his technical advice in the preparation of this manuscript.

REFERENCES

1. Altshuler, G. L., Dziewulski, D. M., Sowek, J. A., and Belfort, G. (1986). Continuous hybridoma growth and monoclonal antibody production in hollow fiber reactors/separators. *Biotechnol. Bioeng.* **28**, 646–658.
2. Andersen, B. G., and Gruenberg, M. L. (1986). Optimization techniques for the production of monoclonal antibodies utilizing hollow fiber technology. *In* "Commercial Production of Monoclonal Antibodies" (S. S. Seaver, ed.). Marcel Dekker, Inc., New York (1987).
3. Baker, M. P., Metzger, L. S., Slaber, P. L., Nevit, K. L., and Boder, G. B. (1986). Large-scale production of monoclonal antibodies in suspension culture. *Ann. Meet. Am. Chem. Soc.* (Abstract).
4. David, G. S., Reisfeld, R. A., and Chino, T. H. (1978). Continuous production of carcinoembryonic antigen in hollow fiber cell culture units: Brief communication. *J. Natl. Cancer Inst.* **60**, 303–306.
5. Hirschel, M., and Gruenberg, M. L. (1986). An automated hollow fiber system for the large scale manufacture of mammalian cell secreted product. *In* "Large Scale Cell Culture Technology" (B. D. Lydersen, ed.). Carl Hanser Publishers, Germany, Macmillan Publishing Company (distributor) (in press).
6. Hu, W.-S., and Wang, D. I. C. (1986). Mammalian cell culture technology: a review from an engineering prospective. *In* "Mammalian Cell Technology" (W. G. Thilly, ed.), pp. 167–197. Butterworth Publishers, Stoneham, Massachusetts.
7. Knazek, R. A., Gullino, P. M., Kohler, P. O., and Dedrick, R. L. (1972). Cell culture on artificial capillaries: an approach to tissue growth *in vitro*. *Science* **178**, 65–66.
8. Knazek, R. A., Kohler, P. O., and Gullino, P. M. (1974). Hormone production by cells grown *in-vitro* on artificial capillaries. *Exp. Cell Res.* **84**, 251–254.

9. Ratner, P. L., Cleary, M. L., and Jones, E. (1978). Production of "rapid-harvest" Moloney murine leukemia virus by continuous cell culture on synthetic capillaries. *J. Virol.* **26**, 536–539.
10. Schratter, P. (1976). Cell culture with synthetic capillaries. *In* "Methods in Cell Biology", vol. 14 (D. M. Prescott, ed.), pp. 95–103. Academic Press, London.
11. Spier, R. E., and McCullough, K. (1987). The large-scale production of monoclonal antibodies *in vitro*. Cambridge University Press (manuscript in progress).
12. Tharakan, J. P., and Chau, P. C. (1986). A radial flow hollow fiber bioreactor for the large-scale culture of mammalian cells. *Biotechnol. Bioeng.* **28**, 329–342.
13. Tharakan, J. P., and Chau, P. C. (1986). Operation and pressure distribution of immobilized cell hollow fiber bioreactors. *Biotechnol. Bioeng.* **28**, 1064–1071.
14. Tyo, M. A., and Spier, R. E. (1987). Dense cultures of animal cells at the industrial scale. *In* "Enzyme and Microbial Technology" **9**(9): 514–520.
15. Tze, W. J., and Chen, L. M. (1977). Long-term survival of adult rat Islets of Langerhans in artificial capillary culture units. *Diabetes* **26**, 185–191.

17

Perfusion Culture Systems for Large-scale Pharmaceutical Production

WILLIAM R. TOLBERT, WILLIAM R. SRIGLEY,
and CHRISTOPHER P. PRIOR
Invitron Corporation,
St. Louis, Missouri, U.S.A.

ANIMAL CELL BIOTECHNOLOGY VOL. 3
ISBN 0-12-657553 3

1. INTRODUCTION

In recent years, researchers have identified many naturally occurring complex human proteins which have potential applications for diagnosis, prevention and treatment of human disease. These advances promise to alter fundamentally the nature of human health care. Such potentially useful proteins include monoclonal antibodies, human growth factors, interleukins, interferons, erythropoietin, plasminogen activators, and the blood-clotting proteins Factor VIIIc and Factor IX. These proteins may be produced by hybridoma cell lines, as in the case of monoclonal antibodies, by cell lines altered through recombinant DNA technology, or by unaltered naturally occurring cell lines.

Many biotechnology and pharmaceutical companies have learned to characterize such proteins, have developed cell lines to produce them, and have made them in limited quantities in the laboratory. However, a significant hurdle to their successful commercialization remains. This is the development of manufacturing processes and capabilities to produce commercial quantities economically, under the conditions consistent with government regulations for the manufacture of pharmaceutical products.

Many, if not most, of these proteins will require mammalian cell-culture manufacturing processes. Mammalian cells have the ability to synthesize these complex human molecules with the proper three-dimensional configurations, correct disulphide bonding, and arrays of sugar side-chains which result in the desired activity of the naturally occurring proteins. Therefore, these proteins derived from mammalian cells are more likely to be efficacious and are less likely to cause a response of the body's immune system which might render such a protein drug ineffective or even endanger the patient.

In order to produce any pharmaceutical product, compliance to extensive government regulations is necessary so that the public is ensured of obtaining safe and effective products made in a reproducible and controlled manner. Pharmaceutical products from mammalian cell culture must meet these same standards. This requires an integration of the cell culture technology used, the documentation and quality assurance methods to ensure product safety and efficacy, and the facility that provides the aseptic environment for manufacturing. This chapter will discuss these three topics.

2. CELL-CULTURE TECHNOLOGY

One of the most fundamental process considerations is the type of production system to be used for large-scale manufacture. Cell-culture systems take many forms and include a number of diverse techniques (*1–6*). A basic distinction between cell-culture systems is whether they utilize batch or perfusion methods.

A manufacturing technology based on batch cell culture typically requires large reactors having a volume of 1000 to 10 000 litres. These batch reactors are generally charged at the beginning of production cycle with seed culture at a density of a few tenths of a millilitre per litre and sufficient medium to sustain the anticipated cell mass during the entire production cycle. As the cell culture grows, the cells utilize the nutrients and produce desired proteins and waste products, both of which remain in the vessel. Under these conditions, most cultured cells are limited to a population density of approximately 1–2 ml of cells per litre of reactor liquid volume. This generally occurs anywhere from 4 to 20 days after the process begins, depending on the cell line's growth characteristics. The contents of the reactor must then be processed to eliminate the cells, waste products, and remaining nutrients. The desired protein product which remains can then be purified. In order to produce additional product, this cyclical growing process must be started again.

In contrast, in a perfusion culture, medium is perfused through the reactor at a rate proportional to the number of cells and their metabolic characteristics. Cell densities from 10 to 100 times those of batch culture can be achieved, that is from 10 to 100 ml of cells per litre of culture. Process parameters are maintained by introducing reagents and gases directly into the vessel or into the medium as it is being fed to the vessel. Perfusion systems generally require a greater number of aseptic connections than batch culture systems. Conditioned medium is removed from contact with the cells on a continuous basis for concentration and purification of desired protein products.

We have developed three, inter-related, large-scale cell-culture manufacturing systems to address the complexities and expense of mammalian cell culture. These culture systems have been designed to simulate the natural environment of cells in which the supply of nutrients to the cells and the removal of waste from the cells is a continuous and carefully regulated process. These systems have been designed to optimize cell density, minimize the required level of animal serum in the nutrient medium and retain the flexibility necessary to accommodate the variable growth characteristics of different mammalian cell lines.

One system has been developed for culturing cells in suspension, another for cells attached to a surface, and a third system for production of desired protein product from cells maintained at very high density in a non-proliferative state. These systems can be used independently or in conjunction with one another to create an optimal environment for each cell type.

Computer automation provides continuous monitoring and control of the manufacturing process. The facility is designed with 37°C incubation rooms adjacent to 4°C product cold rooms, so that cell-conditioned medium containing the product of interest may be rapidly removed from the high-temperature reactor environment to a refrigerated storage vessel. By storing product at a temperature too low for most enzymatic activity, the opportunity for product degradation by proteolytic enzymes in the culture medium is greatly reduced. This may be of significant importance for enhancing the stability and specific activity of certain human pharmaceutical proteins.

Another very important feature of this production method is the ability to operate completely without the use of antibiotics. Antibiotics are almost always used in large-scale applications because of the difficulty of preventing a single microorganism from contaminating an entire production run. This major economic risk increases exponentially with the size of the culture. As the systems are modular and operated at high density (and so relatively small), they can be used without antibiotics. In addition, the production reactors, proprietary aseptic connection systems, and the facility were specifically designed to maintain the highest level of sterility in our production process.

The disadvantages associated with the use of antibiotics impact negatively the safety of the product. Use of antibiotics encourages the growth of low levels of resistant organisms which are difficult to detect and can result in adulterated, unsafe drug products. Antibiotics also tend to permit less than optimum operational standards, which could lead to other types of contamination or loss of product integrity. The absence of antibiotics guarantees very high quality in all aspects of the production process. Therefore, we feel that these substances should not be used in pharmaceutical systems.

The following is a description of the three perfusion-based cell culture manufacturing systems.

2.1. Perfusion Culture System

This system has been designed for the growth of mammalian cells in suspension, either as single cells or as small aggregates of cells (see Fig. 1). A constant environment is maintained in the growth vessel by continuously adding fresh nutrients and removing waste products at a rate proportional to the number of cells in the vessel. Under these conditions, the system can operate for prolonged periods with cell densities from 10 to 30 times the maximum cell density in a typical batch reactor. A proprietary gentle "sail" system is used to prevent cell damage at these high cell densities. The filtration unit separates the cells from the product-containing cell-conditioned medium

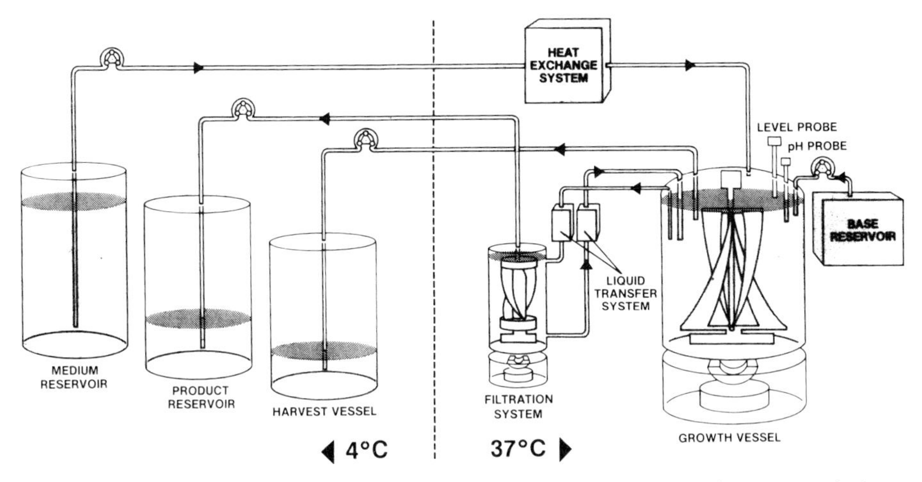

Fig. 1. Perfusion culture system: liquid flow rates, dissolved O_2, CO_2, and N_2 gases, and other process parameters are controlled by proprietary perfusion culture system computer software.

and returns the cells to the growth vessel as fresh medium is added. Dissolved oxygen and carbon dioxide levels, as well as pH levels, are continually monitored and maintained in pre-set ranges throughout the production run. The cell-conditioned medium removed from the growth vessel, or reactor, contains waste products and the desired protein products secreted by the cells. It is transferred from the 37°C warm room to the product reservoir in the refrigerated 4°C cold room to reduce the proteolytic enzyme activity which could degrade the protein product. Product reservoirs may be removed periodically for further processing by concentration and other purification techniques, as will be described below.

When prolonged cell growth is required for production of desired protein products, the system maintains an optimal cell density in the growth vessel by transferring excess cells to the harvest vessel. After the optimal cell density has been attained in the growth vessel, the serum concentration in the fresh medium can be significantly reduced, thereby simplifying the subsequent purification required to obtain the desired product. This perfusion culture system (PC) requires substantially less serum in all stages of the production process than the traditional batch method. The system currently utilizes a growth vessel of 100 litres capacity and has been successfully used to manufacture products from a number of cell lines (see Table I). Figure 2 shows operation of the 100-litre perfusion culture system within the pharmaceutical manufacturing facility.

TABLE I Productivity Comparison

	Small batch vessels		Large perfusion systems				
Cell type	Product concentration ($mg\,l^{-1}$)	Serum concentration (%)	Reactor type*	Product concentration ($mg\,l^{-1}$)	Serum concentration (%)	Total volume (litre)	Total product (g)
Hybridomas							
Line A	50	6	PC	80	1	3400	270
Line B	10	10	PC	32	10	2100	67
Line C	10	6	PC	16	2	3300	52.5
Line D	20	6	PC	36.5	2	2600	95
rDNA							
Line E	10	7.5	PC	13	7.5	8100	105
Line F	0.2	10	MR	1.0	0	680	0.66
Line G	30	10	SMR	47	0	1400	66.2
Line H	10	7.5	SMR	12	0	2600	30

* PC perfusion culture system; MR, microcarrier reactor; SMR, static maintenance reactor.

Fig. 2. The perfusion culture system has the daily equivalent production capacity of 1000 litres of conventional batch suspension culture. A typical perfusion culture run averages 90 days.

2.2. Microcarrier Perfusion System

Many mammalian cells require attachment to a surface for growth, for optimal production of proteins, or for survival. The microcarrier method first introduced by A. L. van Wezel (*7*) is used with a reactor similar to the one described above (see Fig. 3). In the microcarrier reactor (MR), the very gentle agitation system provided by the flexible sail agitators is even more important. During collisions between microcarriers, or between microcarriers and the vessel wall or agitator of the vessel, the force of the collision is transmitted directly through the cell layer. Owing to the much greater mass of the microcarrier as compared to individual cells, these collisions are much more severe for attached cells on microcarriers than for suspension cells. By providing the necessary agitation energy through as large a surface area as possible with an agitator rotating between 8 and 12 r.p.m., much higher amounts of microcarriers, and therefore surface area, may be used. Up to one-quarter of the settled volume of the reactor can be microcarriers. For a 100-litre reactor, this corresponds to a bead surface area of between 500 and 1000 m^2, depending on the type of microcarrier.

Cells and microcarriers are prevented from entering the higher-shear region of the filtration system by a settling bottle which also allows close contact between the microcarriers. This enhances the formation of cell–bead aggregates, thereby further increasing the effective surface area within the growth vessel. In a recent production run, 1.3×10^{12} cells of a recombinant DNA cell

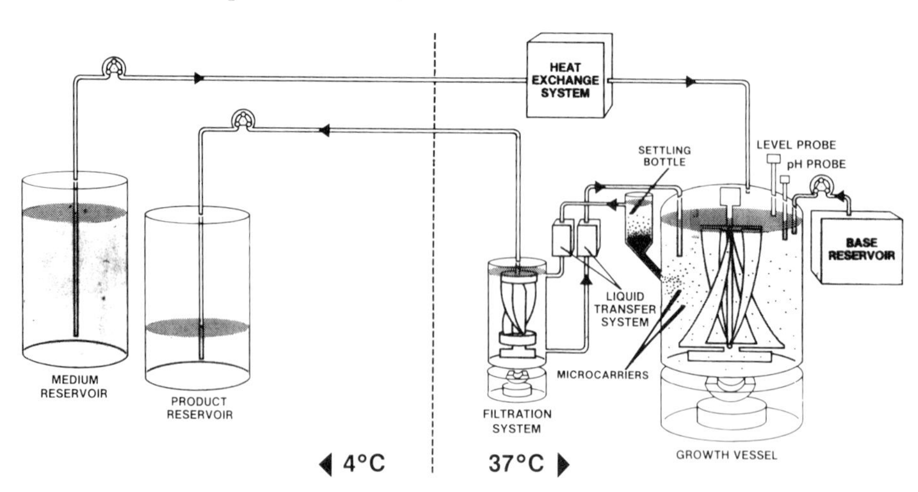

Fig. 3. Microcarrier reactor system: liquid flow rates, dissolved O_2, CO_2, and N_2 gases, and other process parameters are controlled by proprietary microcarrier perfusion culture system computer software.

line were grown to produce 660 mg of a rare human protein which had previously only been available in microgram quantities from natural sources (Table I).

2.3. Static Maintenance Reactor

The third culture system involves immobilizing cells at high density in a matrix material so that they may be maintained productively for long periods of time.

Traditionally, cell-culture and cell-derived protein production have focused on cell growth and maximizing cell proliferation. However, many valuable proteins produced by cultured cell lines are made more efficiently and abundantly during a non-proliferative phase. In the body, most cells are not actively growing or proliferating, but are instead expending their energy making and secreting proteins or performing other functions.

By simulating an *in vivo* environment, the static maintenance reactor (SMR) enables cells which are maintained at high-density in a non-proliferative state to expend less energy on growth and more on production of desired cell-derived products. The reduction of proliferative activity in the static maintenance reactor also lowers the risk of mutation as a result of cell division, which could otherwise occur in growing systems. Moreover, by removing cells from an agitated-medium environment and immobilizing them, the risk of cell damage due to collisions between cells is eliminated. Because cell growth is not desired, the need to supplement the culture medium with expensive serum, or other mitosis-stimulating factors is greatly reduced. The static maintenance reactor can be used to maintain both suspension and anchorage-dependent cells at densities approximately 100 times those obtained in a conventional batch system (see Fig. 4).

One to two kilograms of cells are grown in either of the perfusion growth systems (PC or MR) and then concentrated, mixed with a finely divided matrix material and then pumped as a thick slurry into the reactor. The reactor is filled until the cells are completely immobilized in the interstices of the matrix material. Nine-tenths of the reactor volume is matrix and one-tenth is evenly distributed cell mass.

The reactor consists of a cylindrical vessel penetrated by an array of porous tubes. One set of tubes allows perfusion of a fresh nutrient medium into the reactor and the second set of adjacent tubes removes cell-conditioned medium containing product. All cells are within 2 cm of a fresh source of nutrient medium. A distributed semi-permeable membrane is also provided to allow diffusion of oxygen to the cells and diffusion out of carbon dioxide. These

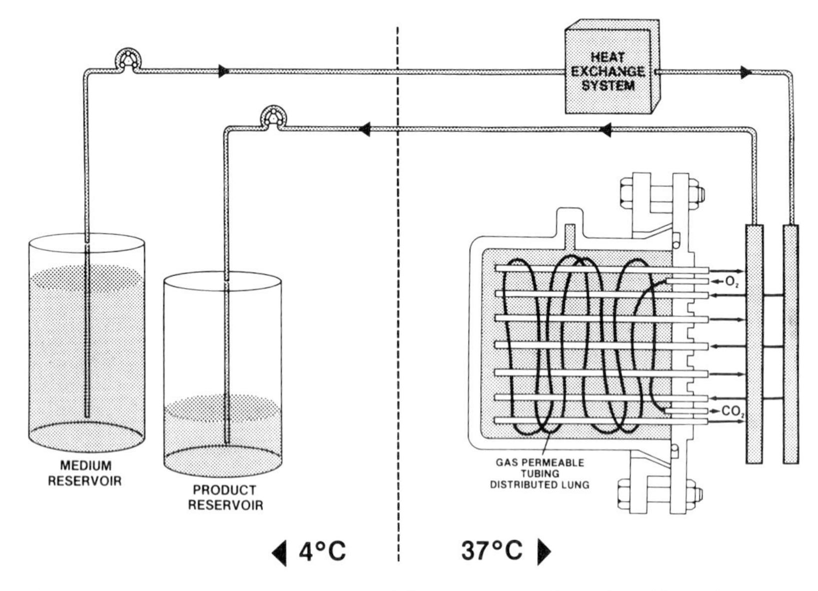

Fig. 4. Static maintenance reactor: liquid flow rates, dissolved O_2, CO_2, and N_2 gases, and other process parameters are controlled by proprietary static maintenance reactor computer software.

reactors have been used for as long as 3½ months, continuously making product (see Fig. 5).

In a recent run, 66 g of a human protein was produced by a recombinant cell line using 1400 litres of serum-free medium (Table I).

2.4. Large-Scale Process Purification

Cell-culture-conditioned medium consists of a wide variety of components that may contaminate the desired protein product. For example, serum proteins and other media constituents such as hormones, vitamins, steroids, lipids, amino acids, nucleic acids, as well as other cellular proteins, will be mixed with the desired protein product. This product will generally be at very low concentration, from 100 mg l^{-1} to as low as 100 μg l^{-1}. From a conditioned medium containing 10% serum, these would require from 50- to 50 000-fold purification respectively.

In addition, the biological activity of the protein product may suffer significantly from any harsh purification conditions such as extremes of pH, oxidizing or reducing conditions, freezing, physical damage under conditions which produce foam, and degradation by proteolytic enzymes. Formation of

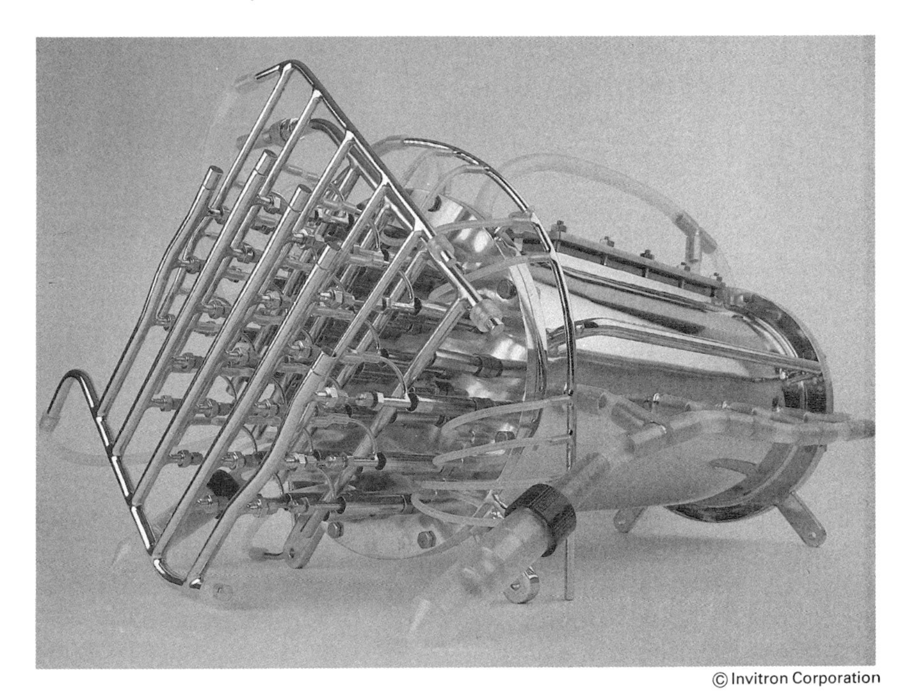

Fig. 5. The static maintenance reactor daily produces gram quantities of medically important proteins. A typical SMR run averages 90 days.

aggregates, or complex formation with other cellular or serum proteins, can also reduce biological activity. Of course, contamination by bacteria or other microorganisms can destroy the product. For these reasons, cell-culture-conditioned medium containing product should be processed as soon as possible by a method that is as gentle as possible to preserve product integrity.

Perfusion systems have major advantages for the downstream process purification. As mentioned above, serum proteins can, in general, be reduced or even eliminated, thereby removing the major background contamination from which the protein product must be isolated. By perfusing the product directly from the 37°C warm room into the 4°C cold room, proteolytic enzyme degradation can be minimized. Also, the relatively high perfusion associated with the static maintenance reactor, in which products secreted by cells are removed from the reactor within a matter of 3–6 hr has significantly reduced aggregate formation and the formation of inhibitor complexes. Perfusion systems also allow product to be collected and processed on a continuous basis, thereby also reducing the opportunity for product degradation.

Fig. 6. Aseptic concentration of product from conditioned cell-culture media.

When handling large volumes of cell-conditioned medium containing a protein product in dilute solution, the first step is a 50- to 100-fold concentration. Generally, an ultrafiltration membrane having a pore size smaller than the desired protein may be used in a number of configurations, e.g. a hollow-fibre system, a plate-and-frame system, or a spiral-wound cartridge. These systems exploit tangential flow of the conditioned medium across the membrane surface to minimize clogging and maintain acceptable flow rates (see Fig. 6).

Once the volume is sufficiently reduced, it is possible to exchange the concentrate with a buffer of lower ionic strength or different pH. This process of diafiltration now places the protein molecules in a buffer environment suitable for subsequent purification fractionation steps. Between uses, the membranes and other surfaces which come into contact with potential product are subjected to chemical disinfection followed by sterilization when applicable.

It should be noted that ultrafiltration serves strictly as a concentration and buffer-exchange step, with minimal or no effect on protein purification; however, if these ultrafiltration systems are scaled-up they will handle large quantities from a few hundred to a thousand litres in 3–4 hours.

The next step in the purification process generally employs ion-exchange chromatography utilizing the electrostatic properties of the molecules to separate them from the bulk of serum and other contaminating proteins. Conditions are generally chosen to bind the desired protein on the column and remove the bulk of contaminating protein in the flow through. This can be followed by opposite conditions in which the desired protein flows through a column and additional contaminants are bound.

With many proteins, it is possible to establish conditions under which the ultrafiltration step can be eliminated and the conditioned medium directly, or with some adjustment of pH or salt concentration, passed over special highly cross-linked ion-exchange resins such as those produced by Pharmacia, Inc. These resins also allow an extremely high flow rate of hundreds of litres per hour and have high capacity for binding proteins. This combines both the concentration and a first purification step in a single operation, thereby eliminating losses associated with ultrafiltration (see Fig. 7). Additional purification steps are determined by the physical characteristics of the desired protein and the purification necessary to achieve isolation. If very high-fold purification is required, immunoaffinity columns can be used with monoclonal antibodies having a specificity against the desired protein bound to the resin.

Gel filtration chromatography provides an excellent finishing step in any pharmaceutical purification process. Acting as a molecular sieve, gel filtration removes potentially toxic residual chemicals and potentially antigenic large protein aggregates. In addition, gel filtration allows a purified protein to be formulated by exchanging the protein into a buffer system that confers the maximum solubility and biological stability. Operational capacity using gel filtration chromatography is limited, as the sample size can be only 1–5% of the total column volume. This means that the protein must generally be concentrated. A method such as ammonium sulphate precipitation or ultrafiltration is used to reduce the sample volume initially, but still very large column

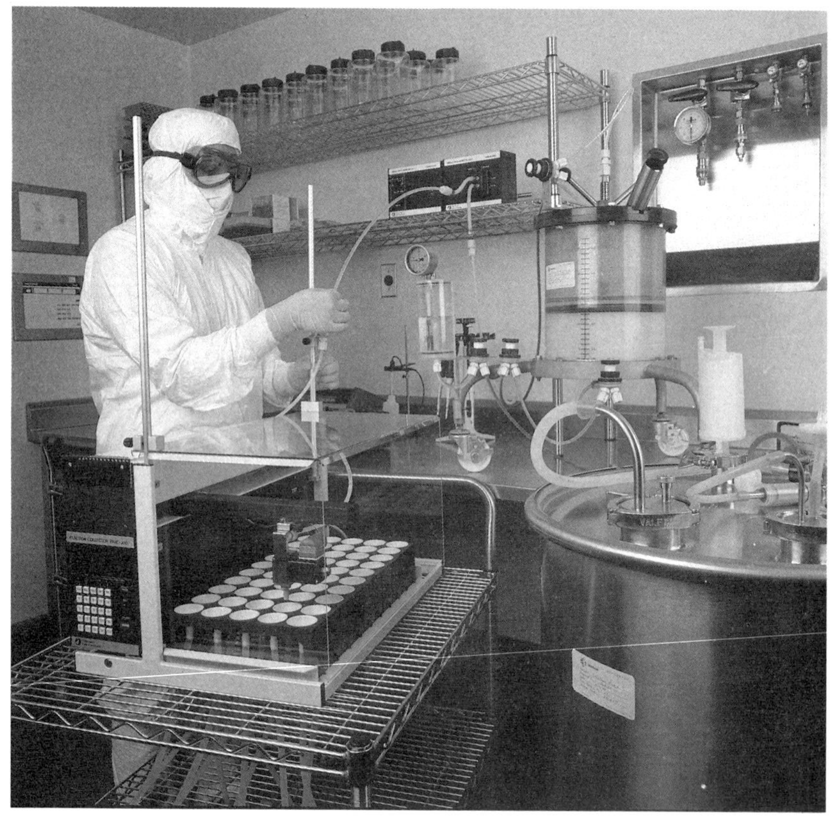

Fig. 7. Chromatographic purification of product is conducted utilizing current good manufacturing practices.

volumes are often needed. Depending on the protein, the resultant purified solution may be kept at 4°C, frozen, or, in some cases, lyophilized.

All stages of the process of purification must be performed under highly aseptic conditions, generally at cold room temperatures, to minimize risk of product degradation or of microbial contamination, formation of endotoxin, and loss of product.

3. QUALITY CONTROL

The production of drug products by any means requires strict adherence to established pharmaceutical manufacturing principles and practices. The fact

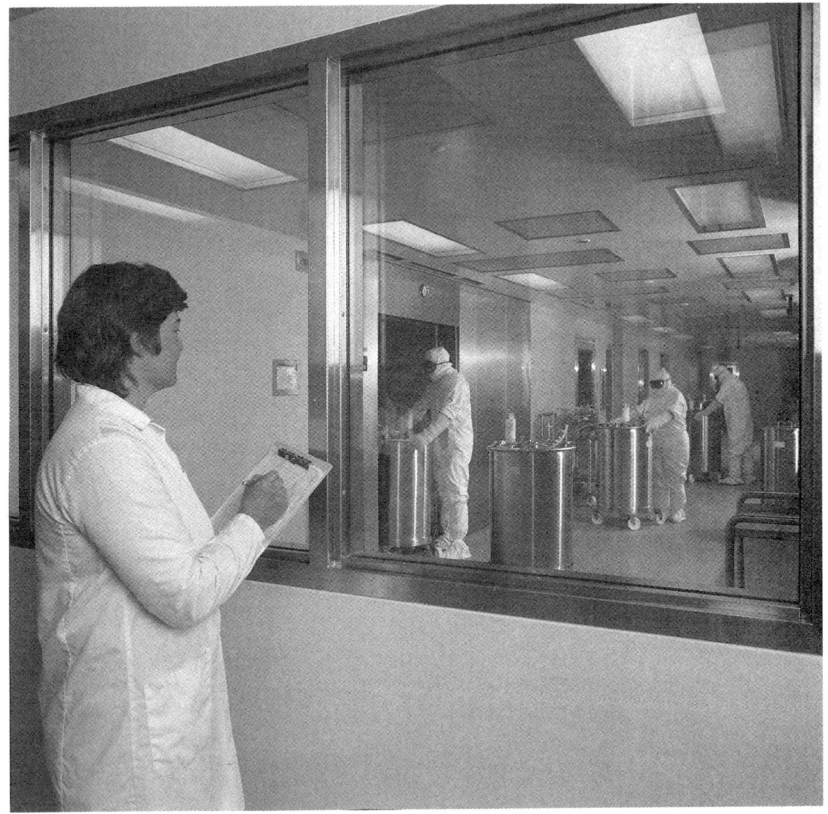

© Invitron Corporation

Fig. 8. Quality assurance person observes the unloading of autoclave-sterilized media containers.

that the products of cell culture are often complex biomolecules whose efficacy may depend upon them retaining their native structure, and whose safety depends on freedom from extraneous contaminants, only makes this all the more essential. These standards apply early in the design of the equipment and extend to the operation of the production facility. While by no means exhaustive, the following discussion will provide a basis for the reader's more detailed research on the subject (see Fig. 8).

3.1. Equipment

The design of equipment used in cell culture is of critical importance in determining the characteristics of the resultant product. All product contact surfaces must be free of process residuals, microorganisms or their by-products. This requires the equipment to be designed in such a manner that it may be sufficiently disassembled to permit every surface which directly or indirectly contacts the product to be cleaned and examined. While this may be more easily accomplished with modular systems which are completely disassembled after every use, it is no less important for systems which are designed to be cleaned in place. The design of clean-in-place systems requires special attention to the location of the spray nozzles, the type of cleaning solutions employed and the duration of the cleaning and rinse cycles. Provisions must be made for examining, or in some way sampling, the product contact surfaces at periodic intervals to evaluate the continuing effectiveness of the cleaning procedures.

Vessels, agitators, pump housings and impellers, sampling tubes, inlets and outlets must be free from cracks and crevices. Threaded fittings should be eliminated. Materials of construction should be as non-reactive as possible. Typically, this is best achieved through the use of 316L stainless-steel polished to at least a 240 grit, electropolished, and passivated just prior to use. Welds should be inspected for imperfections and tested for leaks.

Provisions must be made for sterilizing equipment. While the introduction of steam at sufficient pressure is of great importance, the simultaneous removal of condensate from the lowest portion of the load and any lengths of tubing is equally critical. Means must be provided for monitoring the effectiveness of steam penetration into all portions of the vessel and its ancillary equipment and connections.

3.2. Facilities

Adequate isolation and spatial separation should exist to preclude mix-up or cross-contamination of cells and products. Wall, floor, and ceiling surfaces should be constructed of materials which are durable, smooth, non-absorbent, and able to withstand repeated cleaning and sanitization. Air supplied to the production areas should be filtered and supplied at a rate which provides individual rooms with the over-pressure necessary to achieve containment and process objectives. The number of air handlers used should be sufficient to permit appropriate isolation of critical production and quality control rooms. Steam used to sterilize process or product contact surfaces should be saturated, clean, and free of boiler additives. Process gas lines should be cleaned prior to use and constructed of non-additive materials. Generally, this precludes the use of soldered joints.

The process water system should be as good as possible. Water is the largest single component of cell-culture-derived product, and high-quality water is very difficult to make. A good pre-treatment system, followed by a multi-effect still which feeds water to a 316L stainless-steel tank vented through a 0.2 μm filter can consistently produce water which meets the requirements of the U.S.P. monograph for Water for Injection. Holding the water at 80°C and circulating it through a properly installed and maintained stainless-steel loop can generally assure that the water which is supplied to the production area is consistently of excellent quality. This is not the only way to achieve that objective, but it is probably the one with the longest track record of success.

3.3. Standard Operating Procedures, Batch Records and Training

Cell culture is a developing science. Nevertheless, its use as a vehicle for the manufacture of pharmaceutical products requires that the process be described in detailed procedures. These procedures should include the cleaning and set-up of equipment, sterilization, cell culture, inoculations, media preparation, culture maintenance, collection and concentration of conditioned media, product purification, and the disposal of cultures and media. Procedures for testing, raw materials receiving, storage, facility maintenance, and so on are also necessary, but are typically more easily reduced to written form.

The design of a documentation system suitable for recording the critical steps of a cell culture process in a manner which complies with CGMP's presents unique challenges. The variability of cells from different species, or in some cases, different clones of the same cell line, makes the advance preparation of a uniform batch record applicable to all cells virtually impossible. At the same time, the prospect of attempting to write a customized master formula and batch record for each individual cell line becomes formidable if any more than a few cell lines are involved. In a similar vein, while the natural variability of biological systems makes it likely that acceptable differences in the production process might be encountered from batch to batch, reasonable process control requires that some boundaries be established which prescribe the allowable latitude for process variability. These judgements are not easy ones, and yet it is imperative that they be made.

Training employees in the proper execution of these procedures is essential (see Fig. 9). Management must continually emphasize the need for comprehensive training programmes and follow-up monitoring to assure that deviations from accepted procedures are corrected.

© Invitron Corporation

Fig. 9. Large volumes of nutrient media are prepared and sterilized through multiple cartridge filters.

3.4. Testing of Raw Materials and Product

Consistency in the manufacture of product is only possible to the extent that the characteristics of the raw materials used in the process are known and maintained within defined limits. Internal testing of purchased materials is an essential element of a quality control. Of no less importance, however, is an active vendor-communication programme. Such a programme might involve not just vendor audits, but also discussions regarding sampling plans and future requirements, as well as changing or additional specifications. Frequent interactions with vendors let them know that you consider them an important part of your business and make it more likely that you will get what you need when you need it.

A well-designed in-process testing programme provides an understanding of the process which is often impossible to obtain by testing final product. Information regarding product characteristics, endotoxin levels, microbial counts, specific activity, concentration of contaminants, or the presence of deleterious enzymes might allow problems to be detected early while they can be corrected, or suggest ways of improving yields. Frequently, the establishment of such a programme must await the development of a process which is at least somewhat routine. The sooner it can be done, however, the more likely it is that the process will be able to be controlled and improved.

3.5. Validation and Monitoring of Process and Facility

The manufacture of pharmaceutical products by any means requires a strict adherence to principles and practices which have been utilized in the industry for many years. The fact that the products of cell culture are often complex biomolecules whose efficacy may depend upon them retaining their native structure, and whose safety depends on freedom from extraneous contaminants only makes the task of manufacturing them all the more difficult. Prior to the mid 1970s, the method by which the quality of drug products was proven was primarily end-item testing. The thesis was that if a lot of final product passed a set of release tests, it was suitable for commercial distribution. In 1976, the U.S. Food and Drug Administration (FDA) published "Proposed Good Manufacturing Practices for Large Volume Parenterals." These proposed regulations set forth a concept which has dramatically changed the way that the quality of drug products is defined. In essence, the Administration said that simply meeting final product specifications was no longer a sufficient basis on which to conclude that a lot of product was able to be released. In addition, the manufacturer must also be able to prove that the production and control systems used to make, hold, and test the product were operating within established limits during the time the product was manufactured, held, and tested in the facility. The basis for such proof is validation—proving that a system process or test consistently works the way it was intended to work.

This entails defining the critical aspects of facility and equipment design and construction, verifying that they were adhered to, establishing that the plant and production systems function within acceptable specifications and documenting the results (see Fig. 8). It also means that once things are working in a controlled fashion to yield the desired results, changes must be evaluated to ascertain that they do not adversely impact the product or production systems. The supposition is that if you have demonstrated that the plant and production systems are consistently capable of producing acceptable product, as long as no changes are made to those systems, it is likely that you will continue to produce acceptable product. End-item testing then becomes a confirmation of that success, rather than a determination of it. This is an extension of the principle that quality cannot be tested into a product; quality is either there or not there as a consequence of the manufacturing process. Validation is the means by which you insure that the quality *is* there.

4. REGULATORY AFFAIRS

An important aspect of the manufacture of pharmaceuticals, including those derived from cell culture, is the conduct of the company's relationship with the Food and Drug Administration. In established drug companies, this function is generally carried out by a staff of regulatory professionals who typically have either a legal or scientific background. Depending on the particular interests of the company, the responsibilities of this group might include FDA contacts: good laboratory practices (GLP), good clinical practices (GCP), and Environmental Protection Agency (EPA) audits within the corporation; management of clinical trials; and the review of labelling and advertising.

In smaller companies, the regulatory function is often more modest in scope and may involve only one or two people. In some instances, particularly when the company is relatively new, it may be limited to handling contacts with the FDA, and delegated to a fairly junior level in the organization. This might suffice for a time, but if the company is successful it may soon become necessary to put in place a more sophisticated regulatory affairs function.

5. CONCLUSIONS

The manufacture of pharmaceutical products by cell culture involves the use of complex biological processes and subtle biochemical operations. That these processes must be carried out in compliance with the FDA's current Good Manufacturing Practices, and in accordance with the drug industry's high standards for process validation, makes the task even more challenging.

In this chapter, the authors have attempted to provide examples and guidance as to how this might be done. In the case of cell culture, a perfusion culture system is described which has been used successfully to manufacture safe and efficacious biomolecules on an industrial scale. Principles of protein purification typical to the manufacture of cell-derived products are reviewed with emphasis on those factors which are critical to product safety. The importance of quality control was discussed; in particular, its role as a means of establishing and maintaining high operational standards for the facility and the process.

REFERENCES

1. Clark, J. (1983). Microcarrier-bound mammalian cells. *In* "Immobilized Cells and Organelles",Vol. 1 (Mattiasson, B., ed.), pp. 57–88. CRC Press, Boca Raton, Florida.
2. Feder, J., and Tolbert, W. R. (eds.) (1985). "Large Scale Mammalian Cell Culture", Academic Press, New York.
3. Fink, D. J., Curran, L. M., and Allen, B. R. (eds.) (1985). "Research Needs in Non-Conventional Bioprocesses", pp. 67–99. Battelle Press, Columbus, OH.
4. Hopkinson, J. (1983). Hollow fiber cell culture: applications in industry. *In* "Immobilized Cells and Organelles", Vol. 1 (Mattiasson, B., ed.), pp. 89–99. CRC Press, Boca Raton, Florida.
5. Thilly, W. G. (ed.) (1986). "Mammalian Cell Technology", p. 223. Butterworth, Stoneham, MA.
6. Tolbert, W. R., and Feder, J. (1983). Large scale cell culture technology. *In* "Annual Reports on Fermentation Processes" (G. T. Tsao, M. C. Flickinger, and R. K. Finn, eds.), pp. 35–74. Academic Press, New York.
7. van Wezel, A. L. (1967). Growth of cells strains and primary cells on microcarriers in homogeneous culture. *Nature (London)* **216**, 64.

PART IV

DOWNSTREAM PROCESSING

18

Downstream Processing of Animal Cell Culture Products — Recent Developments

ALAN ROSEVEAR and CHRISTOPHER LAMBE
Biotechnology Group, Harwell Labs,
Didcot, U.K.

1. INTRODUCTION

Downstream processing encompasses all those operations required to recover a material from the production vessel and present it to the end-user in an acceptable form. Except in those rare cases in which the biomass or the culture fluid itself can be used directly, the costs of recovering and purifying a culture product will generally make a significant contribution to the overall production cost. This is particularly true for the new, high-purity products of mammalian cell culture, for which downstream stages may represent an even higher proportion of operating costs than the 67% reported for purification of microbial enzymes (*87*).

Product recovery is the last stage of production: it is often also the last stage to be considered during the move to full-scale operation. This status is understandable, since the actual generation of product is the *sine qua non* of the whole process and so demands priority attention. Furthermore, with a totally new product, especially one which commands a high price, enormous increases in cellular yield often far outweigh the differences arising from optimizing primitive recovery steps.

Many animal cell products may still be in this early stage of development. The volume of material is often small compared to more traditional biotechnological processes. Therefore almost any purification technique can still be employed, irrespective of cost or complexity. Products may also become outdated more rapidly than in more well-established areas, so that getting something to work quickly is more important than achieving the highest theoretical yield.

However, a few mammalian cell culture products are entering a more mature phase. The new uses of monoclonal antibodies in therapy, imaging and processing, mean that markets of over $1 billion per year are projected by 1990 (*63*). As the scale of production increases, the cost of processing will make a more significant impact on final cost (*40*) while, without improvements in yield, the total value of product lost during purification will rise substantially. At the same time purification protocols become fixed by registration with regulatory authorities, so that any future modifications will require expensive retesting and re-registration. Consequently, it is becoming more important to get the downstream processing "right" at the development stage. In a mature business area, reaching the market first becomes less important, as cellular yield begins to approach an upper limit which can be achieved by several competitors. It is here that the difference between commercial success and failure will depend markedly on the quality of the downstream processing.

The main aim of this chapter is to review the recovery of products from mammalian cell cultures. Nevertheless, some reference will be made to those methods which are already used in the purification of material derived from whole animal tissue, since this technology provides valuable pointers to the scaling-up of the methods required to prepare cell-culture products. In addition, limited mention will be made of those mammalian cell products which have been produced by genetically engineered microbial cells. In many respects these processes only differ in the primary solid/liquid separation stage and in the quantity rather than the kind of contaminants present in the process fluid. It should also be recalled that the culture fluids will contain animal serum or compounds such as insulin and transferrin, derived from animals, so that the purification requirements will be similar to those already used on a large scale in fractionation of human blood serum. Finally, the cell lines which are used to produce large amounts of protein are generally "abnormal" and so their products must be freed of all genetic material in the same way as microbial fusion products.

The subject of downstream processing of animal cell products was reviewed by Cartwright in an earlier volume of this series (*96*). It is a mark of the industrial interest in this subject that after only two years a further review is necessary. The basic principles of main purification processes will not be dealt with in detail and the reader is referred to the book by Scopes (*91*) for a fundamental description of the methods involved, particularly in purification of proteins.

2. PURIFICATION

Downstream operations are required to purify the product from contaminants which might impair its function or acceptability and to increase the specific activity, so that an effective dose can be delivered in a convenient quantity of material. It must be remembered that no new product is generated in downstream operations; all we do is add costs through processing and cut overall yield by product loss and deactivation.

Table I illustrates that large quantities of contaminants must be removed in order to purify to homogeneity important products such as interleukins. This can be achieved in several steps but at the cost of loss of well over half the product originally synthesized in the culture. Thus, the processor must be constantly aware of how much purification is actually needed to satisfy the end user, and to what extent this purity justifies expensive, multistep purification. If the product is to be used for diagnostic and analytical purposes or as a

TABLE I **Purification of Biologically Active Compounds from Culture Fluid: Interleukins 1, 2, and 3** (*25, 73, and 45*)

	Protein (mg)	Purification (×)	Cumulative yield (%)
Interleukin 1			
Start	31 800	1	100
Concentration by Amicon YM10 ultrafilter } Gel permeation on Ultragel ACA54 }		10	44
Immunosorption	0.25	3 500	18
Interleukin 2			
Start	15	1	100
Concentration on Pellicon and diafilter }		1.4	54
Gel permeation on Sephadex G100 }		5.6	54
Phenylsepharose			39
DEAE-Sephacell		11.1	39
Reverse phase HPLC	0.05	84.2	27
Interleukin 3			
Start	72 000	1	100
50–80% ammonium sulphate		1.5	63
DEAE-cellulose		180	40
Hydroxyapatite		419	20
Gel permeation Sephadex G-75		6 792	16
Reverse phase HPLC	0.003	1 800 000	8

process aid, the unique biological function of the product may reduce the need for absolute purity. For instance, many immunologically based assay kits will work satisfactorily with ascites fluid which contains many contaminating proteins. However, for therapeutic use, a product must be free of antigenic contaminants and genetic material if it is to be widely accepted. Food and Drug Administration (FDA) guidelines recommend that for mammalian cell products destined for use *in vivo*, there should be less than 35 ng of endotoxin and less than 10 pg of DNA in each dose received by a patient (*31*). The market for such injectable products is growing rapidly and, although it places extra demands on the purification process, it also offers greater financial rewards than the diagnostics area.

For the purpose of this review, downstream processing is taken to include all aspects of product recovery, whether in the culture vessel itself or after the process fluids have left this vessel. The use of this term also implies that significant quantities of the product are to be isolated at a cost which the end-user can bear. In the case of animal cell products, this might not be on the large scale of the fermentation industry. The high value and potency of mammalian products means that often no more than a few grams of product are needed to satisfy an economically viable market. For example, a single dose of a viral antigen vaccine may only be fractions of a milligram, so that as little as 10–100 g of product could supply a million doses. This contrasts with typical antibiotics, where almost a ton of product would be required to provide the same number of doses. Affinity-purification columns of only 500 ml capacity of Protein A Sepharose are currently quite adequate for the production of monoclonal antibodies, for instance. However, production for sale does impose the requirement of reliability and reproducibility on the methods selected, so that quality and yield can be guaranteed. Furthermore, the same techniques will be required on a larger scale if the predicted needs for kilogram amounts of product for therapy (*8*) are to be met.

3. AN OVERVIEW OF DOWNSTREAM OPERATIONS

During downstream processing the product must be freed of the major contaminants associated with its production. At the same time, further contamination during the purification operations has to be minimized.

The many components of the culture medium and the cell metabolites released by cells are the important gross impurities from which the product must be separated. In addition, there may be undetected viral, mycoplasmal, fungal, or bacterial infections present in the process stream. Process aids such as antifoaming agents and antibiotics might also have been added to assist in the earlier stages of production. During downstream processing, the fluid may leach compounds from process aids such as adsorbent resins and membranes and from equipment surfaces. Even if the product itself can be kept sterile, pyrogenic materials such as microbial cell wall may be introduced in the water supply or through build-up of a biological load on adsorbents. Finally, the product itself may be degraded to give compounds which impair activity but are physically similar to the original product, thus making purification particularly difficult. For instance, in the recovery of insulin from pancreas, several degraded forms become of concern in determining the acceptability of the product for injection. Careful analysis of immunglobulins eluted from Protein A columns has shown that, even with such a specific adsorbent, breakdown products generated at low pH or by prolonged processing are co-purified with the monoclonal antibody (*65*).

The sequence of downstream processing should take account of these factors. For instance, the process fluid is usually concentrated first and sub-fractionated later. This reduces the volume of liquid which must be stored and minimizes problems if there is a hold-up in the batch operations downstream. Since the quantity of undesirable material leached from adsorbents is probably related to the volume of liquid passing through them, pre-concentration has the further advantage of reducing the total amount of contaminant in the end product. More importantly, concentration of leachate-contaminated

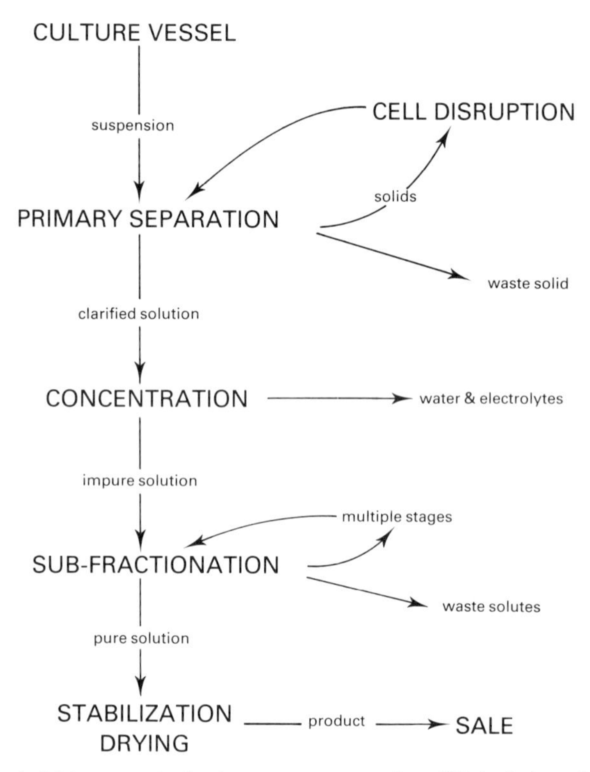

Fig. 1. Major stages in the downstream processing of biological products.

product late in the process might lead to even a low level of unwanted material reaching unacceptable levels in the final product.

During the concentration step, the membrane acts as a subsidiary barrier, retaining infective cells, so that it might reduce subsequent spoilage of product. This is a further point favouring concentration before fractionation. A final advantage of this approach is that the concentration of product during membrane processing can improve the performance of adsorbents used downstream, since they will be operating at higher points on the adsorption isotherm.

In these respects, the recovery of mammalian cell products follows a similar pattern to that for other biologicals and can be considered as four sequential stages (Fig. 1). It is first necessary to separate the cells and cell debris from the process liquors. This primary separation, or clarification, usually requires the removal of a discontinuous solid phase from a continuous aqueous phase.

In those few instances in which the product is held in the cells, this step serves to concentrate the material and the technologist must then extract or leach a solute from this solid residue, or lyse the cells to release an intracellular product. Where the material is adsorbed on the cell or is a part of the cell structure (e.g. a viral antigen) it can be used directly as a product. Alternatively, it can be solubilized with a detergent or other solvent. Thus, irrespective of the intermediate steps, the process fluid going forward from this stage for further purification, is a cell-free solution or stabilized colloid (e.g. virus particles).

The more usual situation is for the mammalian cell product to be found in the original clarified fluid. The active material is generally present in a dilute solution, so a concentration stage normally follows primary separation. The impure concentrate is then passed on for subfractionation before the pure solution is stabilized in a form which can be formulated for sale.

Each of these stages may involve more than one unit operation, particularly at the subfractionation step. Even when yields are high at each individual stage, overall recoveries from multistage processing are inevitably lower. Consequently, there are commercial advantages in recovering the product directly from the culture broth by one- or two-stage processes. This approach has been used successfully for microorganisms at production scale (*6*). Products such as ethanol can be selectively taken into the gas phase (*86*) and for plant cell cultures and microbial fermentations, secondary metabolites have been adsorbed into immiscible liquids or onto solid adsorbents (*34*).

However, animal cell products provide a particularly difficult challenge for this approach. Firstly, the product is at very low concentration so that even adsorbents with a high binding constant are working on an inefficient portion of the isotherm and capacities are lower than normal. Secondly, the proteinaceous product is contaminated with a large number of similar compounds so that very high specificity and low levels of non-specific binding are required. When highly specific immunosorbents have been used, it has been found that the low solute concentration has led to significant non-specific binding to adsorbents (*88*). Finally, the system must be operated aseptically even when recycle of the spent medium is not required.

Despite these difficulties, Nilsson *et al.* (*71*) have successfully used an immobilized Protein A column in an extracorporeal shunt to remove unwanted antibodies circulating in the blood of human patients.

4. UNIQUE FEATURES OF PROCESSING ANIMAL CELL PRODUCTS

Although animal cells make products with many unique biological properties, at the molecular level they differ only in detail rather than in kind from other biological materials. Thus, downstream processing of mammalian products has not created totally new types of separation techniques but has

TABLE II Unique Features of Animal Cell-culture Products

Factors relating to source of product	
Mammalian cells	are large;
	are easily broken;
	release proteases;
	are possible hosts for abnormal genes.
Culture medium	is nutrient rich;
	contains molecules with wide range of sizes;
	contains additives (antibiotics, polyols).
The culture is	usually at low cell density;
	slow in product formation.
The product is	generally extracellular;
	high-molecular-weight;
	mainly proteinaceous.
Factors relating to end-use	
The product is	very biologically active;
	immunologically active;
	often glycosylated;
	labile.
The process may need to guarantee that	no non-specific immunological response occurs;
	no gene sequences are present (DNA <10 pg per dose);
	no pyrogens are present (<35ng/dose).

required the separation scientist to put a different emphasis on particular methods and has placed special restraints on the use of familiar techniques.

Table II summarizes those features which make the recovery of materials from mammalian cells different from those routinely found in the more well-established area of microbial products. These features can be divided into characteristics which result from the nature of the cells and ones which arise from the use to which the product is finally put.

4.1. Differences in the Cells

Animal cells are much larger than microbial cells and lack the robust extracellular wall present in yeast and plant cells. This makes cell breakage extremely easy: an advantage if one wishes to release an intracellular product, but a serious problem in the more usual situation in which the product is secreted into the medium. The cell debris is more difficult to sediment and may be an important source of fouling in subsequent membrane processes. In addition, compounds released from within the cell during primary separation will increase unnecessarily the degree to which the desired product is contaminated. Furthermore, any lysosomal enzymes released during cell death can cause rapid degradation and loss of the product.

Mammalian cells require a much richer nutrient medium than prokaryotic cells (*40*). Thus, the product stream not only contains complex biological molecules and waste products synthesized by the cells but may also include a large number of proteins, peptides, glycoproteins, carbohydrates, lipids, and organic acids which must be added to the basic salts media for successful growth of the cells. A number of these culture components have similar physicochemical characteristics to the products, greatly increasing the demands placed on the sub-fractionation stages. The move away from serum-containing media is as much to simplify the problems of product recovery as to reduce the industry's reliance on costly foetal calf serum. A more expensive serum-free medium may be preferred over a more traditional complete medium if the former avoids the need for extra sub-fractionation stages. For instance, the recovery and purity of recombinant tissue plasminogen activator is substantially higher when the product is recovered from a serum-free medium rather than from one containing 10% calf serum (*17*).

Nevertheless, it should be noted that "serum-free" does not mean "free of macromolecular components". Such media still contain proteins and peptides as well as polymeric additives. The fouling of membranes caused by antifoams added to microbial fermentations should be taken as an indication of how the solving of problems upstream can complicate subsequent processing. In this respect, hydrophobic polymers added as albumin substitutes in mammalian cultures, could cause similar problems by adsorbing to plastic surfaces during downstream processing.

The general principle of avoiding problems downstream by taking appropriate action upstream should always be considered (*17*). For instance, if the later stages of the culture can be sustained by perfusion with a simple salts medium (*100*), great benefits can result downstream. In this regard, immobilized cell systems are at a particular advantage and are given special consideration in Section 5.1.2.

For those familiar with the fermentation of microorganisms, the low density of cells in a mammalian culture is very striking (typically up to 10^6 cells ml^{-1} compared to 10^{10} cells ml^{-1} for fermentation). In addition, the product concentration is also low, being typically fractions of a gram per litre for monoclonal antibodies (*8*), compared to many grams of antibiotic per litre from a microbial fermentation. Thus a concentration or de-watering step assumes even greater importance for culture products than in simple fermentation processes. Fortunately, mammalian cell products are generally of high molecular weight, reducing the demands placed on membrane-based systems to discriminate between product and major impurities in the bulk solution.

4.2. Differences in End-use

In downstream processing of animal cell culture products, one is more likely to be concerned with very large proteins and proteins which have been glycosylated or have lipid substituents. These features are important in the recognition by the host immune system of compounds which are injected into patients. Even small differences compared with the natural product can provoke an adverse immune response. In the case of vaccines, cell wall constituents exhibiting antigenic determinants in the correct configuration are an important product group. As a result of these factors, purification of animal cell products makes greater use of the more sophisticated forms of sub-fractionation than is normal for biotechnology as a whole. Consequently, processes thought of only as laboratory curiosities for microbial products are worthy of consideration for the sophisticated needs of the mammalian cell product. Affinity-type separations based on the use of macromolecular ligands such as Protein A, concanavalin A and immunoglobulin (itself a major animal cell product) are of great importance in this sector. The interaction of solutes with these ligands is not only very specific but also extremely tight with dissociation constants ranging from 10^{-4} M for lectins to 10^{-12} M for certain monoclonal antibodies (*62*). Thus, particular attention must be paid to inexpensive methods for the desorption of products under conditions which do not deactivate the ligand or the product.

A common feature of many mammalian products is that they possess a very high biological activity. This is often the main justification for incurring the substantial cost of using such a delicate eukaryotic cell. Consequently, protection of operators from accidental contact with the concentrated product in the later stages of purification is essential. This restriction is further compounded by the fact that mammalian extracts, particularly those acting on an immune response, may have a more direct effect on humans than do fermentation products. Where the cells have been virally transformed to facilitate cell culture, or have been specifically infected to produce vaccines, there are further difficulties owing to the risk of infective agents remaining in the process fluid. The concern over viral infection and the transfer of oncogenes places further demands on therapeutic products in which the total absence of polynucleotides may be a requirement for general use. Close monitoring of the working environment and the product places great emphasis on powerful analytical techniques during all stages of product recovery.

A large proportion of products such as vaccines, interferons and the higher-value monoclonals are destined for therapeutic use. They must therefore be processed under conditions which prevent the build-up of pyrogens and which must certainly give a sterile end-product. Totally aseptic operation is difficult to achieve, but hygienic operation and the regular sterilization of equipment with reagents acceptable to regulatory authorities is essential. Running immunosorbent columns in a clean-room environment may have a further advantage in extending the operational life-time of the adsorbent (*46*).

Wherever possible, metal junctions should be smoothly welded and dead spaces in which liquid may remain stagnant should be avoided. Steam is the preferred sterilization agent, but not all components will tolerate such high temperatures. Treating equipment with 0.5 M caustic soda solution will kill many vegetative microbes, bacterial spores, and viral contaminants in a matter of hours at room temperature (*81*). Furthermore, it will remove lipids, though these might also be removed from equipment by raising the temperature to 40–60°C.

5. INDIVIDUAL UNIT OPERATIONS

A limited number of unit operations have become established for the separation of biological materials (*10*). These will be dealt with individually, following the general sequence outlined in Fig. 1. Special attention will be given to recent developments and new perspectives which have been reported since this topic was reviewed in Volume 2 of this series (*96*). In Section 6, the way in which these novel techniques are linked to more traditional methods will be exemplified by considering the total separation process actually used for recovery of a number of important products.

5.1. Primary Separation

Animal cell products can be generated in one of two very different types of culture system. The older of the techniques is restricted to cells such as fibroblasts which grow while attached to a surface. The more recent technology uses selected cells, often transformed in some way, which are capable of growth while suspended in the culture medium (see Vol. 1 of this series (*95*)).

Cells grown in suspension culture present similar problems to those found in typical fermentation processes. The main requirement is to remove a finely dispersed, near-neutral buoyancy solid from a larger volume of aqueous solution. However, cells grown while attached to surfaces are an important feature of animal cell culture, particularly for vaccine production and for some of the newer genetically engineered cells. In many respects, these cultures can be regarded as immobilized cell systems, a technology which is only recently being considered as a novel approach by the larger fermentation industry. In principle, retention of the cells results directly in a clarified process stream. Thus, although immobilization is normally considered an upstream activity, it has such a profound effect on downstream processing that it will be considered briefly in this section.

No special consideration will be given to methods of cell disruption. Although this is an important unit operation in the recovery of intracellular products from microbial cells, the sensitivity of mammalian cells to mechanical shear means that cell disruption is rarely a problem for mammalian culture products. The problems are more frequently those of preventing cell breakdown when it is not wanted.

5.1.1. Clarification

The most important methods for the primary separation of biological products are summarized in Table III. Although several of these might be suitable for separating animal cells and cell debris, only sedimentation and microfiltration have been exploited to a significant extent. The applications of these methods to animal cell products were reviewed in Vol. 2 of this series (*96*).

Techniques for debris removal are of greatest interest in this area of product recovery. Disc-stack centrifuges are preferred when a large quantity of solids must be removed and continuous operation is desirable. Tubular bowl centrifuges achieve a higher g-force and give better de-watering but have a limited capacity and operate discontinuously. Both types of machinery can be adapted or contained so as to operate in a sterile manner without generating aerosols.

TABLE III Unit Operations for Clarification of Biological Fluids

Operation	Comments
Pressure-driven	
Dead-end filtration	Rapid blockage of filters
Cross-flow microfiltration	Increasingly important particularly for intermediate scale
Gravity-driven	
Centrifugation	Traditional method of choice for very small and very large scale
Flocculation	Useful if polymers acceptable
Unit-gravity sedimentation	Good for isolating intact cells but little suitable equipment
Others	
Flotation	Proteins denatured at interface
Dielectrophoresis	Potentially mild—no suitable equipment
Two-phase aqueous	Carry over of polymers Unpredictable
Magnetic methods	Promising for clarification or *in situ* recovery

The performance of most clarification techniques is improved by the addition of flocculants. Since the majority of high-value mammalian products could be destined for therapeutic applications, the use of general flocculants such as polyacrylates is unacceptable owing to concern over toxicity. Hence, the use of natural materials such as sulphonated polysaccharides (*20, 29*) or polyuronic acids (*53*) is of particular interest for removing cells and debris from culture media. Flocculation will aggregate debris, thus increasing the particle size and density of the precipitate, and facilitating sedimentation and recovery of the solid by centrifugation. It may have further, less apparent, benefits in removing other materials which may promote fouling downstream.

Although the newer animal cell-culture products are generally soluble, there is also a need to recover intact cells. The methods available are restricted by the risk of cell breakage resulting from high shear forces. The relatively large size of mammalian cells makes sedimentation under normal gravity feasible, especially if this is done in a specially designed vessel in which sedimentation path lengths are short. Unit-gravity separators are likely to be of increasing interest in this area. Animal cells needed for vaccine production have been isolated in a device in which a series of closely spaced, slightly inclined trays are connected to a central harvesting point (*106*). This facilitates rapid collection of the solids from a large volume of culture fluid, over a period of hours.

An alternative approach is the elutriation of a cell suspension in a conical chamber (*1*). This operates in a similar manner to a fluidized bed, with the cells sedimenting into zones, against the upward flow of media. It has been used to separate tumour from smaller lymphoid cells or thymic nurse cells from thymocytes in a thymus digest.

Microfiltration is being used increasingly as an alternative to centrifugation, especially for the smaller scale of production common with culture products. The equipment is less costly and less energy-demanding than centrifugation. Furthermore, it is easier to employ diafiltration (washing the solid in place with fresh liquid), while shear may not be so severe as with the present design of centrifuges. However, microfiltration is, like all membrane methods, subject to fouling, and as the flux falls process times may increase unacceptably. To an extent, cross-flow filtration reduces the build-up of debris on the active face of the membrane by shearing the surface at a flow rate ten times that of the transmembrane flux. However, adsorption still causes some fouling, necessitating regular backwashing and cleaning of the membranes.

5.1.2. Immobilized Animal Cells

Retaining cells within the production vessel is a very effective way of obtaining a clarified culture fluid. The main methods for immobilizing animal cells have been reviewed recently (*102*). The most important factors which impinge on the downstream processing of the arising products are that

1. The product is always dissolved in a cell-free solution.
2. The process may be continuous or involve some form of cell perfusion to remove products from live but non-dividing cells. Aseptic recycle of this partially depleted bulk fluid would be an advantage, since it reduces the volume of fresh, expensive medium required by the culture.
3. The product is often more concentrated and contains a lower level of contaminating nutrients as a result of the higher process intensity and possibility of using maintenance medium rather than growth medium.

Although it does not conform to all the features of a classical immobilization system, the Damon Encapcel system illustrates many of the advantages that can result from holding cells at high packing density within an immobilization matrix (*26*). Typically, this approach is used to culture hybridoma cells in simple tank reactors of up to 40-litre capacity. It is claimed that these produce 5–20 g of antibody every 2–3 weeks, equivalent to the output of a 500-litre vessel using a suspension culture of cells. In this process, an inoculum of animal cells is entrapped within small semi-permeable capsules made by casting a polymer shell around a calcium alginate gel bead containing cells. The preliminary gel is then dissolved and the cells are grown on as a high-density culture within the capsule. Careful choice of the membrane porosity ensures that key nutrients such as insulin and transferrin move freely in from the surrounding medium. Likewise, waste products such as lactate and proteases diffuse out freely through the capsular wall.

However, larger molecules such as immunoglobulins are retained and concentrated in the beads during the culture. The capsules are robust and settle easily so that harvesting the solid phase by sedimentation is rapid and simple. Furthermore, the capsules can be washed free of any residual nutrient solution before they are broken open to release the cells and the concentrated product. The resulting solution contains up to 10 g l^{-1} of immunoglobulin with an initial purity of 80% and can be purified to an acceptable quality by ammonium sulphate precipitation and ion-exchange. This avoids the losses normally associated with the use of highly specific immunosorbents (*26*). This contrasts with a normal suspension culture fluid, which, following clarification, contains 0.01–0.5 g l^{-1} of monoclonal with a purity of only 10% (*72*) and requires pre-concentration before being sub-fractionated in a multistep process.

Animal cells can be immobilized at high packing density in stable, three-dimensional gels such as agarose and other marine polysaccharides (*58*) or in weighted, collagen microbeads (*50*). This facilitates perfusion with maintenance medium to maximize cell productivity. Alternatively, hybridoma cells have been immobilized in membrane reactors using hollow-fibre ultrafiltration units or sheets of semi-permeable polymer to separate compartments containing the cells from those through which fresh nutrient solution can rapidly perfuse. Though differing in detail, the MBR and Endotronics systems exploit this approach (*102*). As with the capsules, an immunoglobulin product accumulates on the cell side of the membrane. However, in this system it can be recovered by washing the cell side at regular intervals, flushing out the product but retaining some of the cells which remain caught in the macroporous matrix of the membrane.

The systems described so far in this section are designed to deal with cells which normally grow in suspension. All can probably be used just as well for attached cells, but this level of complexity is often unnecessary for cells which will spontaneously bind to a solid surface. For many years, vaccine production depended exclusively on attached cells such as fibroblasts. Initially these were grown on the walls of vessels so that medium could be changed by carefully pouring the liquid out of the flask or roller bottle. Various methods have been suggested for increasing the surface-to-volume ratio of these cultures, but the most successful has been the microcarrier concept, pioneered by the late Toon van Wezel (*95*). This has prompted a number of manufacturers to supply specially prepared beads such as Cytodex to which fibroblasts will adhere strongly. Although the specific gravity of these particles must not be too high, in order to maintain them in suspension during culture, they do sediment over a period of minutes under normal gravity once agitation is stopped. Fluid can then be drawn easily from the vessel, though it has proved possible to perfuse these cultures through a rotating steel-mesh cage fixed to the impeller (*95*). Such an arrangement gives a cell-free solution directly and is obviously a convenient method for rapid isolation of extracellular products. However, for production of lytic viruses such as polio or foot-and-mouth disease, some release of debris is to be expected.

5.2. Concentration

The commonly used methods for the concentration of culture products are precipitation, batch adsorption and ultrafiltration. These are each reviewed in Vol. 2 of this series (*96*).

Precipitation is the simplest means of concentration and has been used routinely in the isolation of monoclonal antibodies at the pre-production stage. Ammonium sulphate is the most acceptable salt for precipitation and concentrations of 50% give adequate recovery of immunoglobulin and achieve some degree of purification at the same time (*74*), though this sometimes leads to some loss of activity. Unlike polymeric precipitants, the salts are easily removed from the high-molecular-weight product in later stages of processing. Nevertheless, the technique is not well suited to large-scale sterile operation and ultrafiltration is now used regularly as the preferred method of concentration (*42, 25*).

The semi-permeable membranes used in ultrafiltration offer a rather more sophisticated means of concentrating the dilute solution arising from primary separation. Water can be considered the major bulk contaminant of the product at this stage, and its removal greatly reduces the size of equipment required in the later stages of downstream processing. Reverse osmosis (RO) membranes, which reject almost all solutes, only result in solute concentration. However, since the majority of animal cell products are of high molecular weight, it is possible to use much more open ultrafiltration membranes. These will permit some of the low/medium-weight contaminants to pass through with the water and will operate at lower pressures than RO systems. In principle, the range of membranes available should make some degree of sub-fractionation possible at this stage. However, the molecular size cut-offs stated by the manufacturers are not absolute and it is not practical to separate molecules which have a size difference of less than five-fold. Molecules much larger than those stated may bleed into the permeate, while smaller molecules may be retained by the concentration polarization layer of proteinaceous gel which rapidly builds up on the active surface of the membrane. Thus, it is normal to choose a membrane with a nominal cut-off of only 10 000 Daltons to ensure quantitative retention of product at an acceptable operating pressure. The larger the product, the more effectively is it retained, so that viral particles can be very effectively concentrated in production-scale devices such as Amicon and Romicon hollow-fibre cartridges.

Cross-flow microfiltration has been employed to concentrate cell suspensions and a number of commercially designed modules are available for small-scale work. Hollow-fibre microfilters appear to be preferable to plate-and-frame membrane units for concentrating large volumes of culture fluids from which intact cells are to be recovered. Following experience in concentrating 64L cultures of rat cells, Shiloach *et al.* (*92*), recommended that a number of short hollow-fibre modules be operated in parallel to give a small axial dimension. This minimizes the pressure drop and hence the shear to which cells are subject, thus avoiding undue cell breakage.

Dynamic microfiltration offers an alternative approach for concentrating delicate mammalian cell suspensions. The feed suspension is introduced into an annulus, bounded by two filter surfaces. Rotation of the inner surface generates Taylorian vortices which keep the active surface of the filter free of solids. In contrast to cross-flow filtration, the residence time of the suspension in the device is independent of the tangential flow, so that multiple passes are unnecessary, while the ordered nature of the vortices is said to reduce shear (*39*). Hybridoma and cancer cells have been concentrated up to nine-fold at a pressure of 0.2 bar in such a device with no damage to the cells. This makes it possible to consider the recycling of concentrated biomass to increase residence time or cell density in a normal suspension culture, effectively giving many of the benefits associated with immobilization.

The possible problems of membrane fouling have already been referred to with respect to cell debris and polymer additives. The soluble proteins themselves may also lead to problems as the concentration in the unit increases and proteins aggregate or precipitate. Furthermore, protein will absorb on to the surfaces in the membrane module. Albumin in particular has been shown to bind to polysulphone membranes, but less adsorption occurs on the more hydrophobic materials such as cellulose acetate (*67*). In the case of certain blood proteins, the surface may also activate clotting reactions. The combined effect of these is to increase back-pressure in the system to an extent that further operation eventually becomes impossible. This not only results in loss of material from the system, but increases process times and necessitates more frequent cleaning and sterilization of closed parts of the system.

Since only the smaller contaminants are removed by membranes, the product may still contain undesirable components such as proteolytic enzymes or infective particles and so must be handled with care. The inner surfaces of a membrane system are a potential site for microbial growth and pyrogen production. Although the newer membranes such as polysulphone and coated ceramics are stable to cleaning with caustic soda and steam sterilization, some of the components of the equipment may not be so robust. Nevertheless, this requirement is now recognized by the manufacturers and sterilizable units are now available from most suppliers.

5.3. Sub-fractionation

This is probably the most challenging topic in downstream processing and is the area in which the greatest progress has occurred since it was reviewed in Vol. 2 (*96*). The high value of animal cell products means that a wide range of the newer separation techniques can be considered. Several of these (e.g. affinity separations) would only be appropriate as laboratory-scale methods in the context of bulk microbial products but are already used as the preferred purification option for processing the multigram quantities of mammalian product currently required in specific market sectors.

These techniques which have been used to purify cell culture products are summarized in Table IV along with the molecular characteristic which forms the basis of the separation. Chromatographic and electrophoretic methods are the most widely used general techniques, and each of these is reviewed in more detail below.

TABLE IV Main Methods for Sub-fractionation of Proteins

Basis of separation	Techniques	Comment
Size	Gel permeation	Dilutes
	Ultrafiltration/microfiltration	Fouls
Density	Sedimentation/centrifugation	Low purification
Charge	Electrophoresis	Dilutes
	Ion-exchange	May dilute
Non-polar residues	Hydrophobic chromatography	Dilutes
	Non-specific adsorbents	
Several spatially-fixed sites	Dye ligand chromatography	Ligand leakage
	Simple affinity chromatography	
Bioaffinity	Concanavalin-A chromatography	Not sterilizable
	Protein-A chromatography	
Immunoaffinity	Antibodies	Not sterilizable
	Antigens	Not sterilizable

5.3.1. Adsorption and Chromatography

This group of techniques depends on the selective adsorption and elution of solutes from a second phase. The chromatographic mode of operation is generally used when the main objective is to sub-fractionate a number of similar solutes. In purification of mammalian cell products, this usually involves solid particulate stationary phase packed in a column through which a number of aqueous solutions are passed sequentially. This approach differs from the batch adsorption methods employed for product concentration in that it presents the developing front of solute with multiple opportunities to interact with fresh adsorbent. In such a plug-flow arrangement, the complete adsorption of selected components is possible, so that high levels of purification can be achieved. This contrasts with batch adsorption, in which a single equilibration will never achieve absolute removal. Nevertheless, if binding constants and selectivity are high, batch adsorption may be practicable, although elution is still better done in a packed bed in order to elute products as distinct fractions in the minimum volume (*66*, *42*).

5.3.1.1. Liquid-phase Systems. Liquid-phase adsorbents have limited applicability in the separation of mammalian cell products, since most water-immiscible liquids, such as organic solvents, denature these biopolymers. Two biocompatible aqueous phases can form when soluble but mildly hydrophobic polymers such as polyethylene glycol (PEG) are mixed with salts or hydrophylic polymers such as dextran. The selectivity of adsorption is often poor and in its simplest form the technique is usually limited to removal of cell debris. When it is used for fractionation, the selective partitioning of cells between the two phases is related to net surface charge. This leads to cells in different phases of the cell cycle showing different partitioning characteristics and to different types of particles being distinguished, so that viruses can be concentrated 100-fold in a pure form using dextran sulphate/PEG/salt systems (*104*).

The partition coefficient of a selected compound into the unfavoured PEG phase can be increased by attaching an affinity ligand such as a lectin, antibody, or dye to the polymer (*62, 104*). For instance, the partitioning of albumin is improved by covalently attaching a blue dye ligand to the polyether, while the affinity of erythrocytes has been improved by using PEG-antibody conjugates (*51, 93*). In this latter case, a countercurrent distribution with 20 transfers gave adequate separation of two components. Nevertheless, this example also illustrates how residual polymer may interfere with subsequent downstream processes as the soluble conjugate did not bind to DEAE ion-exchangers during its purification.

Animal cell products are generally larger than microbial products, so that accessibility to binding sites deep in the resin can only be assured with the most porous types of matrix. Furthermore, desorption may be so slow that the flow rate through the bed may have to be extremely slow to prevent tailing of the product peak. In this respect, free diffusion in liquid/liquid systems has great advantage, though its exploitation is limited by problems in getting several ligands onto each polymer molecule and obtaining high distribution coefficients (*62*).

5.3.1.2. Process Considerations. The separating power of adsorbents is restricted by non-specific adsorption of other components which alter the binding characteristics of the product, interfere with the reversibility of the interaction, and may also leach into the product fraction. When the proportion of product in solution is small and the ionic strength of the solution low, non-specific binding appears to be at its most serious (*88*), and may negate any specificity arising from the intended interaction. Non-specific binding may occur on ion-exchange sites introduced during coupling procedures (*46, 105*) or may be due to unfortunate interactions with particular groups on the specific ligand or on other adsorbed molecules. The impact of non-specifically adsorbed material may be reduced by washing the resin before elution, but some carry-over of contaminants may still occur.

Another important general problem with adsorbents is the leaching of ligand. This not only reduces capacity, but also contaminates the product. It probably takes place in most materials but is most apparent where the ligand is coloured (e.g. blue dye). It is of most concern where the technique is being used in the latter stages of purification, when there are fewer opportunities to remove the contaminant. The amount of ligand lost may be small (<10 ng ml^{-1}

of bed/run for Affigel immunosorbents (*75*)) and may not affect the capacity of the columns, some of which are reported to give unchanged performance over many hundred adsorption cycles and a year's use (*4*). However, some users have noted a slow change in elution patterns with repeated use (*54*).

Some of the more important ligands such as Protein A and protease inhibitors could be dangerous contaminants in the product, and so their elimination is essential. Removal of free ligand is particularly difficult with affinity ligands which remain tightly bound to the soluble complement. Wherever possible, stable ether or amide bonds should be used in preference to esters and carbamates (*75, 105*), though this does not ensure ligand stability if the resin itself deteriorates. A complementary ligand to that which has leached may be used to polish a product, but this itself may introduce contamination and will lead to some product loss. A simple, general method such as ion-exchange purification has been recommended for removing residual Protein A ligand from therapeutic products (*54*). Though this may also remove other general contaminants such as endotoxins, the residual salts may have to be removed later.

The main chromatographic techniques will be reviewed below in order of increasing power of separation. However, this is also the order of increasing cost and decreasing stability to severe conditions. Unfortunately, the more sophisticated methods use proteinaceous ligands which are deactivated if sterilized by steam or chemicals, so their use for purification of therapeutic products is limited. Nevertheless, immobilized ligands are more stable than the free protein, so that a 2% aqueous solution of propylene oxide has been used to store immunosorbents for up to 3 months with only slight loss of adsorbent properties (*89*). Even where sterility of an individual batch is maintained by immobilization of the ligand under aseptic conditions (*71*), it is impossible to verify or record this and so is unlikely to be acceptable to regulatory authorities under normal circumstances. Thus, the less-specific techniques, used in sequence, are likely to remain of particular importance in this area despite the undoubted superiority of affinity methods in recovering extremely pure products from complex mixtures.

5.3.2. Specific Types of Adsorption

The principles behind most of the common types of adsorption will be reviewed briefly in the order of increasing sophistication.

5.3.2.1. Gel Permeation. This separates compounds on the basis of differences in the apparent size. It is well suited to separations when size differences are very clear, e.g. large virus particles from proteins, proteins from salts, or recovery of small molecules such as insulin and insulin-like growth factors (*94*). However, it often results in dilution of the product and so is at a disadvantage with respect to other size-discrimination methods such as ultrafiltration. Consequently, its recent use in the cell product area is limited to those instances where a change in buffer salts is needed between other treatments. It is also of use when eluents must be removed from the product prior to final formulation (*78*) or to remove a protease inhibitor which has been added to prevent product loss in earlier stages of processing (*27*).

5.3.2.2. Simple Adsorption. The level of discrimination achieved is low if no specific interaction is involved. Thus, methods such as adsorption of Factor VIII on to aluminium hydroxide (*47, 70*) or controlled pore glass (CPG) (*66*), and IgG on calcium phosphate (*48*), are primarily intended for concentration rather than fractionation. The porous structure of the adsorbent may provide an additional selectivity by excluding large products from the adsorptive surface within the beads. This approach made it possible to remove proteinaceous contaminants from FMD vaccine using Macrosorb T (a macroporous titania) and to purify Factor VIII on CPG (*66*). A number of inorganic oxides, salts, and natural earths show some fortuitous interactions with proteins. The most important of these is hydroxyapatite, which adsorbs nucleic acids as well as immunoglobulins. Purified fractions can be eluted from this type of calcium phosphate using a phosphate salt gradient. Although the absolute level of purification may not be good (*74*), the capacity is high (*48*) and the nature of the interaction differs from the more powerful affinity methods. Thus it is able to distinguish sub-groups within samples that appear homogeneous by more specific methods (*49*).

5.3.2.3. Ion-exchange. This is probably the most widely used chromatographic method, since almost all the important products from cell culture carry a charge which varies with pH. Selecting the best pH for maximum loading on to a particular resin and for greatest separation from contaminants is crucial to success with ion-exchangers. It is now possible to predict these conditions quite accurately by electrophoretic titration (*79*). The isoelectric point (pI) of most proteins is such that they bind well to anion exchangers. DEAE ligands have been used extensively, though the quaternary amines are of increasing importance since they may have a higher capacity and remain ionized over a much wider pH range. However, for many mammalian cultures, the desired product is a minor component of the process fluid. Consequently, it may be advisable to choose conditions which minimize the degree to which contaminating proteins take up the capacity of the resin (*72*) and may reduce problems from co-purified proteases. Hence cationic exchangers such as the traditional CM-cellulose or the newer, strong-acid polymers such as S-Sepharoses are of use. Small ions compete for the exchange sites and this is the basis of most methods of selectively eluting particular fractions from resins. However, it also means that the operational capacity of resins in rich culture media is much lower than that stated by the manufacturers.

5.3.2.4. Hydrophobic. This complements ion-exchange in relying on non-polar interactions. In contrast to ion-exchangers, the capacity of these resins is greatest at high salt concentrations and desorption is generally achieved at low ionic strength. This method is of increasing importance since many of the new products which must be recovered from cultures (e.g. interferon, human growth hormone, plasminogen activator) appear to be more hydrophobic than the main contaminants. Unfortunately, certain proteases are co-purified particularly well on hydrophobic resins (*94*), and may attack the protein, though proteases may also be deactivated fortuitously by these resins. Hydrophylic beads with non-polar ligands such as phenyl and octyl chains are used routinely for the larger proteins, but with smaller molecules or polypeptides which do not rely on delicate three-dimensional structure for their activity it is possible to exploit reverse-phase techniques with octadecyl ligands on a fully hydrophobic silica (*73, 45*). The importance of the link between the ligand and support is emphasized by a recent review of new hydrophobic resins for the sub-fractionation of blood proteins (*84*). Thio-octyl groups separate albumin better than octyl ligands with the corresponding ether link. In the case of thioaromatic ligands, there is a marked thiophilic effect compared to the ether equivalent, leading to the conclusion that charge transfer mechanisms are involved.

5.3.2.5. Metal Chelate. A number of important mammalian proteins bind selectively to divalent metals held on resins by chelation. Zinc is the most commonly used metal, interacting with urokinase (*28*), α-macroglobulin, and haemopexin (*84*) among others. The adsorbed proteins are easily eluted with a soluble chelating agent such as EDTA. Only a limited number of proteins exhibit this type of interaction, so metal chelation chromatography may have good powers of discrimination between chosen products and contaminants.

5.3.2.6. Dye Ligand. This is often referred to as pseudo-affinity chromatography, since it is capable of discriminating between proteins on the basis of quite subtle differences in structure. It is the most powerful of the techniques in which the resin can be sterilized routinely. It relies on the fortuitous combination of ion-exchange and hydrophobic interactions between anthroquinone dyes and particular proteins (*62*). The blue dye F3GA appears to have a surface topology which mimics the co-factor NAD so that many dehydrogenases and kinases bind to it strongly. However, the interaction is not unique, since the interaction of the dye with human albumin is thought to be with the bilirubin binding site, while that with bovine albumin is with the fatty-acid binding site (*59*). The interactions are probably the result of both ionic and hydrophobic forces, and elution may be affected by use of choatropic salts such as thiocyanate or hydrophobic polymers such as PEG, as well as analogues of the affinity ligand. This type of chromatography covers a range of ligands which are very different in their mode of adsorption. Thus it is not possible to predict performance or specificity from first principles and although α-interferon will adsorb to blue dye ligands, γ-interferon will not but goes preferentially onto a Procion Red ligand (*24*).

5.3.2.7. Affinity. The specific interaction between one biopolymer and another is one of the most powerful methods of separation available. The ligand attracting the most interest for mammalian cell products is Protein A, a soluble protein made by mutant strains of the microorganism *Staphylococcus aureus*. It interacts with several important immunoglobulins, binding up to 20 mg of protein per ml of adsorbent, and is the principal method of purification for monoclonal antibodies on the sub-kilogram scale. Product can be eluted repeatedly (*79*) in weak acid at a pH which varies with the immunoglobulin (*48, 79*), though pH 3.5 is recommended to guarantee desorption on all occasions (*98*). Other bacterial proteins, such as Protein G (*30*), with wider specificity for immunoglobulins, are being studied and may be used in a similar way to Protein A. An elegant extension of the interaction of Protein A with immunoglobulins is the incorporation of a Protein A chain in recombinant proteins. For instance, a gene fusion of insulin-like growth factor and Protein A makes it possible to recover the product on immobilized IgG (*80*). The conjugate can be cleaved enzymatically while adsorbed on the support, so as to release the pure growth factor which is eluted first and the Protein A is desorbed later. A similar approach has been employed (*46*) in forming a peptide chain on the fusion product. The conjugate can then be recovered on a general immunosorbent and the peptide subsequently cleaved.

A second proteinaceous ligand of importance for mammalian products is concanavalin A. This lectin is extracted from Jack beans and binds the non-reducing end of glucose and mannose sugar residues. The interaction is only sufficiently strong to bind a compound when several sites on the molecule interact, but this is commonly the case with the large, glycosylated products which eukaryotic cells are uniquely capable of synthesizing. The product is normally desorbed using a competitive hapten such as methyl mannoside (*28, 60*), though complexing agents such as borate could be employed. A number of other lectins specific to different carbohydrate residues are commercially available.

Other specific interactions which have been exploited for purification of mammalian products are the interaction of urokinase with a number of inhibitors and pseudo-substrates such as benzamidine, fibrin, gelatin (*94*) or lysine (*27*). Anticlotting factors have been purified on a heparin-agarose conjugate (*94*) and fibrinectin can be fractionated on gelatin adsorbents (*62*).

It is essential that the link between the ligand and the matrix is stable and does not introduce sites for non-specific adsorption. Immobilization of the ligands onto pre-activated gels is obviously very convenient and cyanogen bromide activated material has been used extensively in this manner (*46*). However, the stability of the supposed iso-urea linkage is not absolute and leakage of around 20 ng of antibody per ml of support per run has been recorded with CNBr-activated cellulose (*75*). This leakage, like that of dye ligands, appears to be at its worst during elution of the product. It is most serious when acetic acid or thiocyanate are employed as eluents and leakage rate seems independent of the number of cycles. Tresyl-chloride-activated beads are a suitable alternative (*46*), while oxidized cellulose or *N*-hydroxy-succinimide-activated Affigel are reported to be stable (*75*). Coupling methods that involve some degree of cross-linking, such as glutaraldehyde or divinylsulphone, appear to give greater stability to tetrameric proteins such as Con A, probably by multipoint attachment (*105*). The problems of non-specific binding to sites introduced on the matrix can be reduced by adding salts or detergents to the process buffers (*105*).

5.3.2.8. Immunosorbents. These are a special sub-set of affinity adsorbents which depend on the binding of an antibody and an antigen. Either component may form the ligand; immobilized antigens are used to adsorb impure antibody and this purified product is then available to separate more pure antigen. The availability of monoclonal antibodies has made the latter approach the preferred method of purifying high-value products such as interferon (*94*).

Immunosorbent columns can be stable for over one year and it is possible to automate eluting and loading cycles (*4*) to make efficient use of what is always an expensive adsorbent. The performance of the immunosorbent can be affected by the concentration of ligand on the support matrix (*33*). The adsorption kinetics for very large molecules, typical of those found in mammalian cell cultures, are far better when there are low levels of substitution of antibody on the support. This leads to the proposal that immunosorbent columns should be large, with a low concentration of ligand, though this concept will be limited by the high cost of the support. The binding can be extremely strong and so it is preferable to use an antibody with a relatively low binding constant (*103*). However, the homogeneous nature of monoclonals makes it possible to define specific desorption conditions and obtain higher yields of pure product than with polyclonals (*27*). When using heterogeneous polyclonal antibodies, complete release of product is dictated by the most strongly binding component and so very disruptive conditions have to be employed (*88*). Desorption of antigens from immobilized monoclonals is said to be achieved more effectively with chaotropic agents such as potassium thiocyanate than by the reduction of pH used frequently for polyclonal antibodies (*7*). However, in the case of antibodies to Factor IX, glycine/ethanediol solution at pH 10 was best (*65*), while changes in the calcium binding characteristics of some clotting factors have been exploited as a simple method of desorbing these from immunosorbents (*60a*).

5.3.3. Large-scale Options

Although the value of most mammalian cell products is such that a wider range of chromatographic methods can be considered, even in this area there are pressures on the separation scientist to take different options as the scale of production increases. On the gram scale, production methods can include most of the techniques mentioned above, and in many cases it is possible to move directly from the type of material used in laboratory-scale work to an equivalent preparative grade. Thus, a process can be developed on an FPLC system using small surface-coated beads and then scaled directly to preparative-scale gel beads with the same ligand substituents (*16, 76*). Most manufacturers of HPLC equipment also provide facilities to move from the microgram scale to the milligram and gram scale using preparative-grade materials with ligands similar to the analytical grade adsorbents on which a technique has been developed (*14,61*). Although some silica-based HPLC particles may present problems owing to the restricted diffusion of the larger proteins to sites within the matrix, preparative HPLC systems have the advantage of high chromatographic efficiency and can probably be modelled more reliably to assist in scale-up. Even the more sophisticated methods of affinity HPLC have been successfully scaled-up to the level where grams of mammalian enzymes can be recovered on each operational cycle of a 3.3-litre bed of silica particles to which Procion Blue MX-R dye had been linked (*23*).

5.3.4. Electrophoresis

This is one of the most powerful analytical techniques available and separates molecules on the basis of mobility in an applied electric field. The mobility of a particle, molecule, or cell is determined by the net charge on its surface and its hydrodynamic size. Until recently, however, electrophoresis has only been appropriate for dealing with biological materials up to milligram quantities. Early attempts to scale-up electrophoresis were restricted by problems of Joule heating owing to the high currents required. This led to convective mixing, particularly when the electrophoresis took place in the absence of a support matrix (free-flow). This problem can be circumvented by running the equipment in the near-zero-gravity environment of space (*68*). However, transportation costs and irregular access to product make this approach of limited value.

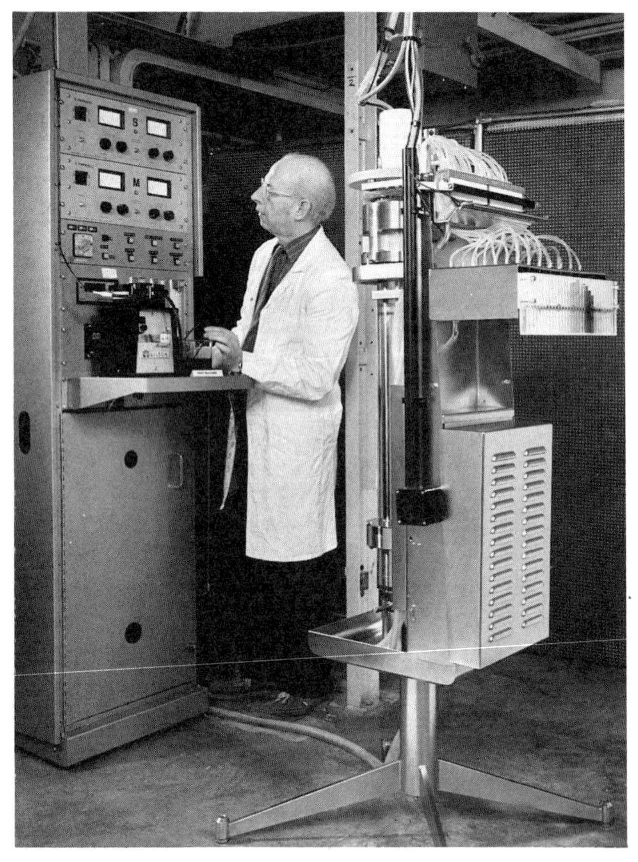

Fig. 2. The Biostream free-flow electrophoretic separator, capable of continuously fractionating up to 1 g min^{-1} of protein.

It is known that fluid flowing in an annulus in which the outer wall rotates is stabilized against convective mixing. This phenomenon has been exploited in the design of the Biostream Electrophoretic separator (Fig. 2), which is capable of continuously sub-fractionating up to 1 g min^{-1} of protein into 29 fractions. This equipment has been used industrially to fractionate blood plasma, giving a high yield of labile compounds such as Factor VIII (Fig. 3). It is also capable of separating other soluble molecules such as immunoglobulins from major contaminants such as albumin and of sub-fractionating cells (*57*). While this technique separates molecules on the basis of their relative velocities in an electric field, another, isoelectric focusing, moves materials until they reach a zone of pH where their net electric charge is zero. Materials are hence

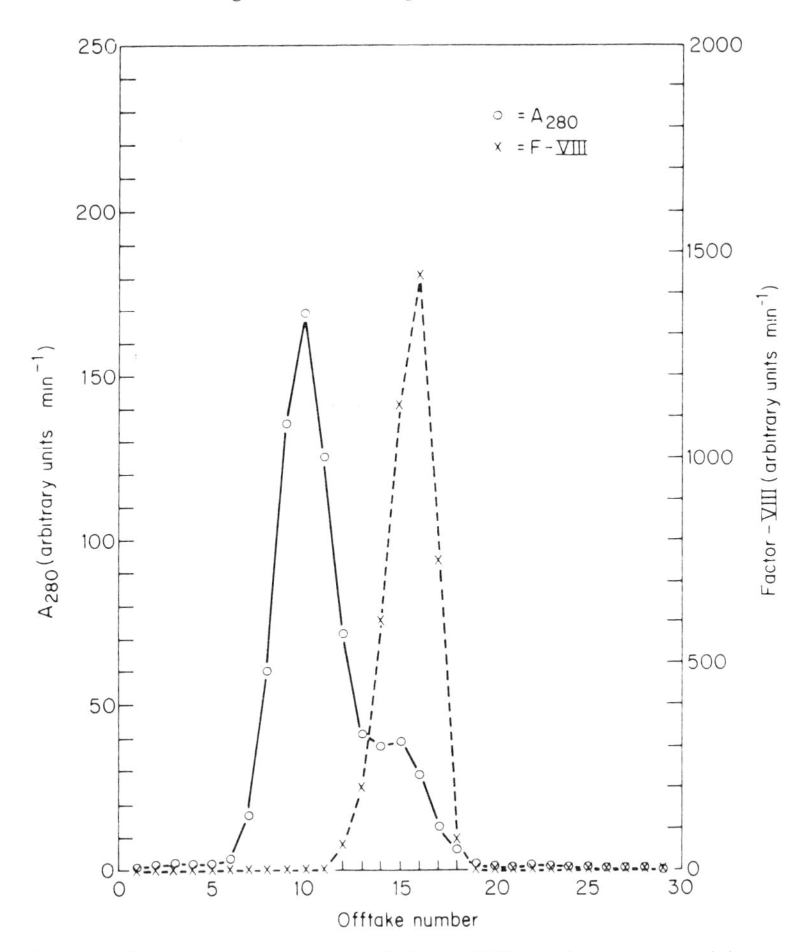

Fig. 3. Purification of the anti-haemophilic Factor VIII from human cryoprecipitate.

fractionated on the basis of their isoelectric point. This technique has been scaled-up close to the point of being applicable to industry, with the pH-controlled compartments into which the charged materials move being separated from one another by membranes (*69*).

5.4. Final Presentation

This subject was dealt with in Vol. 2 (*96*) and will not be expanded on here. Nevertheless, the stabilization, sterilization, and storage of the final product is essential if it is to be commercially successful. Freeze-drying of biological materials is widely used but ill-understood, and will require considerable attention if mammalian cell products are to reach their full potential.

5.5. Analytical Procedures

The rapid improvement in techniques for sub-fractionating mammalian cell products has been paralleled by a similar increase in the range of analytical methods available for process control. Electrophoretic and immunologically based methods are of great importance, while chromatographic methods in particular have advanced to a stage where rapid analysis of the product of interest is possible using an HPLC (*74*) or FPLC (*72*) method which closely matches the production-scale method. Furthermore, major products of animal cells such as monoclonal antibodies are themselves contributing to the sophistication and breadth of immunologically based assays which can be used to monitor the progress of a purification process.

6. PRODUCTS OF INTEREST

It is important to view product recovery as an integrated process in which the unit operations covered above are combined in various ways depending on the product and its end use. This concept is illustrated below by detailed consideration of six groups of mammalian cell products, which are of commercial interest for clinical and diagnostic uses. These compounds are serum proteins, blood coagulation factors, plasminogen activators, monoclonal antibodies, interferons, interleukins, and cell fragments. This selection reflects a greater interest in soluble proteins than has been the case in the traditional animal cell culture industry where particulates such as cells and viruses are major products. Although the chief concern of this review is the separation of culture products, the majority of these compounds were first isolated from body fluids such as blood or urine or ascites fluid. Consequently, improvements in these areas are relevant and illustrate how the older techniques are used in conjunction with recent developments to improve the overall process.

Typical purification processes for each of the compounds under consideration is summarized in Table V using the generalized view of separation policy outlined in Fig. 1. The pattern which emerges is similar to that found in a more general survey of biological separations (*10*) and re-emphasizes the similarity in philosophy for selecting purification processes.

TABLE V **Purification Processes for Selected Products**

Serum proteins	
Source	Blood (human or bovine)
Primary separation	Centrifuge removes intact cells
Concentration	Ethanol precipitation (*32*)
Sub-fractionation	Anion and cation exchange (*35, 99*) for albumin and IgG Blue dye ligand (*62*)
	Albumin immunosorbent (*18*)
Coagulation factors	
Source	Cryoprecipitate or genetically engineered cells
Primary separation	Not needed or cell attachment
Concentration	Adsorption by alumina (*47*)
	Precipitation by PEG/salts (*47, 70*)
	Precipitation by natural polymers (*20, 29*)
Sub-fractionation	Controlled pore glass (*66*) } Factor VIIIc
	Gel permeation (*41*) } Factor VIIIc
	Electrophoresis (*57*) } Factor VIIIc
	Polyelectrolyte (*36*) } Factor VIIIc
	Anion exchange (*32, 35, 97*)
	Immunosorbents (*7*)
Plasminogen activation	
Source	Cultured cells
Primary separation	Cell attachment/centrifugation
Concentration	Hollow-fibre membrane
Sub-fractionation	Zinc chelate ligand (*28*)
	Concanavalin A ligand (*28*)
	Gel permeation (*28, 27*)
	Immunosorbents (*27*)
	Small-affinity ligand (*27*)
Monoclonal antibodies	
Source	Ascites suspension culture
	Immobilized hydridoma
Primary separation	Centrifugation, microfiltration
Concentration	Ultrafiltration, ammonium sulphate precipitation (*74, 3*)
Sub-fractionation	Cation and anion exchange (*72, 61, 90, 99*)
	Gel permeation (*13*)
	Hydroxyapatite (*48, 74, 49*)
	Hydrophobic resin (*74*)
	Dye/ion-exchange (*12*)
	Antigen ligand (*78*)
	Protein A (*72, 48, 79*)
	Immunosorbents (*3*)
	Electrophoresis (*57*)
Interferon	
Source	Blood, suspension cultures
Primary separation	Centrifugation
	Filtration (*42*)

(*continued*)

TABLE V *(continued)*

Concentration	Precipitation with ethanol or potassium thiocyanate (*52*)
	Adsorption to controlled pore glass (*42*) or silica (*60*)
	Two-phase aqueous (*15*)
Sub-fractionation	Immunosorbent (*42*)
	Ion-exchange (*60*)
	Con-A ligand (*60*)
	Poly-U ligand (*60*)
	Blue-Dye ligand (*60*)
Interleukins	
Source	Suspension cultures
Primary separation	Centrifugation
Concentration	Ultrafiltration (*25, 73*)
	Ammonium sulphate precipitation (*45*)
Sub-fractionation	Gel permeation (*25, 73, 45*)
	Hydrophobic ligand (*73*)
	Hydroxyapatite (*45*)
	Ion-exchange (*45*)
	Immunosorbent, R-P HPLC (*73, 45*)
Cell membranes	
Source	Tissue, attached and suspension cultures
Primary separation	Gradient centrifugation in detergent (*37*)
Concentration	Membranes
Sub-fractionation	Ion-exchange (*37*)
	Gel permeation (*37*)
	Immunosorbents (*101*)
	Protein-A ligand (*36*)
	Con-A ligand (*9*)
	Electrophoresis (*57*)

6.1. Serum Proteins

Fractionation of human blood plasma into its major constituents has been a well established process since Cohn defined the basic operational parameters during World War II. The process involves the fractional precipitation of proteins from cell-free plasma using ethanol at low temperature and defined pH. The first component, cryoprecipitate, is the solid left when the frozen plasma is thawed and this contains Factor VIII. The first precipitate is obtained with 21% ethanol and contains mainly immunoglobulins, while subsequent fractions using 30% ethanol, contain mainly albumin. Intermediate fractions will be mixtures of both proteins along with transferrin. The technique is best viewed as principally a concentration method, since the degree of purification is low. However, the products have an established market and the impurities in each fraction are of little concern since, after appropriate screening, the starting material, plasma, is already accepted as safe. Furthermore, the high concentration of solvent reduces the risk of microbial contamination of the product during processing. Precipitation has the advantage that it requires far less process water than chromatography, which is generally reserved for purification of the sub-fractions. In this respect, blood fractionation closely resembles the purification of products from serum containing culture fluids. Although most of this experience relates to human plasma, the same general consideration should apply to the bovine and horse sera used in tissue culture.

The important advances in this sector have been the better control of the precipitation and the development of chromatographic techniques to deal with albumin. The sub-fractionation to purify coagulation factors and progress in the de-activation of viral contamination are dealt with in Section 6.2.

Although in blood fractionation it is necessary to deal with thousands of litres of fluid, control of the precipitation has been improved by scaling-down the volume at the critical stage. The continuous precipitation method developed by the Scottish Blood Transfusion Service (*32*), leads to greater control of temperature, pH, and ethanol concentration by bringing the fluid streams together in a static mixer device. Although acoustic methods were used to enhance coalescence of the precipitate, a simple holding section is now known to be adequate. Microprocessor control of flow prevents overshoot of conditions, reducing pre-precipitation of albumin in Fraction IV, and hence improving quality and yield.

Chromatographic methods based on the sequential use of cationic and anionic adsorbents have been developed to replace the traditional fractional precipitation method (*35*). These separate albumin from immunoglobulins and will be discussed in more detail with respect to monoclonal antibody purification (see Section 6.4). There is industrial interest in the more specific purification of albumin by dye ligand chromatography and the use of a 16-litre column capable of recovering up to 300 g of 95% pure albumin with a 6-hr cycle time has been reported (*62*). Human albumin, and to a lesser extent bovine albumin, show a fortuitous affinity for the dye Cibacron F3GA (*59*). However, there is concern that ligand leaching from the support may not only contaminate the albumin product but may be carried into other products which must be recovered from the supernatant. The capital cost of chromatography is higher than of precipitation for a bulk product such as albumin and so the more specific techniques may be better suited to removal of trace amounts of albumin from other materials. In the case of very high-value culture products, the use of immunosorbents to remove trace contamination may be justified. For example, contaminating levels of bovine serum albumin may be reduced 200-fold in a single pass through a column of Sepharose–anti BSA (*18*).

6.2. Coagulation Factors

Although Factor VIIIc has been produced in genetically engineered mammalian cells, most separation methods are based on the cryoprecipitation of human plasma. The main aim of the purification is to reduce the protein load in the doses given to haemophiliacs and to remove fibrinogen and fibrinectin, which cause aggregation and problems during heat treatment and subsequent resolubilization of the product for injection.

It is possible to precipitate these contaminating macromolecules at low temperatures, but more recently polyethylene glycol alone (*47*) or this polymer along with carefully controlled amounts of glycine and salt (*70*) have been used to improve purification. More specific purification from contaminants is achieved with sulphonated polysaccharides such as heparin (*20*) which have the advantage of ready acceptability but are rather costly (*29*).

Alternatively, the Factor VIIIc can be adsorbed, either on alumina (*47*), or, more recently, on a synthetic ethylene maleic anhydride copolymer electrolyte (*36*). The great size of Factor VIII has been exploited by treating solutions with controlled pore glass (500 Å pores) such that contaminants entered the pores and adsorbed to the high internal pore surface (*66*). Herring *et al.* (*41*) reported an elegant use of gel permeation to purify Factor VIIIc by altering the size of the soluble conjugate. At low salt concentrations the material associates with von Willebrands factor to form a very large aggregate which passes through the column in the excluded volume. It is then re-chromatographed at high salt concentration, when the complex dissociates and Factor VIII is found in the included volume (*41*). Unfortunately, Factor VIII is deactivated on contact with many surfaces and so purification by free-flow electrophoresis has proved particularly suitable (Fig. 3).

Factor IX has been cloned in rat hepatoma cells and the product has been purified by adsorption to freshly precipitated barium citrate followed by immunopurification (*2*). On the large scale, prothrombin is recovered from the supernatant in thawing plasma, by batch adsorption to DEAE ion-exchangers (*32, 35*). Although DEAE cellulose gives slightly better fractionation than the Sephadex equivalent, it is reported that the eluent from the cellulose matrix occasionally gives a pyrogenic response (*97*). The specificity of this separation can be improved by the use of immunosorbents carrying monoclonal antibodies to the protein (*7*). An elegant extension of this concept is the use of monoclonal antibodies against the Factor IX–calcium complex. In this method, the bound product is eluted under mild conditions involving the use of EDTA to chelate the metal and has also been applied to a recombinant γ-carboxylated Factor IX molecule (*60a*).

Factor VIII is traditionally produced from the pooled plasma, taken from up to 1000 individual donations and the risk of viral contamination has been recognized for some time. Consequently, the handling of this product provides valuable guidance on how viral contamination might be dealt with in culture products. Fortunately, methods developed to remove the more robust, infectious human viruses such as hepatitis are already in place and have been adapted to deal with recent concern over contamination of products by AIDS virus. Certain sub-fractionation techniques, such as adsorption onto polyelectrolyte (*36*) do fortuitously separate hepatitis virus from Factor VIIIc with no loss of yield. However, deactivation of the virus is probably essential if a product is to be widely accepted. Some process conditions reduce viral infectivity (*36*), but heat inactivation is the only acceptable method at present. This inevitably carries a penalty in lost product and so conditions have to be set empirically. The differential in stability is probably greatest under conditions of dry heat and it is reported that freeze-dried Factor VIIIc can tolerate 80°C for 72 hr, sufficient substantially to kill the hepatitis virus (*29*). The virus may be further destabilized by addition of detergent, while the Factor VIIIc is reported to be stabilized by use of salts/amino acid/PEG mixtures (*70*) or glycine/sugar (*5*), such that it can be pasteurized in solution at 60°C for 10 hr to inactivate viruses.

6.3. Plasminogen Activators

Thrombolytic compounds such as urokinase have been purified from urine for some time using affinity-type interactions with gelatin or fibrin to concentrate the protein and gel permeation to sub-fractionate (*43*). However, the clinically more useful tissue plasminogen activator has routinely been recovered from tissue culture fluid. The product is a specific protease zymogen, so care must be exercised not to use processes which irreversibly activate the proteolytic activity. It is therefore normal to include protease inhibitors such as aprotinin (*27*) during the initial stages of purification, and to remove them in the final gel-permeation stage.

The classic process is a typical multistage scheme involving simple centrifugation to remove cells, a membrane concentration in a hollow-fibre system with a 10 000 Dalton cut-off, followed by three chromatographic steps (*28*). The first two adsorbents bear group specific ligands, the zinc chelate relying on the formation of an ill-defined metal complex, the immobilized concanavalin A being considered specific for the carbohydrate side-chains present in glycosylated proteins. The plasminogen activator is eluted from the Con A with methyl mannoside, a small hapten which binds competitively to the ligand. This glycoside is a more expensive reagent than a simple salt and must be removed subsequently. This is easily achieved by both membrane concentration and gel permeation, since it differs greatly in size from tPA. Although there are minor differences between the product from epithelial cells and melanoma cells (e.g. a higher concentration of mannoside is required to elute the former) the separation process seems sufficiently robust to be applied generally to this group of products. A recently reported alternative purification process uses an immunosorbent to give primary purification, and lysine-Sepharose to distinguish the two sub-forms of the enzyme (*27*).

6.4. Monoclonal Antibodies

This is a particularly interesting group of tissue culture products since they are themselves valuable as process aids in purification of high-value compounds. The large-scale purification of this group of products has recently been reviewed by Ostlund (*72*). Biochemically they are considered to be immunoglobulins ranging in molecular weight from the four-sub-unit IgG at around 150 000 Daltons to IgM's which are hexamers of this basic unit at just under 10^6 Daltons.

The purification of monoclonal antibodies from hybridoma cells in tissue culture is based on methods developed for the recovery of the same product from ascites fluid or the even older processes of recovering immunoglobulins from serum. The clarification and concentration procedure employed in recovering monoclonal antibodies is typical of those used for other macromolecules. Although ammonium sulphate precipitation may be used on a small scale (*3, 74*) and in one general production process (*26*), it can cause some loss of activity (*22*). Membrane concentration in hollow-fibre or tangential-flow units now seems to be preferred on the production scale. Nevertheless, membranes can suffer from serious problems of fouling during prolonged operation, necessitating regular cleaning of the active surface despite the high cross-membrane fluid velocities.

Pre-treatment of the crude culture fluid may reduce some of the fouling downstream. For instance, cellulose powder with a low charge density has been employed as a guard column to remove acidic contaminants (*90*) and improve performance of anionic exchangers downstream. Pre-treatment of samples with poly(vinylpyrolidone) is also recommended as a prelude to chromatography on ion-exchangers (*74*). The successful use of fluocculents to remove all debris makes it unnecessary to use microfiltration prior to concentration by ultrafiltration and may reduce membrane fouling problems. It is reported that polygalacturonic acid is a particularly useful flocculating agent for use in purifying monoclonal antibodies (*53*).

The most interesting aspect in the separation of monoclonal antibodies is the sub-fractionation. The source of the antibodies may affect the degree of pre-concentration required and the level of impurities, but does not materially affect the separation methods selected. It should be appreciated, however, that the composition of the fluid presented for downstream processing can vary widely. In ascites there may be 3–15 mg of Mab per litre, but this constitutes only 16–30% of the protein. Meanwhile the culture fluid from hybridoma cells may vary from 0.01 to 10 g l^{-1} and up to 80% purity, depending on the amount of serum used in the culture and type of reactor system chosen (*72*). Sub-fractionation may be based on differences in charge, low-specificity interaction, or affinity binding; the degree of purification increases along with cost through this series. Although at the present sub-kilogram level of production most of the techniques discussed below are considered, the move to kilogram-scale is focusing attention on the low-cost, generalized methods such as ion-exchange followed by gel filtration (*16*). These give higher yields than the affinity methods, but do not yet reach the levels of purity demanded of a therapeutic material.

Ion-exchange is the simplest chromatographic method for recovery of monoclonal antibodies, giving yields of over 80% with purities in the range 50–90%. However, the monoclonals should not be considered as a homogeneous group, since isoelectric points ranging from 4.9 to 8.3 have been reported (*78*). Most proteins in the culture fluid bind to anionic exchangers such as the tertiary amino group DEAE (*11*) and stronger quaternary amines (*61, 90*). The various proteins are eluted sequentially on a salt gradient. The semi-pure eluent is then further fractionated on a cationic exchanger (*90*) and by gel permeation (*13*).

Repeated use of anion-exchangers for sub-fractionation of complex solutions such as serum, can result in the build-up of tightly bound compounds on the column. This may necessitate scrubbing the bed with alkali as often as every three cycles. The costs in lost production time and re-equilibration buffers can be significant compared to the value of the product. Composite ion-exchangers fabricated from very pure agarose appear less prone to non-specific adsorption and can be used up to 13 times for serum fractionation without alkaline scrubbing being needed (*99*).

The elution of proteins from the anionic resin dilutes the material and the large amount of contaminating protein and components of culture media, such as phenol red, which also binds to the resin, makes poor use of the total capacity of the column (*72*). Consequently, the use of cationic resins to which only the immunoglobulins and a limited number of other proteins bind is recommended for large-scale operation where capacity is important. The stronger cationic exchangers such as S-Sepharose, which remain ionized over a wider pH range, are favoured for this approach irrespective of whether the product is in serum-free (*16*) or serum-containing medium (*64*). A pH of about 5.5 and low salt concentration ensures that albumin (pI 4.9) does not bind. Contaminating transferrin (a heterogeneous pI of 5.2–6.1) and bovine immunoglobulins (pI 5.8–7.3) are selectively eluted at higher salt concentrations with the murine immunoglobulin eluting at 75–200 mM salt depending on its form (*64*). Endotoxins and other undesirable contaminants bind strongly to ion-exchangers and so some degree of secondary purification also takes place, the columns being cleared of residual material with 1 M salt or dilute alkali.

The transition from small-scale FPLC or HPLC to larger columns is facilitated by the availability of ion-exchangers which perform in a similar way in both systems (*61, 77*). At the small scale, HPLC methods anionic exchangers still seem to be the norm (*74*). In the case of the large IgM molecule, a tandem purification using anionic Mono Q beads in an FPLC system has been followed by gel permeation on Sepharose 6 column to yield a pure high-activity product (*5*). Alternative low-affinity adsorbents such as hydrophobic resins and hydroxyapatite have been used successfully for monoclonal antibody purification. When used in conjunction with ion-exchange and gel-permeation, these may offer the degree of purification at low cost which will be required to prepare kilogram quantities of material for therapeutic applications.

Hydroxyapatite has been used to pre-concentrate material prior to loading on to Protein A columns (*48*) but is said to give poor resolution with crude mixtures (*74*). However, it has been used successfully to distinguish three idiotypes in a monoclonal antibody which appeared homogeneous on a Protein A column (*49*). This could be particularly important if hybrid idiotypes are used to overcome immunological problems in therapy. Hydrophobic resins give a lower product yield than ion-exchangers (*74*), but mixed-function resins appear to have some advantages. For instance DEAE-Affigel, which has both ion-exchange and blue dye substituents, gives easy elution of monoclonal antibodies under mild conditions and removes protease contaminants from culture fluid (*12*). More specifically, important contaminants such as esterases and proteases can be removed from immunoglobulin preparations with a range of Procion dye ligands.

The high-affinity binding of antibodies to an anti-IgG (*3*) or to its immobilized antigen (*78*) may be used to provide high degrees of product purity. However, the specific interaction between the Fc region of most of the economically important monoclonal antibodies and immobilized Protein A is the most versatile affinity method available. Although the adsorbent is more expensive than ion-exchangers, it gives a highly purified product which can be eluted under fairly mild, acidic conditions and is used extensively at current production scales. It is claimed that a 5-litre column of immobilized protein could produce up to 100 g of product in a 3-hr operational cycle (*72*). In practice, the expense of the columns makes it important that they are used efficiently, and Celltech report the use of simple mathematical models to predict changes in performance with use and facilitate automated cycling of columns on a semi-continuous basis. Such columns, typically of only 500 ml volume, are reported to be of use for at least 100 cycles without detectable decrease in capacity (*54*). Under ideal conditions the columns can be used for prolonged periods to give 50–80% yields of products with almost no contaminants, although the recovery of activity may not be as high as that achieved with ion-exchangers (*26*).

It was thought for some time that IgG_1 did not bind well to Protein A and it was recommended that antibodies of this sub-class be produced in serum-free medium and purified on cationic SP Sephadex G-50 followed by gel permeation (*16*). However, this low binding can now be overcome (*48*, *79*). Work reported by Pharmacia indicates that the binding of all sub-groups depends on both pH and ionic strength. IgG_1 monoclonal antibodies are unusual in requiring 3 M salt to achieve 98% binding, presumably owing to an increased hydrophobic interaction. Thus it appears that Protein A can be used almost universally. Desorption is also dependent on the same parameters, and it was often recommended that monoclonal antibodies be desorbed at pH of about 3, samples being dropped directly into alkaline buffers if the product was labile (*79*). It now appears that even the tightly bound IgG_2 can be desorbed in 0.1 M citrate at pH 4 and IgG_1 at pH 6 (*48*).

The main concern with the use of immobilized Protein A is that the ligand leaks from the cyanogen-bromide-activated Sepharose routinely used as a support. Thus, although it is considered satisfactory for production of diagnostic reagents, further purification is usually required for therapeutic applications. To this end it is claimed that any contamination in the final product can be removed by an ion-exchange treatment which also adsorbs pyrogenic materials (*54*).

6.5. Interferon

The pharmaceutical industry has displayed intense interest in this group of products. This has ensured that almost every significant method for separating biopolymers has been used in attempts to purify interferon from human blood or lymphoblastoid cells (*83*).

Most batches of trial material were partially purified by ethanol precipitation under acidic conditions, but potassium thiocyanate is now seen as an alternative (*52*). Horowitz (*42*) reports the use of a Zeta cartridge to remove cell debris prior to concentration on a Pellicon ultrafilter and precipitation with 0.5 M thiocyanate at pH 3.5. Adsorption on controlled pore glass followed by immunopurification resulted in an overall purification of over 7000-fold with a yield of 65%. Partial purification of human interferon has been achieved in a two-phase aqueous system formed with 31% ammonium sulphate and 30% polyethylene glycol. The product was found in the lower-salt phase, giving a 1.5-fold purification (*15*). Precipitation with 45% PEG, ion-exchange on CM-Sephadex, and concentration with an Amicon PM10 membrane increased the purification factor to 117-fold with an overall yield of 79% (*15*). Ley *et al.* (*60*) report the use of most of the principal chromatographic matrixes (silica, DEAE-Sepharose, Con A, Poly U, and blue–Sepharose). However, none of these achieves the impressive purification power of over 4000-fold for a single step claimed by the suppliers of the NK2 and YOK monoclonal antibodies which are now being used for preparative-scale immunosorbents (*19*).

6.6. Interleukins

These polypeptides are found in low concentration in macrophage cultures but have been expressed as acid-soluble inclusion bodies in E. coli (*56*). Interleukins are very hydrophobic and adsorb strongly to equipment as well as to adsorbents such as Phenyl Sepharose and Procion Red ligands (*55, 56*). Recently reported purification procedures for the three major interleukins have already been illustrated in Table I, to emphasize the degree of purification required for such compounds. Although most of the more common techniques for protein purification are included in this list, it is interesting to note the successful use of reverse phase HPLC (*45, 73*). This technique, which presents an extremely hydrophobic polymethylene chain to the solute, is further evidence of the hydrophobic nature of these peptides, while the small size of the product makes it practicable to use a non-aqueous solvent, acetonitrile, as eluent. The trifluoroacetic acid employed as an ion-pairing reagent is, in principle, sufficiently volatile to be removed from the final product. Less-toxic acids would make this technique more acceptable, but residual impurities from acetonitrile might be of more concern to regulatory authorities.

6.7. Cells and Membranes

The sub-fractionation of cells and cell membranes is currently limited to fundamental studies of cell function but is potentially of importance for preparing enriched cell cultures for lymphocyte-replacement therapy. It provides an interesting extension of the concepts already outlined for the soluble proteins reviewed above.

Membrane components are generally separated from disrupted cells by differential centrifugation on a sucrose gradient. In the purification of mouse cell beta-adrenergic receptors, it was possible to dissolve the crude preparation in 1.5% digitonin to facilitate further purification of the fragments as a homogeneous solution using conventional chromatography (*37*). The mixture was subjected to affinity purification on an immobilized antagonist, (−)alprenolol, separated on DEAE-Sepharose and finally passed down a series of gel-permeation columns to give an overall purification of 16 000-fold. Membrane components may show high non-specific binding, and in the preparation of a virus membrane antigen it was only possible to reach 50% purity on an immunosorbent, even in the presence of a detergent (*85*). It has been suggested that affinity purification of membrane antigens is best achieved with the antibody held on a spacer arm. The antibody is first treated with succinimidobiotin and the biotinylated protein then bound to a Sepharose–streptavidin conjugate (*101*).

Affinity binding to the cell surface markers of intact cells has also been used in cell sorting. In "panning" techniques a selective surface is created on a plastic culture dish. This approach is illustrated by the use of a monolayer of *Staphylococcus aureus* cells to which antibodies to the chosen cell are bound and cross-linked (*36*). Cells can actually grow on this substratum and can be separated on the basis of viral markers expressed on the cell surface. Alternatively, the Protein A from the *Staphylococcus* can be immobilized on Sepharose and used to bind target cells expressing IgG on their surface or to bind soluble antibodies to cell surface markers (*38*). Thiolated concanavalin A bound to trisacryl activated with an organomercury compound was used to adsorb mouse thymocytes, while a similar method using thiolated antibodies was able to bind sheep erythrocytes (*9*). The column-packing approach presents a higher surface area for adsorption but does carry the risk of unbound cells becoming physically trapped in the interstices, so purification factors may be low (e.g. 2-fold (*38*)). The binding of B cells to immobilized anti-Ig, and T cells to antibodies against Thy-1 antigen has been reported as a practical method of sub-fractionating white cells (*44*), while *Helix pomatia* lectin bound to Sepharose is capable of giving 87% pure fractions of these lymphocytes for tissue-typing purposes (*82*).

The different settling characteristics of cells are exploited in zonal unit gravity elutriation, in which large cells can be separated from smaller particles using a device similar to a fluidized bed to wash mixed populations (*1*). Free-flow electrophoresis is able to sub-fractionate particulate material, such as cells and sub-cellular particles, without disrupting cellular function. Recovery of the sample is generally quantitative and the technique has been used successfully to enrich blood fractions in key components such as platelets and sub-populations of leukocytes on the basis of their different mobilities in an electric field (*21*).

7. CONCLUSIONS

In the two years since it was reviewed in Vol. 2, there has been a dramatic increase in the amount of information available on the downstream processing of mammalian cell products. The fact that development work has been underway is not surprising in view of the huge commercial market now assured for such products. It is the freedom with which this information is disseminated which is encouraging. It is an area in which innovative companies have moved new separation methods from laboratory to pilot plant very rapidly, refining the techniques as they go. As the industry moves into a less frenetic era, it is to be hoped that the same success can be achieved in consolidating and improving the many novel methods now being applied at production scale.

Acknowledgements

The authors are involved in research projects covering mammalian cell technology and the separation of biological materials. They wish to acknowledge the financial support of the Biotechnology Unit of the U.K. Department of Trade and Industry in these projects and the help given by colleagues and BIOSEP industrial members in preparing the review.

REFERENCES

1. Andrews, P., and Shortman, K. (1985). Zonal unit gravity elutriation. *Cell Biophys.* **7**, 251–266.
2. Anson, D. S., Austen, D. E. G., and Brownlee, G. G. (1985). Expression of human clotting factor IX. *Nature (London)* **315**, 683–685.
3. Bazin, H., Cormont, F., and De Clercq, L. (1986). Purification of rat monoclonal antibodies. *Method. Enzymol.* **121**, 638–652.
4. Bazin, H., and Malache, J. M. (1986). Rat and mouse monoclonal antibodies V. *J. Immunol. Method.* **88**, 19–24.
5. Behringerwerke AG. (1966). Stabilizing coagulation factors against heat. EPO. 80-81025C/46.
6. Belter, P. A., Cunningham, F. L., and Chen, J. W. (1973). Development of a recovery process for novobiocin. *Biotechnol. Bioeng.* **15**, 533–549.
7. Bessos, H., and Powse, C. V. (1986). Immunopurification of human coagulation factor IX. *Thromb. Haemost.* **56**, 86–89.
8. Birch, J. R., Boraston, R., and Wood, L. (1985). Bulk production of monoclonal antibodies in fermenters. *Tibtech.* **3**, 162–166.
9. Bonnafous, J. C., Dornand, J., Favero, J., Sizes, M., Boschetti, E., and Mani, J. C. (1983). Cell affinity chromatography using cleavable mercury sulphur bonds. *J. Immunol. Method.* **58**, 93–107.

10. Bonnerjea, J., Oh, S., Hoare, M., and Dunnill, P. (1986). Protein Purification—The right step at the right time. *Biotechnology* **4**, 954–958.
11. Boschett, E., and Sene, C. (1986). Production et purification de composes biologique. *Biosciences* **5**, 80–86.
12. Bruck, C., Debrin, J. A., Glineur, C., and Portetelle, E. (1986). Purification of mouse monoclonal antibodies by DEAE Affi-gel Blue chromatography. *Method. Enzymol.* **121**, 587–596.
13. Burchiel, S. W. (1986). Purification and analysis of monoclonal antibodies by HPLC. *Method. Enzymol.* **121**, 596–615.
14. Burgoyne, R. F. (1985). HPLC techniques advance monoclonal antibody isolation. *Res. Dev.* 82–85.
15. Cantel, K., Hirvonen, S., and Kaupinin, H. L. (1986). Production and partial purification of human interferon. *Method. Enzymol.* **119**, 54–63.
16. Carlsson, M., Hedin, A., Inganas, M., Harfast, B., and Blomberg, F. (1985). Purification of *in vitro* produced monoclonal antibodies. *J. Immunol. Method.* **79**, 89–98.
17. Cartwright, T. (1987). Isolation and purification of products from animal cells. *Tibtech.* **5**, 25–30.
18. Celltech. (1986). Reselute BSA, leaflet. Slough, U.K.
19. Celltech. (1986). Anti-interferon data sheet. Slough, U.K.
20. Central Blood Laboratory Authorit, (1986). Purification of coagulation factor VIII. GB 2172-000-A.
21. CJB (1987). Biostream: Electrophoresis for production. Brochure. Portsmouth, U.K.
22. Clezardin, P., Bourgro, P., and Macgregor, J. L. (1986). Tandem purification of IgM monoclonal antibodies from mouse ascites. *J. Chromatogr.* **354**, 425–433.
23. Clonis, Y. D., Jones, K., and Lowe, C. R. (1986). Process scale high performance affinity chromatography. *J. Chromatogr.* **363**, 31–36.
24. Coppenhaver, D. H. (1985). Interaction of human interferon with immobilized dye ligands. *IRCS Med. Sci.* **13**, 809–810.
25. Dukovich, M., and Mizel, S. B. (1985). Murine interleukin I. *Method. Enzymol.* **116**, 480–492.
26. Duff, R. G. (1985). Microencapsulation technology. *Tibtech.* **3**, 167–170.
27. Einarsson, M., Brandt, J., and Kaplan, L. (1985). Large scale purification of tissue plasminogen activator using monoclonal antibodies. *Biochim. Biophys. Acta.* **830**, 1–10.
28. Electicwala, A., and Atkinson, T. (1985). Purification of plasminogen activator from epithelial cells. *Eur. J. Biochem.* **147**, 511–516.
29. Evans, D. R., Robinson, A. E., and Smith, J. K. (1986). Plasma fractionation and machine plasmaphoresis. *Plasma Ther. Transfus. Technol.* **7**, 33–40.
30. Fahnestock, S. R. (1987). Cloned streptococcal G protein. *Tibtech.* **5**, 79–83.
31. Federal Drug Administration (1985). Draft Report on Points to consider in the manufacture of injectable monoclonal antibody products intended for human use in vivo. Office of Biologics Research and Review Centre for Drugs and Biologics, Bethesda, MD, U.S.A.
32. Foster, P. R., Dickson, A. J., Stenhouse, A., and Walker, E. P. (1986). Process control system for the fractional precipitation of human plasma proteins. *J. Chem. Tech. Biotechnol.* **36**, 461–466.
33. Fowell, S. L., and Chase, H. A. (1986). Variation of immunosorbent performance with the amount of immobilized antibody. *J. Biotechnol.* **4**, 1–13.
34. Frej, D. K. (1984). Recovery of lactase by adsorption from unclarified *E. Coli* homogenate. "Proc. Third European Congress on Biotechnology Munich," Vol. 1, pp. 655–663.
35. Friesen, A. D. (1985). Column ion exchange chromatographic production of albumin IV, ISG and factor IX. *Dev. Haematol. Immunol.* **13**, 97–103.

36. Galpin, S. A., Karayiannis, P., Middleton, S. M., and Thomas, H. C. (1984). The removal of hepatitis B virus from factor VIII. *J. Med. Virol.* **14**, 229–233.
37. George, S. T., and Malbon, C. C. (1985). Large scale purification of beta andronergic receptors. *Prep. Biochem.* **15**, 349–366.
38. Ghetie, V., and Sjoquist, J. (1984). Separation of cells by affinity chromatography on protein A gels. *Method. Enzymol.* **108**, 132–138.
39. Goldinger, W., Rebsamen, E., Brandli, E., and Zeigler, H. (1986). Dynamic micro and ultra filtration. *Sulzer Tech. Rev.* **3**, 10–12.
40. Griffiths, B. (1986). Can cell culture costs be reduced? *Tibtech.* **4**, 268–272.
41. Herring, S. W., Shitanishi, K. T., Moody, K. E., and Enns, R. K. (1985). Isolation of human factor VIIIc. *J. Chromatogr.* **326**, 217–224.
42. Horowitz, B. (1986). Large scale production and recovery of human leukocyte interferon. *Method. Enzymol.* **119**, 39–47.
43. Husain, S., Gurewich, V., and Lipinski, B. (1983). Purification and partial characterization of a single chain form of urokinase from human urine. *Arch. Biochem. Biophys.* **220**, 31–38.
44. Hubbard, R. A., Schluter, S. F., and Marchalonis, J. J. (1984). Separation of lymphoid cells using immunosorbents. *Method. Enzymol.* **108**, 139–147.
45. Ihle, J. N., Weinstein, Y., Keller, J., Henderson, L., and Palaszynski, E. (1985). Interleukin 3. *Method. Enzymol.* **116**, 540–552.
46. Jacks, G. W., and Wade, H. E. (1987). Immunoaffinity chromatography of clinical products. *Tibtech.* **5**, 91–95.
47. Johnson, A. J., Mathews, R. W., and Fulton, A. J. (1984). Fractionation of factor VIII and IX, an overview. *Scand. J. Haematol. Suppl.* **40**(*33*), 513–524.
48. Juarez-Salinas, H., Bigbee, W. L., Lamotte, G. B., and Ott, G. S. (1986). New procedures for the analysis and purification of IgG murine monoclonal antibodies. *Int. Biol. Lab.* **4**, 20–27.
49. Juarez-Salinas, H., Ott, G. S., Chen, J. C., Brooks, T. L., and Stanker, L. H. (1986). Separation of IgG idiotypes by high performance hydroxyapatite chromatography. *Method. Enzymol.* **121**, 615–622.
50. Karkare, S. B., Phillips, P. G., Burke, D. H., and Dean, R. C. J. (1985). Continuous production of monoclonal antibodies by chemostatic and immobilized hybridoma culture. *In* "Large Scale Mammalian Cell Culture." (J. Feder and W. R. Tolbert, eds.), pp. 127–149. Academic Press, London.
51. Karr, L. J., Shafer, S. G., Harris, J. M., van Alastine, J. M., and Snyder, R. S. (1986). Immunoaffinity partition of cells in aqueous polymer two phase systems. *J. Chromatogr.* **354**, 269–282.
52. Kauppinen, H. L., Hirvonen, S., and Cantell, K. (1986). Effect of purification procedures on composition of human leukocyte interferon preparations. *Method. Enzymol.* **119**, 27–35.
53. Kenny, A. C. (1985). A process for separating animal cells from a liquid culture. EPO 160 520 A2.
54. Kenny, A. C. (1986). Production and recovery of monoclonal antibodies produced in airlift reactors. Proceedings of Conference "Large scale production of monoclonal antibodies" Dec. 1986. Society of Chemical Industry, London.
55. Krakauer, T. (1986). Human interleukin 1. *CRC Crit. Rev. Immunol.* **6**, 213–244.
56. Kronheim, S. R. (1986). Purification and characterisation of human interleukin 1 expressed in *E. Coli. Biotechnology* **4**, 1078–1082.
57. Lambe, C. A. (1986). Continuous electrophoresis for production scale purification. *In* "Bioactive Microbial Products", Vol. 3 (J. D. Stowell, P. J. Bailey, and D. J. Winstanley, eds.), pp. 191–207. Academic Press, London.

58. Lambe, C. A., and Walker, A. G. (1987). Reactor requirements for animal cells. In "Plant and Animal Cells, process possibilities", (C. Webb and F. Mavituna, eds.), pp. 116–124. Ellis Horwood, Chichester.
59. Leatherbarrow, R. J., and Dean, P. D. G. (1980). Studies on the mechanism of binding of serum albumins to immobilized cibacron blue F3GA. *Biochem J.* **189**, 27–34.
60. de Ley, M., van Damme, J., and Billiau, A. (1986). Production and partial purification of human interferon. *Method. Enzymol.* **119**, 88–92.
60a. Liebman, H. A., Limentani, S. A., Furie, B. C., and Furie, B. (1985). Immunopurification of Factor IX by using conformational antibodies directed against factor IX-metal complex. *Proc. Natl. Acad. Sci. U.S.A.* **82**, 3879–3883.
61. LKB. (1986). Grams of monoclonal antibody in hours. Technical leaflet. Bromma, Sweden.
62. Lowe, C. R. (1984). New developments in downstream processing. *J. Biotechnol.* **1**, 3–12.
63. McCormick, D. (1987). Pharmaceutical markets for the 1990s. *Biotechnology* **5**, 27.
64. Malm, B. (1987). A method suitable for the isolation of monoclonal antibodies from large volumes of serum containing hybridoma cell culture supernatants. Pharmacia technical report. Uppsala, Sweden.
65. Manil, L., Motte, P., Pernas, P., Troalen, F., Bohuon, C., and Bellet, D. (1986). Evaluation of protocols for mouse monoclonal antibodies. *J. Immunol. Method.* **90**, 25–37.
66. Margolis, J., Gallovich, C. M., and Rhoades, P. (1984). A process for the purification of high purity factor VIII. *Vox Sang.* **46**, 341–348.
67. Matthiasson, E. (1983). The role of macromolecular adsorption in fouling of ultrafiltration membranes. *J. Membrane Sci.* **16**, 23.
68. Morrison, D. R. (1984). Bioprocessing in space—an overview. *In* "World Biotech Report 1984," Vol. 2, pp. 557–571. Online Publications, Pinner.
69. Nagabhushan, T. L., Sharma, B., and Trotta, P. P. (1986). Application of recycling isoelectric focusing for purification of recombinant human leukocyte interferons. *Electrophoresis* 7(12), 552–557.
70. Ng, P. K., Eguizabal, H. C., and Mitra, G. (1986). Preparation of high purity factor VIII concentrates. *Thromb. Res.* **42**, 825–834.
71. Nilsson, I. M., Sundquist, S. B., and Freiburghaus, C. (1984). Extracorporeal protein A–Sepharose for removal of antibodies. *Progr. Clin. Biol. Res.* **150**, 225–241.
72. Ostlund, C. (1986). Large scale purification of monoclonal antibodies. *Tibtech.* **4**, 288–293.
73. Paetkau, V., Riendeau, D., and Bleackley, R. C. (1985). Murine interleukin 2. *Method. Enzymol.* **116**, 526–539.
74. Pavlu, B., Johansson, U., Nyhlen, C., and Wichman, A. (1986). Rapid purification of monoclonal antibodies by HPLC. *J. Chromatogr.* **359**, 449–466.
75. Peng, L., Calton, G. J., and Burnett, J. W. (1986). Stability of antibody attachment in immunosorbent chromatography. *Enzyme Microb. Technol.* **8**, 681–685.
76. Pharmacia. (1985). Q and S Sepharose fast flow. *Downstream* **1**, 2–4.
77. Pharmacia. (1986). Good news for immunologists. *Pharmacia News* **2**(3), 1.
78. Pharmacia. (1986). Monoclonal antibody purification. *Pharmacia Sep. News.* **13**(4), 1–4.
79. Pharmacia. (1986). Specific monoclonal antibody purification techniques. *Pharmacia Sep. News.* **13**(5), 1–4.
80. Pharmacia. (1986). Protein A gene fusion. *Analects* **14**(2), 3–6.
81. Pharmacia. (1986). Process hygiene in industrial chromatography. *Downstream* **2**, 1–3.
82. Pharmacia. (1980). Cell affinity chromatography. Pharmacia Fine Chemicals, Uppsala, Sweden.
83. Phillips, A. W., Finter, N. B., Burman, C. J., and Ball, G. D. (1986). Large scale production of interferon. *Method. Enzymol.* **119**, 35–38.

84. Porath, J. (1986). Salt-promoted adsorption. *J. Chromatogr.* **376**, 331–341.
85. Randle, B. J., Morgan, A. J., Stripp, S. A., and Epstein, M. A. (1985). Large scale purification of Epstein–Barr virus membrane antigen. *J. Immunol. Method.* **77**, 25–36.
86. Roffler, S. R., Blanch, H. W., and Wilky, C. R. (1984). *In situ* recovery of fermentation products. *Tibtech.* **2**, 129–136.
87. Rossen, C. G., and Datar, R. (1983). Primary separation steps in fermentation processes. *In Biotech 83*, pp. 201–224. Online Publications, Pinner.
88. Sada, E., Katoh, S., and Sukai, K. (1986). Adsorption equilibrium in immunoaffinity chromatography with polyclonal and monoclonal antibodies. *Biotechnol. Bioeng.* **28**, 1497–1502.
89. Sato, H., Kidaka, T., and Hori, M. (1985). Sterilization of therapeutic immunosorbents with aqueous propylene oxide. *Int. J. Artif. Organs.* **8**, 109–114.
90. Schwartz, W. E., Clark, F. M., and Sabran, I. B. (1986). Process scale isolation and purification of IgG. *LC-GC.* **4**(5).
91. Scopes, R. K. (1982). "Protein Purification—Principles and Practice". Springer Verlag, New York.
92. Shiloach, J., Kaufman, J. B., and Kelly, R. M. (1986). Hollow fibre microfiltration methods. *Biotechnol. Progr.* **2**, 230–233.
93. Sharp, K. A., Yalpani, M., Howard, S. J., and Brooks, D. E. (1986). Synthesis and application of PEG-antibody affinity ligands for cell separation. *Anal. Biochem.* **154**, 110–117.
94. Sofer, G. K. (1986). Current applications of chromatography in biotechnology. *Biotechnology* **4**, 712–715.
95. Spier, R. E., and Griffiths, J. B. (eds.) (1985). "Animal Cell Biotechnology," Vol. 1. Academic Press, London.
96. Spier, R. E., and Griffiths, J. B. (eds.) (1985). "Animal Cell Biotechnology," Vol. 2. Academic Press, London.
97. Stampe, D., Wieland, B., and Kohle, A. (1986). Isolation of factor IX concentrates. *J. Chromatogr.* **363**, 101–103.
98. Stephenson, J. R., Lee, J. M., and Wilton-Smith, P. D. (1984). Production and purification of murine monoclonal antibodies. *Anal. Biochem.* **142**, 189–195.
99. Sterling Organics. (1987). Macrosorb KAX Composites, Information Bulletin, Newcastle, U.K.
100. Tharakan, J. P., and Chau, P. C. (1986). Serum free fed batch production of IgM. *Biotechnol. Lett.* **8**, 457–462.
101. Updike, T. V., and Nicolson, G. L. (1986). Immunoaffinity isolation of membrane antigens. *Method. Enzymol.* **121**, 717–725.
102. Van Brunt, J. (1986). Immobilized mammalian cells. *Biotechnology* **4**, 505–510.
103. Van Heyningen, V. (1986). A simple method for ranking the affinities of monoclonal antibodies. *Method. Enzymol.* **121**, 472–481.
104. Walter, H., and Johansson, G. (1986). Partitioning in aqueous two phase systems. *Anal. Biochem.* **155**, 215–242.
105. Zhoa, Y. J., and Belew, M. (1986). Effect of coupling methods on the adsorption of serum proteins by immobilized Con A. *Biotechnol. Appl. Biochem.* **8**, 75–82.
106. Zoletto, R. (1985). A multiplate settling vessel to improve the separation speed of cells from growth media. *Dev. Biol. Stand.* **60**, 313–316.

Index